Construction Delays

Second Edition

Construction Delays

Understanding Them Clearly, Analyzing Them Correctly

Second Edition

Theodore J. Trauner, Jr., P.E., P.P.

William A. Manginelli

J. Scott Lowe, P.E.

Mark F. Nagata

Brian J. Furniss

Trauner Consulting Services, Inc., Orlando, Florida

ELSEVIER

AMSTERDAM • BOSTON • HEIDELBERG • LONDON
NEW YORK • OXFORD • PARIS • SAN DIEGO
SAN FRANCISCO • SINGAPORE • SYDNEY • TOKYO

Butterworth-Heinemann is an imprint of Elsevier

Elsevier Butterworth–Heinemann
30 Corporate Drive, Suite 400, Burlington, MA 01803, USA
525 B Street, Suite 1900, San Diego, California 92101-4495, USA
84 Theobald's Road, London WC1X 8RR, UK

This book is printed on acid-free paper.

Library of Congress Cataloging-in-Publication Data

Trauner, Theodore J.
Construction delays : documenting causes, winning claims, and recovering costs / Theodore J. Trauner. – 2nd ed.
p. cm.
Includes bibliographical references and index.
ISBN 978-1-85617-677-4 (hardcover : alk. paper)
1. Construction industry–Management. 2. Production scheduling. I. Title.
TH438.T727 2009
692–dc22 2009008998

British Library Cataloguing in Publication Data
A catalogue record for this book is available from the British Library

ISBN 13: 978-1-85617-677-4

For all information on all Elsevier Butterworth-Heinemann publications
visit our Web site at www.elsevierdirect.com

Printed in the United States of America

Transferred to Digital Printing, 2011

Dedication

This book is dedicated to the staff of Trauner Consulting Services, Inc. Their efforts have been an essential part of the research required for this book. Over the years, the staff of Trauner Consulting Services has invested countless hours studying construction delays, learning the nuances of scheduling software, and continuously questioning the approach to a delay analysis with the goal of providing analytically correct and objective assessment of construction delays.

Contents

The companion Web site which contains an image collection of all of the figures in the text can be found at http://www.elsevierdirect.com/companions/9781856176774

Foreword

This book was first written and published in 1990. Over the past 19 years, we have received significant positive feedback on the contents of the book. As part of that feedback process, readers have asked questions and made suggestions concerning the content. For that reason, we decided to prepare this second edition. We trust that many of those questions will be answered in the new edition and that we will provide more examples of the proper approach to analyzing delays. Also, scheduling software has become far more powerful. As a consequence, some of the scheduling "rules" are no longer sacrosanct. The power of the software has allowed schedules to expand far beyond the basic "forward and backward pass" days when Critical Path Method (CPM) scheduling was created. For this second reason, we decided to update the information to reflect the subtle and significant changes that one may see when reviewing or analyzing a schedule. We will note the areas where the software may have an effect on how one assesses a schedule and determines the critical path. We have also incorporated more examples and more complex examples. Our first edition kept the information as simple as possible because we wanted our audience to be as broad as possible and still allow everyone to gain a clear understanding of delay analysis. But as the software has grown in power, so, too, has the understanding of our readers. CPM scheduling is far more commonplace in the industry and much better understood.

When a construction Project is delayed beyond the Contract completion date or beyond the Contractor's scheduled completion date, significant additional costs can be experienced by the Contractor, the Owner, or both. Because Contract schedules are so important and delays can be so costly, more and more projects end up in arbitration, litigation, or some form of dispute concerning time-related questions. A judge, jurors, or arbitrators are then faced with the task of sorting out who is to blame from a complex collection of facts and dates. Oftentimes, experts are required both to perform an analysis of the delays that occurred and to provide testimony to explain the analysis. One of the most difficult tasks of the expert is to educate the parties involved so that an understanding can be reached concerning the delays that occurred and who is responsible for them.

This book provides the background information necessary to understand delays. This understanding is not geared solely to the context of disputes but rather provides a framework to help prevent disputes from occurring and to resolve questions of time as they arise during the Project.

Chapter 1, "Project Scheduling," provides an overview and definitions of basic scheduling concepts and terms that will be referred to throughout the book. It is not intended as a CPM scheduling primer. Rather, it addresses important basic concepts required for using Project schedules. Key elements include float, reviewing and approving schedules, the critical path, and early completion schedules.

Chapter 2, "Types of Construction Delays," explains the basic categories of excusable and nonexcusable delays and the subcategories of compensable and noncompensable delays. It addresses the concept of concurrency and also noncritical delays. This primer in delays prepares the reader for the specific issues covered in succeeding chapters.

Chapter 3, "Measuring Delays—the Basics," explains how to approach the analysis, including the starting points of as-planned schedules and as-built diagrams and how one must compare the two in order to quantify the delays that have occurred. The question of liability is addressed separately, since this determination is made most expeditiously after the specific delays have been identified. Chapter 4, 5 and 6 travel through the actual process of analyzing delays with bar charts, CPMs, and no schedule.

Recognizing that there are numerous approaches used in analyzing delays, Chapter 7 comments on some of the more common approaches used and the strengths and weaknesses associated with them.

Damages to the Owner and Contractor are addressed in Chapters 8 through 13. Since inefficiency and acceleration costs are often time-related issues associated with delay, they have been addressed separately in the hopes that some of the myth and magic that surrounds them may be cleared away. Similarly, the topic of costs associated with noncritical delays has been given special attention, since many projects experience these with little or no recognition of the problem.

Chapter 14, "Determining Responsibility for Delay," explains the process used to assess the party who caused the delay. The responsibility for delays is addressed separately from the delay analysis because we believe that this is the proper approach to use: first determine the activities that are delayed and the magnitude of the delay and then address responsibility or liability.

Chapter 15, "Risk Management," could also be called "Prevention of Time-Related Problems," since it focuses on the delay-related risks of the various parties in a construction Project. By maintaining this focus, each of the parties has a tendency to better control time and resolve delay problems as they occur.

This book has been written with the hope that a better understanding of delays, time extensions, and delay costs will help to prevent problems rather than foster and fuel the already litigious atmosphere that exists in construction.

Bear in mind that the methodology described herein can be applied to any type of Project that (1) has a time constraint and (2) is amenable to scheduling and the monitoring and control of time. This category could include supply contracts, manufacturing projects, and research and development projects, as well as traditional construction projects. The approach will be the same for all situations, given a logical and reasoned application within the context of the existing facts.

Acknowledgments

The authors wish to acknowledge the following individuals whose writing efforts are reflected throughout this book and whose research and dedication brought this revised edition to fruition: John Crane, Sid Scott, Rocco Vespe, Linda Konrath, Anthony Ardito, and Geoff Page. Special recognition goes to Janet Montgomery for creating the graphics and coordinating this book.

Introduction to Second Edition

Construction is risky business. And while today's construction projects may be safer than they were in the past, the financial risks continue to be great. Today's construction projects are bid under fierce competition with little margin and require the coordination of many trades under demanding conditions and challenging time frames. Often, everything does not go according to plan, and the parties to the construction Contract find themselves at odds.

Many of the risks that Owners and Contractors face can affect construction time. And the cost of a lost day on a construction Project may be staggering. Unfortunately, the effects on construction time can be difficult to isolate, identify, and quantify. This is true despite the fact that the construction process has employed "modern" scheduling techniques for nearly half a century. More surprisingly, even though the power and capabilities of scheduling software have increased considerably in recent years, identifying and accurately quantifying construction delays continues to challenge even the best Project Managers.

While the parties' management teams are able to analyze and assess most factors related to a change, the effect on construction time remains difficult to understand and accurately measure. And so, even though the parties can often reach compromises related to most aspects of a change, it is the delay component that often prevents settlement. As a result, most construction claims include a component related to delays. Because the expertise required to reliably and convincingly assess delays and delay damages often goes beyond that of the participants, experts are hired to analyze delays for mediation and trial.

This book addresses the topic of construction delays, the resulting effects, and damages. This is a timely subject, since the failure to meet schedules can result in serious consequences with unprecedented cost implications. The financial significance of delays demands that the Project Owner, General Contractor, Construction Manager, Designer, and Subcontractors educate themselves regarding delays and the associated added costs. This book is designed to serve as a primer for that education process. Too many texts on this subject focus on the legal perspective, using legal language. This book is intended as a practical, hands-on guide to an area of construction that is not well understood.

All construction industry professionals should know the basic types of delays and understand the situations that give rise to entitlement to additional compensation. Most important, they should understand how a Project schedule and Project documentation can be used to determine whether a delay occurred, quantify the delay, and assess the cause of the delay. Furthermore, construction professionals should be able to assess the delay's effects on the Project and quantify any costs or damages.

Many techniques are used to analyze delays. Some of these methods have inherent weaknesses and should be avoided. This book points out the shortcomings of these faulty methods and explains how a delay analysis should be performed. It then describes—specifically—how the analysis is done with CPM schedules. The discussion will cover the subtleties of the process, such as shifts in the critical path and noncritical delays.

The subject of damages is covered in detail, including the major categories of extended field overhead and unabsorbed home office overhead costs. Likewise, the damages suffered by the Owner, either actual or liquidated, are also explained.

Finally, a chapter is devoted to managing the risk of delays and time extensions from the viewpoints of the various parties to a construction Project. A discussion of early completion schedules and constructive acceleration is also included.

The authors' substantial experience analyzing delays and quantifying damages provides the readers with numerous benefits, including the following:

- A clear, concise definition of the major types of delays
- A simple, practical explanation of how delays must be analyzed
- A detailed explanation of how delays are defined and quantified for projects with CPM schedules, bar charts, or no schedule at all
- A glimpse of some of the less obvious problems associated with delays, such as delays to noncritical activities
- An understanding of the shortcomings of some delay analysis methods that may not provide reliable results
- A detailed understanding of the various areas where costs can increase and how to calculate these costs
- An understanding of the risks that delays present to various parties to the Project, and how each of those parties can manage those risks

An explanation of delays and delay damages, presented in a straightforward, accessible manner, should be useful to public and private Owners, Construction Managers, General Contractors, Subcontractors, Designers, suppliers, and attorneys whose work involves them in the construction industry.

Chapter | one

Project Scheduling

THE PROJECT SCHEDULE

If we were to ask a Contractor, a construction Owner, or an Architect if they plan their construction projects, undoubtedly they would respond affirmatively. At the inception of a Project, everyone has some form of a plan as to how the work will be executed. That plan will incorporate many different elements such as the number of workers, the types of trades and Subcontractors, the physical aspects of the Project that affect the sequence of the work, the availability of materials, and the time required to perform the different tasks. While this list of factors that should be incorporated into a plan can be expanded, the concept is pretty straightforward: All elements that relate to the execution of a construction Project should be considered during the planning stage.

A Project schedule is a written or graphical representation of the Contractor's plan for completing a construction Project that emphasizes the elements of time and sequence. The plan will typically identify the major work items (activities) and depict the sequence (logic) in which these work items will be constructed to complete the Project. At its most basic level, a Project schedule will visually illustrate the intended timing of the major work items necessary to demonstrate how and when the Contractor will construct the Project.

The Project schedule should include every element of the Project sequenced in a logical order from the beginning of the Project through completion. In addition, the schedule should define specific time periods for each activity in the schedule. The sequencing and summation of the individual time elements will define the overall Project duration. The level of detail shown in a construction schedule will vary, depending on a number of different factors. Those factors include, but are not limited to, the type of schedule used, the Contract requirements, the nature of the work, the Contractor's practices, and so on.

Overall, the Project schedule should portray in a clear fashion the construction tasks that must be performed, the time allocated to each task, and the sequence of the tasks.

THE PURPOSE OF A PROJECT SCHEDULE

Just as a bid is an estimate of the costs required to construct a Project, the Project schedule is an estimate of the time required to construct the Project. A Project schedule is a valuable Project control tool that is used by successful Project Managers to effectively manage construction projects. As noted earlier, the Project schedule should include every element of the construction Project. As such, it is the primary document in the Project record that can provide a detailed picture of the Project's planned construction sequence. If a Project schedule is properly developed and updated throughout the duration of the Project, then it will provide periodic snapshots of the plan to complete the Project, as it may change over time. If this Project control tool is used properly, it will depict the construction plan to the Project participants, allow management to control and measure the pace of the work, and provide the participants with the information to make timely decisions.

Effectively Depicting and Communicating the Construction Plan

Successful Contractors use Project schedules to depict and communicate the construction plan among the Subcontractors and other Project participants. The development of the construction plan should be a collaborative process that includes the General Contractor and its Subcontractors. Involving the Subcontractors in the development of the construction plan/schedule, will significantly facilitate acceptance by the Subcontractors of the overall approach to build the Project. Additionally, incorporation of the Subcontractors' means and methods will strengthen the validity of the Project schedule as a tool that accurately depicts the planned construction sequence.

Once the General Contractor has a Project schedule that it believes is an accurate representation of the construction plan, the General Contractor should share the schedule with the Owner to demonstrate its plan and to inform the Owner when it will need to perform its obligations, which may include the review and approval of shop drawings and submittals and inspection of the work. Effectively communicating the work plan to all parties involved is not only a sound Project management practice, but it also promotes a culture of cooperation and partnering.

Additionally, a properly updated Project schedule will also document changes in the Contractor's plan to complete the Project. Successful Contractors and Owners know that, as a Project progresses, they may encounter unexpected problems or issues. In response to these, the Contractor may need to alter portions of its construction plan, such as its work sequence, crew sizes, and operating hours. Project schedules should be periodically updated to reflect the Contractor's then current construction plan. These updates will provide snapshots of the Contractor's plan as it changes during the course of the Project.

Control and Measure the Work

A Project schedule that is properly and periodically updated throughout the life of the Project will enable the General Contractor and Owner to accurately track and measure the Project's progress. Controlling and measuring the work happen at different levels. If the General Contractor has included its Subcontractors in the planning process, then it will be in a better position to track and enforce the agreed upon sequencing and work durations depicted in the schedule. In the same vein, the Owner should also use the schedule to track the Contractor's progress and keep the stakeholders informed of the Project's status.

Timely Decisions

In addition to tracking and measuring the Project's progress, a properly maintained Project schedule will also enable the parties to identify and deal with unexpected issues as they arise. When a problem is encountered that may delay some element of the Project, the Project participants can use the Project schedule as a tool to predict the effect of the delay on the completion of the overall Project. In addition to predicting the effect of the problem, they can also decide on an appropriate course of action to deal with the problem, which may include accelerating the work, relaxing Contract restrictions to more quickly advance the Project, or deleting work items. This ability to predict and deal with a problem that may delay the Project before it actually does so is perhaps a Project schedule's most valuable attribute. Most Project Managers can see and deal with problems as they occur. However, good Project Managers can also predict how problems today will affect the Project in a month, in six months, and even farther in the future. Relying on the Project schedule as a planning, scheduling, and management tool will enable Project Managers to more competently and reliably control and manage their projects.

TYPES OF PROJECT SCHEDULES

A Contractor can use many different types of schedules to depict its construction plan. Selection of the most appropriate scheduling technique depends on the size and complexity of the construction Project, the preferences of the entity preparing the schedule, and the scheduling requirements of the Contract. The most common scheduling techniques used for construction projects are narrative schedules, Gantt Charts or bar charts, linear schedules, and Critical Path Method (CPM) schedules.

Narrative Schedules

Narrative schedules are typically used on very small construction projects that have very few activities. A narrative schedule consists of a narrative description of the Contractor's planned construction sequence and is typically submitted prior to the start of work. For example, a narrative schedule may tell the Owner that the Contractor plans to work across the Project site in an east to west fashion. In addition, in the narrative the Contractor should also identify the number of crews it plans to use and how long it believes the work will take. The degree of detail in a narrative schedule will vary from Project to Project, but most narrative schedules range between one paragraph and two pages. When a narrative schedule is requested, the Contract scheduling specification should identify the level of detail required. An example of a narrative schedule is shown in Figure 1.1. The physical Project as described in the narrative schedule is shown in Figure 1.2.

Gantt Charts

A Gantt chart, also called a bar chart schedule, visually depicts the Project's major work activities in relation to time. A Gantt chart or bar chart schedule is a simple and straightforward depiction of the construction plan, showing the duration and timing of the work activities with the sequence of tasks implied. The major work items or activities are identified along the vertical axis, and time is tracked along the horizontal axis. The chart contains columns along the left-hand side of the page that identify the number and title of the major work activities, activity durations,

RE: Bridge Project
Narrative Schedule

Dear Engineer,

The bridge construction project consists of two abutments (Abutment #1 and Abutment #2), three pile-supported piers (labeled Pier #1, Pier #2, and Pier #3 from west to east), and four spans with steel girders and concrete decking (labeled Span 1, Span 2, Span 3, and Span 4 from west to east).

I will mobilize to the site on April 1, 2008, and plan to construct this bridge by moving from west to east. I plan to construct the bridge abutments and piers concurrently. One crew will construct Abutment #1 and then Abutment #2 and at the same time Piers #1 through #3 will be constructed starting at Pier #1 through Pier #3. The construction of the spans will follow the installation of the piers.

I expect the construction of each abutment to take approximately 6 weeks and the construction of each pier to take approximately 8 weeks. I expect that all of the abutments and piers will be in place and ready to accept the steel girders for the spans by the end of July.

From the end of July until the middle of October, I plan to erect the steel for the four spans, form and pour the decks, and form and pour the parapets and sidewalks. After completing any punchlist work, I plan to open the bridge to traffic on October 20, 2008.

Please feel free to contact me with any questions.

Sincerely,
Bob Smith
Project Manager

FIGURE 1.1

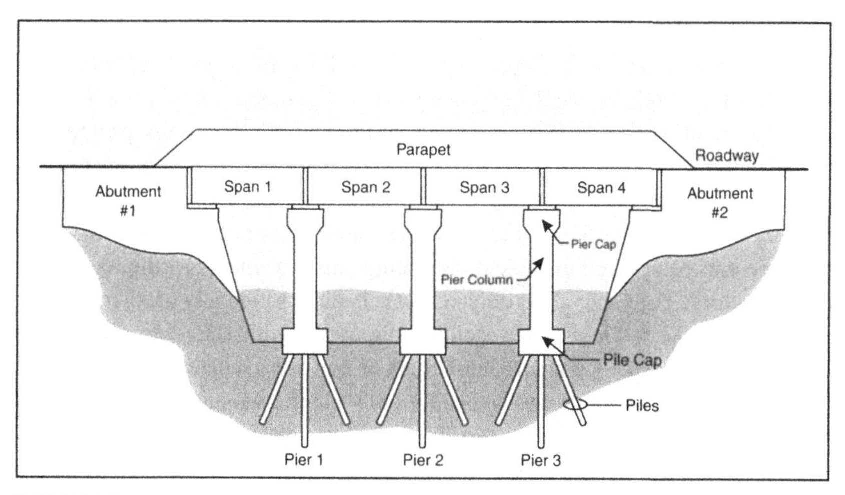

FIGURE 1.2

and the activity start and finish dates. To the right of the columns are horizontal bars that represent the work activities described in the columns. The work activities are typically organized in descending chronological order, with the earliest work item in the first row, the next earliest work item in the second row, and so on. In addition to organizing the work items chronologically, they can also be grouped according to similar locations, phases, and so forth. A compilation of all of the horizontal bars should provide a visual representation of the work for the entire construction Project. An example of a bar chart for a construction Project is shown in Figure 1.3.

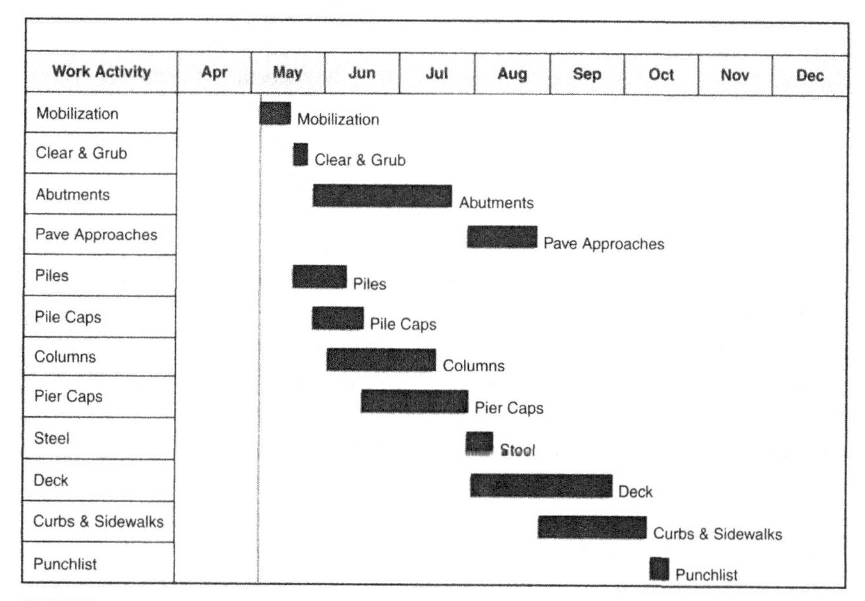

FIGURE 1.3

Once work begins, the actual performance of the Project can be tracked using the same chart. As the Project progresses forward in time, the horizontal work bars that originally depicted when work was planned to begin can be compared to actual work bars that are drawn into the schedule as work progresses. These actual work bars can reflect progress and the actual start and finish dates of activities. If required, Gantt charts or bar chart schedules can be updated periodically throughout the Project to show the status or progress of the work as of that time and be updated to project the timing and sequencing of the remaining Contract work. The Gantt chart or bar chart schedule is typically used on smaller to midsized projects for which the relationships among the activities are obvious or easily recognized. A typical Gantt chart or bar chart schedule only summarizes the major work items and is usually only one to two pages long. It should be noted that bar charts can be very large and very detailed. If done correctly, they can be an acceptable method for scheduling the work, depending on the nature of the Project.

Linear Schedules

The Linear Scheduling Method is also referred to as Line of Balance scheduling. It is most effectively used to plan and manage construction projects that are repetitive or linear in nature, such as highway construction and pipeline and power line construction. A linear construction schedule is usually depicted as a graph with an X- and Y-axis. The entire Project duration (time) starting from day 0 to the end of the Project is plotted along the X-axis and the measure of production that is usually common to all schedule activities is plotted along the Y-axis from one end of the Project to the opposite end in ascending order from the lower limit of the X-axis to its upper limit. The measure of production is typically represented as specific locations along the Project, and, depending on the type of construction Project being depicted, the measure of production can be defined as stationing for highway and pipeline projects or floors for high-rise building construction. The schedule activities are usually represented as a line that starts at the X-axis and that extends upward toward the upper limit of the Y-axis that depicts performance of the schedule activity along the entire length of the Project. The slope of the schedule activity line will represent the production rate of the schedule activity's operation. Figure 1.4 is an example of a linear schedule for a roadway Project.

The major advantage that the linear scheduling method has over other scheduling techniques—for example, a narrative schedule, Gantt chart, and CPM schedule—is that a linear schedule allows the user to easily track planned and actual production rates of individual schedule activities. When used on linear projects, such as roadway and pipeline projects, productivity is often an important factor in measuring efficiency, profitability, and, ultimately, success. However, the major weakness of this technique is that it does not identify the Project's critical path. It should be noted that whether or not a Project has a schedule, the Project will always have a critical path. The inability of the linear

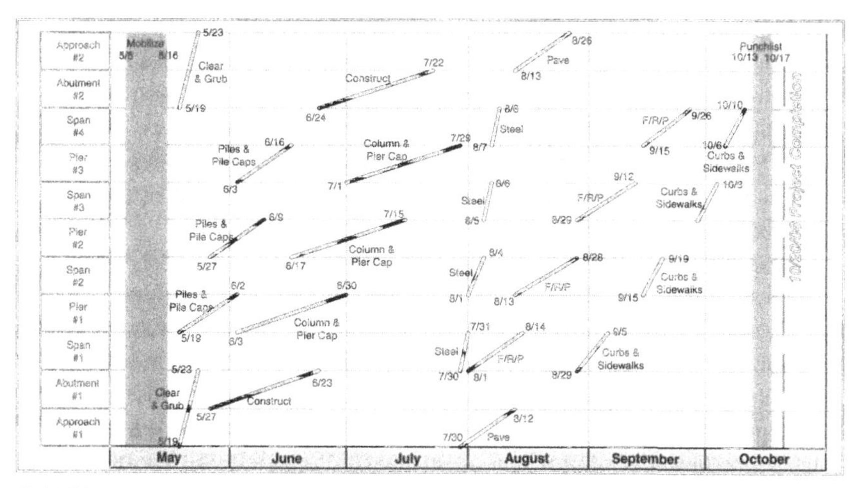

FIGURE 1.4

scheduling method to identify the Project's critical path stems from the fact that the schedule activities are not linked to one another, and thus the schedule cannot accurately correlate delays to the completion of the Project to specific schedule activities in a cause-and-effect manner. One of the major reasons that many Owners and Contractors have chosen the CPM scheduling technique over the linear scheduling methods is the CPM scheduling technique's ability to identify the critical path of the Project. With the critical path defined, the Owner and Contractor are better able to manage the Project with respect to time. Along with this is the benefit of being able to assign Project delays to specific schedule activities and determine whether the Contractor is entitled to additional Contract time.

Critical Path Method Schedules

A Critical Path Method (CPM) schedule is similar to a bar chart schedule in that it contains work activities that represent the Project. However, the CPM schedule usually contains all of the Project work items and connects or links those work activities to one another according to their planned sequence. The result is a network of interrelated activities that defines the various paths of work necessary to construct the Project. A CPM schedule is simply an arrow diagram or logic network of the work activities that graphically or visually represents the construction plan. The linking or interdependency of the work items is a major strength of CPM scheduling because it enables the identification of the critical path or the longest path of work through the network. The critical path predicts the earliest date that the Project can be completed.

CPM schedules are the most frequently used scheduling technique for the planning and scheduling of construction projects, both large and small. Whereas Gantt charts and bar chart schedules usually only depict the major items of work, properly developed and updated CPM schedules can include virtually every item of work in the Contract and may contain between 10 and 30,000 interrelated work activities. CPM schedules have been and continue to be used as planning tools for simple to complex construction projects that require the integration of many components and incorporation of phasing and coordination. CPM schedules can be used on any size and type of construction Project. If constructed properly and updated correctly, they are the most effective form of schedule for a construction Project.

Most CPM schedules are updated with progress on a monthly basis, and in some instances the schedule may even be updated more frequently. When the schedule update records the Project's progress, the timing and order of the remaining work items will automatically be calculated to reflect the actual performance of the completed work items. The strength of the CPM scheduling technique is that it is a dynamic modeling tool that can identify issues and problems *before* they arise. If problems arise, the CPM schedule will identify potential areas of delay and provide a measure of the magnitude of the delay. The CPM schedule's ability to accurately reflect Project progress and the effect it has on the overall plan helps the Project management staff to identify and deal with issues in a real-time setting. An example of a relatively simple CPM schedule is shown in Figure 1.5.

WHAT IS THE CONTEMPORANEOUS SCHEDULE?

Merriam-Webster's Dictionary defines *contemporaneous* as "existing, occurring, or originating during the same time." However, when the term *contemporaneous* is applied to a construction schedule as an adjective, it has a very specific meaning. Then it distinguishes between the schedules that existed during the construction of the Project and the schedules created after the fact or altered after the fact. All too often, analysts will create a schedule well after the Project is completed or, alternatively, will modify the Project schedule, asserting there were errors that needed correction. While some limited adjustments to a schedule may be appropriate, making significant changes or creating a new schedule is very subjective and not an accepted practice for an analysis of delays to a Project.

The contemporaneous schedule is the Project schedule, which typically consists of the baseline schedule and schedule updates that were used to manage and construct the Project. Since the contemporaneous schedules were used by the Project participants to manage and construct the Project, they are the most reliable representation of the construction plan and the status of the work throughout the Project.

WHAT IS THE CRITICAL PATH?

A CPM schedule contains many interrelated activities that are required to construct the Project. All of these activities are combined logically to create the network diagram. The critical path is defined as the longest path of work activities through the network diagram. That longest path determines the earliest date that the Project will finish.

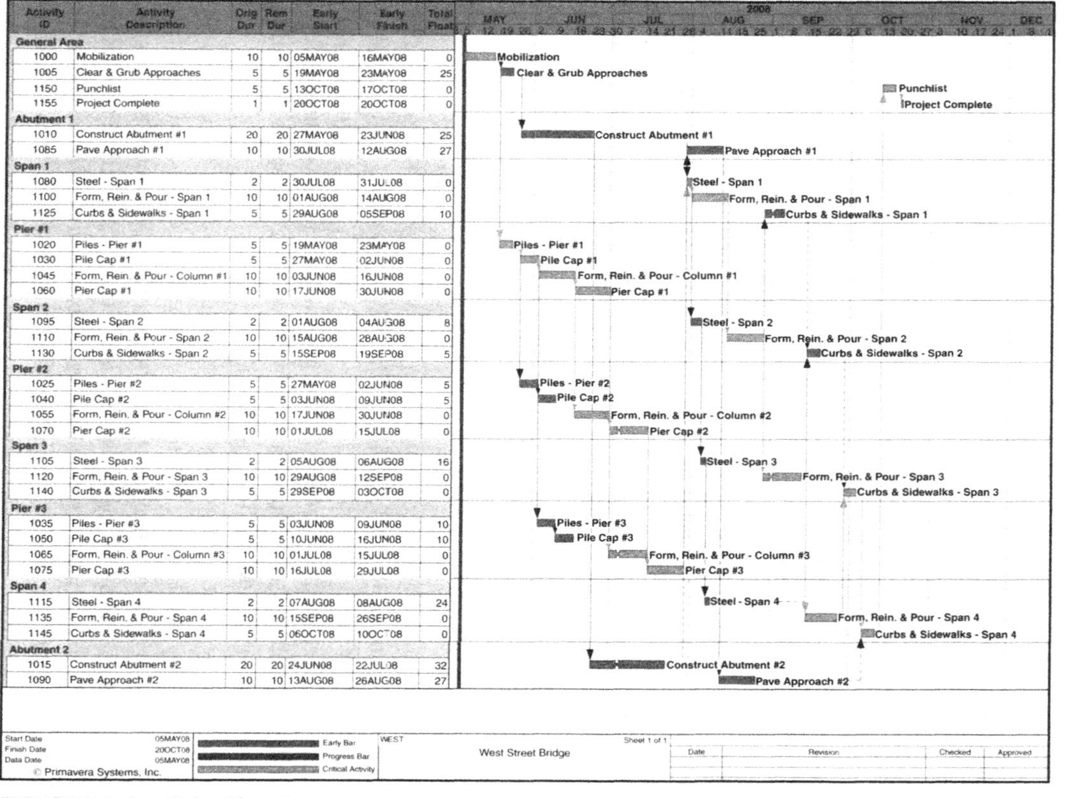

Activity ID	Activity Description	Orig Dur	Rem Dur	Early Start	Early Finish	Total Float
General Area						
1000	Mobilization	10	10	05MAY08	16MAY08	0
1005	Clear & Grub Approaches	5	5	19MAY08	23MAY08	25
1150	Punchlist	5	5	13OCT08	17OCT08	0
1155	Project Complete	1	1	20OCT08	20OCT08	0
Abutment 1						
1010	Construct Abutment #1	20	20	27MAY08	23JUN08	25
1085	Pave Approach #1	10	10	30JUL08	12AUG08	27
Span 1						
1080	Steel - Span 1	2	2	30JUL08	31JUL08	0
1100	Form, Rein. & Pour - Span 1	10	10	01AUG08	14AUG08	0
1125	Curbs & Sidewalks - Span 1	5	5	29AUG08	05SEP08	10
Pier #1						
1020	Piles - Pier #1	5	5	19MAY08	23MAY08	0
1030	Pile Cap #1	5	5	27MAY08	02JUN08	0
1045	Form, Rein. & Pour - Column #1	10	10	03JUN08	16JUN08	0
1060	Pier Cap #1	10	10	17JUN08	30JUN08	0
Span 2						
1095	Steel - Span 2	2	2	01AUG08	04AUG08	8
1110	Form, Rein. & Pour - Span 2	10	10	15AUG08	28AUG08	0
1130	Curbs & Sidewalks - Span 2	5	5	15SEP08	19SEP08	5
Pier #2						
1025	Piles - Pier #2	5	5	27MAY08	02JUN08	5
1040	Pile Cap #2	5	5	03JUN08	09JUN08	5
1055	Form, Rein. & Pour - Column #2	10	10	17JUN08	30JUN08	0
1070	Pier Cap #2	10	10	01JUL08	15JUL08	0
Span 3						
1105	Steel - Span 3	2	2	05AUG08	06AUG08	16
1120	Form, Rein. & Pour - Span 3	10	10	29AUG08	12SEP08	0
1140	Curbs & Sidewalks - Span 3	5	5	29SEP08	03OCT08	0
Pier #3						
1035	Piles - Pier #3	5	5	03JUN08	09JUN08	10
1050	Pile Cap #3	5	5	10JUN08	16JUN08	10
1065	Form, Rein. & Pour - Column #3	10	10	01JUL08	15JUL08	0
1075	Pier Cap #3	10	10	16JUL08	29JUL08	0
Span 4						
1115	Steel - Span 4	2	2	07AUG08	08AUG08	24
1135	Form, Rein. & Pour - Span 4	10	10	15SEP08	26SEP08	0
1145	Curbs & Sidewalks - Span 4	5	5	06OCT08	10OCT08	0
Abutment 2						
1015	Construct Abutment #2	20	20	24JUN08	22JUL08	32
1090	Pave Approach #2	10	10	13AUG08	26AUG08	27

FIGURE 1.5 See Color Plate 1.

Historically, the critical path was the path of activities that had zero total float. However, CPM scheduling software has gotten far more sophisticated. As a consequence, the critical path of a Project in a CPM schedule may not be the activities that are identified as having zero float or the least float. The scheduler can now use multiple work calendars and many different types of constraints that can affect the float calculation. The calculation of float is still arithmetically correct, but the calendars and constraints require that the calculation be based on the calendar and constraint requirements. The result is that the float that is calculated does not have the same meaning that it used to. Zero float or the least amount of float does not necessarily mean that those activities are on the Project critical path. This is difficult and confusing for many of us in the construction industry. If we learned CPM scheduling more than a few years ago, we became confident of our understanding of how it worked, and we learned to trust that the float information would tell us all we needed to know. Those days are gone. Float can still determine the critical path but only in schedules that have limited calendars and limited constraints.

In order to determine the critical path of a Project in today's construction environment, the analyst must look at the longest path regardless of what the float calculations say. That is the most accurate way to determine the Project critical path. Keep in mind that the critical path is the path of activities that if delayed will delay the completion date of the Project. For many years we would refer to noncritical activities as activities that had float. In the technical sense, this characterization may no longer apply.

WHAT IS FLOAT?

Float is a term of art that was conceived when the Critical Path Method was first developed. It is unique to CPM scheduling, but, more important, its significance has changed over the past few years. And it is this change in the significance of *float* that has created many problems in the analysis and understanding of CPM schedules and delays. Let's start with the conceptual basis of float as it was first structured when the CPM method originated.

Conceptually, float is the amount of time an activity can be delayed before it begins to delay the Project. When an activity consumes its available float, it will become critical and may delay the Project. In a CPM schedule, float is the difference between the activity's early and late start and finish dates. The early start and finish dates represent the earliest that an activity can start or finish based on the activity durations and logic relationships in the network diagram. An activity's early start and finish dates are determined during the arithmetical process of the forward pass. The forward pass is a basic step in the algorithm of the CPM process. The forward pass begins from the schedule's data date, and the early dates are calculated by adding the durations of each of the successive activities according to the logic relationships in the network diagram. Additionally, the forward pass calculates the earliest date that the Project can finish.

Conversely, an activity's late start and finish dates represent the latest date that the activity must start or finish before it will delay the Project. An activity's late dates are identified during the backward pass, which is also a basic step in the algorithm of the CPM process. The backward pass begins from the arithmetically calculated completion date and works backward through each path of the network, subtracting the durations of each preceding activity.

The term *float* refers to the difference in workdays between the early and late start and finish dates of an activity. Noncritical activities have float or slack time between their early and late dates. As mentioned earlier, when noncritical activities experience delay, they will consume their float and may ultimately become critical.

The preceding description of float accurately describes the term based on how the Critical Path Method was originally developed. But that definition may no longer be accurate depending on how the CPM schedule is constructed. As we noted in the preceding discussion of the critical path, the generally understood concept of float may not be appropriate depending on the calendars used, the constraints included, and so on. Because of that, the definition of float can most accurately be described as the difference between the arithmetically calculated early and late start and finish dates. The use or consumption of an activity's float may not make the activity critical and may not delay the Project. Float now is applicable only to particular activities or chains of activities. It is not necessarily applicable to the entire schedule or network diagram.

To those of us who have used CPM for many years, this creates a problem. We are conditioned to look at the float values and draw conclusions based on those. For example, if we look at a list of activities and see that the least float is –20 for a chain of activities, we instinctively conclude that the negative-20 path of float is the critical path and that the Project is 20 work days behind schedule. This may not be the case at all. Therefore, when reviewing any CPM schedule, the analyst must develop a clear understanding of how the network was constructed and ascertain exactly what meaning any float values may have.

REDEFINING THE CRITICAL PATH

Ask any construction professional to define the "critical path" and 99 percent of the time the answer you will receive is that the critical path is "the zero-float path." This belief is based on the concept that the critical path should have no float and that a delay to any critical activity on the critical path will result is a corresponding delay to the completion of the Project. Although this definition of the critical path is conceptually correct, the use of advanced scheduling features that more accurately depict the construction plan forces us to reevaluate this traditional definition.

In its book *Construction Planning & Scheduling*, the AGC provides the following definition of the critical path:

> The critical path of a Project is the longest path through the network that establishes the minimum overall Project duration. The critical path is composed of a continuous chain of activities through the network

schedule with zero total float. All activities on the critical path must start and finish on the planned early start and finish times. Failure of a critical path activity to start or finish at the planned early and late finish times will result in the overall Project duration being extended. For the classic schedule calculation it is both necessary and sufficient for an activity to be on the critical path if the activity's total float is zero.

Additionally, *Primavera Project Planner* (P3) defines critical path as "one or more continuous chains of zero or negative float activities running from the start event to the finish event in the schedule." The AGC's definition can be broken down into the following three criteria:

- The critical path is a continuous chain of activities through the network schedule.
- A delay to a critical path activity will result in a corresponding extension to the overall Project duration.
- Zero or negative float values are predictors of the critical path.

However, due to the advanced capabilities of current scheduling software, these three criteria will not always accurately define the critical path. The following discussion will explain why and propose a new definition.

The Critical Path as a Continuous Chain of Activities

This first component of the AGC's definition of critical path states that the critical path should consist of a continuous chain of activities. This component has long been a readily accepted and logical requirement of the critical path. However, due to the use of advanced scheduling software features, such as multiple calendars, the critical path may include time periods during which no critical work is planned. Although the idea sounds counterintuitive, consider the following example. In the northern states, it is common for heavy highway and road construction projects to experience a winter shutdown, during which no Project work is allowed or expected to occur. Additionally, most State Departments of Transportation (DOTs) include standard Contract language that places limitations on the placement of bituminous pavement mixtures. These limitations prohibit the Contractor's ability to place asphalt during winter months or when surface or air temperatures are below a specified level.

Most Contractors incorporate this restriction and other contractually mandated nonwork periods, such as for wildlife and environmental reasons, into the Project schedule through the use of multiple calendars. In instances when the Project contains restrictions such as these, the schedule will typically include a separate calendar that will be assigned to the restricted work activities that prohibit restricted work from occurring during specific time periods. It should be noted that the use of multiple calendars in this manner enables the Project schedule to more accurately model the actual Project conditions. This more accurate tool can assist the Project participants in planning and constructing the Project and dealing with problems as they arise.

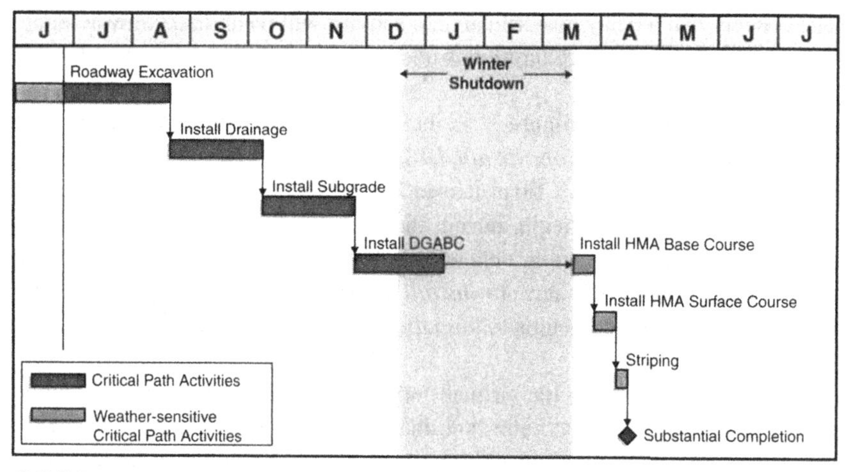

FIGURE 1.6

For instance, as just stated, in northern states that limit the placement of asphalt during the winter months, Contractors often choose to block out the winter months as nonwork days. Additionally, on multiphase, multiyear projects, it is common to find Project schedules that contain critical paths with at least one nonwork period. Figure 1.6 is a simplified example of a critical path with a nonwork period.

Although the example in Figure 1.6 is an oversimplification, many Contractors have chosen to use multiple calendars to manage their risk when planning weather-sensitive work. As you can see, there is a gap of time between the finish of the *Install DGABC* activity and the start of the *Install HMA Base Course* activity that accurately portrays the actual Project conditions. Therefore, as depicted in Figure 1.6 and contrary to both the AGC's and *Primavera Project Planner*'s definition, the critical path will not always be a continuous chain of activities through the network, due to the use of advanced scheduling features, such as multiple calendars.

After establishing that the critical path will not always be a continuous chain of activities through the network due to nonwork periods, such as a winter shutdown, then it should be easy to comprehend how the second requirement will also not always be accurate when defining the critical path.

Delay to a Critical Path Activity

AGC's definition of the critical path states the following:

> All activities on the critical path must start and finish on the planned early start and finish times. Failure of a critical path activity to start or finish at the planned early and late finish times will result in the overall Project duration being extended.

The concept that a delay to a critical path activity will result in a corresponding delay to the overall Project duration is no longer a steadfast requirement for the critical path.

As can be seen in Figure 1.6, the first four activities on the critical path—*Roadway Excavation, Install Drainage, Install Subgrade,* and *Install DGABC*—contain positive float caused by the inability of the *Install HMA Base Course* activity to begin during the winter nonwork period. Therefore, the first four activities can be delayed for about two months and not cause a delay to the Project until activity *Install DGABC* is delayed beyond the winter nonwork period and begins to directly delay the *Install HMA Base Course* activity.

Some may argue that the critical path depicted in Figure 1.6 really only consists of the last four activities and that the first four activities are not critical. Their reasoning would most likely be based on the fact that the first four activities have positive float and that critical path activities cannot have positive float, so the first activities are not critical. However, although the first four critical activities have positive float, all eight of the activities depicted in Figure 1.6 are, in fact, "critical" because the predecessor to the *Install HMA Base Course, Install DGABC,* is planned to finish during the winter shutdown, and its current finish date is effectively restricting the ability of activity *Install HMA Base Course* to begin until after the winter shutdown. Therefore, the start of the *Install HMA Base Course* activity is predicated on two things: first, that the finish of the *Install DGABC* activity is planned to occur during the winter shutdown, and second, the *Install HMA Base Course* activity cannot occur during the winter shutdown period.

In summary, the example depicted in Figure 1.6 demonstrates that the critical path will not necessarily always be a continuous chain of activities and that a delay to a critical activity will not always result in a corresponding delay to the overall Project duration. The third and last criteria of the AGC's definition of the critical path requires the first two to be true. Since we have established that they are not, then the third criterion cannot be true. The next section explains why.

Zero or Negative Float Values as Predictors

As just demonstrated, activities on the critical path can have positive float and can be delayed without extending the overall Project duration due to the use of scheduling features, such as multiple calendars. In the past, when construction schedules used only one work calendar for all of the work activities, you could easily identify the critical path by running a total float/early start sort of the work activities. Unfortunately, the use of multiple calendars has created a situation where work activities along the critical path will have varying float values, thereby making it impossible to use float values as the only predictor of the critical path.

Therefore, it is reasonable that the definition of the critical path as set forth by both the AGC and Primavera needs to be reexamined. To more accurately define which path of work is in fact the critical path, given the preceding discussion, the following definition of the critical path is clearer and more accurate:

> The critical path is the longest path of work through the network.

This definition eliminates the need to use float as a determiner of criticality and takes into consideration the fact that the critical path may contain gaps or periods when no work is planned to occur.

WHO OWNS FLOAT?

For many years, construction professionals have argued the question of "who owns the float" in a CPM schedule. It emerged as a contentious issue for different reasons. One was that Contractors would have activities with float delayed by an Owner. There was no delay to the Project, so no time extension was due. But there may well have been economic ramifications to the Contractor. For example, the Contractor desires to perform a noncritical activity early while he or she has certain equipment available. But the Owner needs to resolve an open design question related to that work. The design question is resolved but only after the Contractor's equipment is no longer available. As a consequence, the work is performed with no delay to the Project but less efficiently and perhaps at increased cost.

Keep in mind that most construction contracts address time delays but normally only in the context of the Project end date and perhaps interim milestone dates. For Owner-caused problems, the Contract may allow time extensions and even additional compensation. But most construction contracts are silent with respect to delays that affect noncritical activities. Because of the historic conceptual meaning of float that was previously discussed, activities with float were always viewed as being noncritical. As a consequence of the lack of direction concerning delays to activities that were noncritical, disputes arose over "who owned the float."

Rather than let the courts decide this question, some creative Contract drafters will actually state that the Owner owns the float. It was believed that by using this wording, the Owner would not have any exposure should it cause delays to noncritical activities. However, the astute Contractor could structure the network diagram such that no float existed at all. What this all meant was that the schedule became a game of positioning rather than a professional tool to manage the Project with respect to time, sequence, and resources. Most construction contracts are still silent with respect to "who owns the float." The authors have always expressed the belief that the Project owns the float, and either party may use float as long as it is not to the financial detriment of the other party.

Recall our earlier discussion of float and the critical path. In today's CPM schedules, the entire argument about float ownership may well be meaningless because float may have no direct correlation to the Project critical path or delays to the overall Project.

The authors still maintain that float is a shared commodity and can be used by either party as long as there is no adverse financial effect to the other party. It is suggested that this description be incorporated into the construction Contract. In line with that philosophy, if an Owner uses float, the Contractor has an obligation to notify the Owner if the use of that float will have an adverse financial affect. The resolution of the issue would then follow the procedures outlined in the Contract normally as described in the changes clauses in the Contract.

REVIEWING AND APPROVING THE AS-PLANNED PROJECT SCHEDULE

As stated earlier in the chapter, the as-planned schedule is usually the only Project document that depicts the entire Project's as-planned construction sequence. This work sequence is tremendously valuable both as a means for gauging how well the Contractor has developed its plan to complete the Project and for identifying and measuring delays that may occur during construction. For these and other reasons, it is incumbent upon the Owner to perform a thorough review of the Contractor's as-planned schedule submission before accepting it as the official Project schedule.

The reviewer of the as-planned schedule should ultimately be able to conclude that the as-planned schedule complies with the Contract requirements, depicts the Contractor's plan to complete the Project, and represents a plan that is constructible, with a critical path that makes sense. The content and quality of schedule submissions will vary greatly according to the Contract provisions, the nature of the work, and even the Contractor's scheduling experience. Although there is no single checklist or foolproof procedure that defines what constitutes a "good" Project schedule, it is recommended that the reviewer should take the following steps when performing the review of the Contractor's as-planned schedule. For the purposes of this discussion, it is assumed that the as-planned schedule is a Critical Path Method (CPM) schedule.

1. UNDERSTAND THE CONTRACT REQUIREMENTS

Before beginning the review of the Contractor's as-planned schedule, the reviewer should have a thorough understanding of the Contract documents. The Contract documents will provide the reviewer with necessary information including, but not limited to, the Project's scope of work, the Project's staging or phasing, Contract milestones, Owner review times, third-party responsibilities, and Contract restrictions.

2. VERIFY THE CONTENTS OF THE SCHEDULE SUBMISSION

After receiving the schedule submission, the first step in the review process is to determine whether or not the Contractor's schedule submittal meets the requirements of the Contract's scheduling specification. Most CPM scheduling specifications require the Contractor to submit a copy of the electronic schedule file along with the schedule narrative and multiple schedule or tabular reports and require the Contractor to submit the schedule within a certain defined time period from Notice to Proceed.

Often, Contractors do not include a schedule narrative or all of the required printouts as part of the schedule submission because they believe the copy of the electronic schedule file provides all of the necessary information. A schedule narrative is similar to the method statement in that it can provide the Owner with a better understanding of the Contractor's sequence of operation and the planned utilization of resources. Although many schedule submissions are not submitted on time or lack the schedule narrative and required schedule/tabular reports, or both, these deficiencies should not in and of themselves constitute reasons for an immediate rejection of the submitted schedule. Instead, in its written review, the reviewer should state whether the submission was submitted in accordance with the time requirement in the Contract, identify the missing reports, require that the missing reports are included in future submissions, and begin the review of the schedule.

3. DOES THE SCHEDULE CONFORM TO THE CONTRACT REQUIREMENTS?

After checking the contents of the schedule submission and with a thorough knowledge of the Contract documents, the reviewer should then determine whether the Project schedule conforms to the Contract requirements. Here is a sample of some relevant questions to ask and recommended items to evaluate when reviewing the schedule's conformance to the Contract requirements.

- **Does the schedule conform to the Contract milestones, phasing, sequencing requirements, and so forth?** The as-planned schedule should conform to the Contract's phasing and sequencing and should not forecast Project milestone dates later than the Contract milestone dates. Additionally, if the Contract identifies Project-specific restrictions, including but not limited to limitations to access, seasonal limitations, or environmental restrictions, the schedule should include consideration of these requirements.
- **Does the schedule conform to the CPM Scheduling & Time Extension Provisions?** The Contract's CPM Scheduling Provision will usually identify specific characteristics that the Project schedule should include, which may do the following:
  —Limit the duration of activities
  —Limit the use of unnecessary or inappropriate constraints
  —Prohibit activities with open ends
  —Prescribe the use of certain logic relationships, lags, and leads
  —Establish the required level of detail in the schedule

More Owners are developing CPM scheduling special provisions or auxiliary CPM manuals that standardize everything from activity-naming conventions, activity ID numbering, abbreviation usage, Work Breakdown Structure (WBS), and even activity durations (based on known quantities and productivity rates).

In addition to the CPM Scheduling Provision, the Contractor should also ensure that the as-planned schedule conforms to the requirements of the Contract's Time Extension Provisions, which often vary greatly from Contract to Contract. For example, if the Time Extension Provision states that the Owner will only grant additional Contract time for unusually severe weather, then the Contractor needs to account for anticipated levels of inclement weather in the as-planned schedule.

- **Include all of the tasks necessary to complete the Scope of Work.** The schedule should include all of the work activities and tasks that are necessary to construct the Project. In addition to activities for the preparation and submission of key submittals, time for the procurement of key equipment and material, and adequate time for certain Owner activities, including submittal reviews and procurement of Owner-furnished materials or equipment, the schedule should include the following types of tasks, if applicable:
  —Time for permitting or third-party/regulatory activities
  —Concrete cure times
  —Surcharge or embankment settlement durations
  —Winter shutdown perlods
  —Permitting restrictions
  —Work area restrictions
  —Coordination with adjacent projects

If the as-planned schedule is deficient in meeting the Contract requirements, Contract Scheduling, or Time Extension Provisions or in properly depicting all the work activities necessary to construct the Project, then it may not provide an accurate representation of a plan to complete the Project under these particular Project circumstances. Deficiencies in these areas will typically provide sufficient justification to require the Contractor to revise and resubmit the schedule.

4. IS THE SCHEDULE CONSTRUCTIBLE AND ARE THE DURATIONS REASONABLE?

After verifying that the as-planned schedule meets the Contract requirements, the next step involves an evaluation of both the schedule's sequencing of the work activities and the activity durations. Evaluating the sequencing of the work activities consists of reviewing the overall schedule from a commonsense standpoint. A schedule's sequence of work must make sense based on the physical and contractual restrictions of the Project. Every Project should follow a generally accepted construction sequence based on the physical characteristics of work, which is sometimes referred to as "mandatory" logic. For example, a building's foundation has to be installed before structural steel can be erected, and the structural steel must be erected before the metal decking can be placed.

Therefore, the construction sequence of almost every Project will have some mandatory sequence that must be followed, and the reviewer must ensure that the Project's mandatory construction sequence or mandatory logic relationships are not violated. If the schedule contains instances that show the Project work occurring in a sequence that violates the Project's mandatory construction logic, then the reviewer should ask the Contractor to revise and resubmit the schedule correcting the mandatory logic.

The reviewer must also evaluate the reasonableness of the activity durations, which is perhaps the most subjective part of a schedule review. An activity's duration can vary based on how and to what level the Contractor chooses to apply resources to complete that specific task and is dependent on the Contractor's means and methods. Some of those factors include the amount of manpower and equipment that the Contractor plans to utilize and the crews' planned productivity. If the Contractor has submitted a schedule narrative as part of its submission, then the reviewer may be able to rely on the manpower and equipment information in the narrative to evaluate the reasonableness of the durations. Unfortunately, in most cases, the reviewer does not have access to the level of information needed to evaluate the reasonableness of the activity durations and has to rely on his or her own experience, which in most cases is sufficient to identify overly optimistic or exaggerated durations. If the reviewer feels that he is not qualified to evaluate the reasonableness of activity durations, then it is recommended that the reviewer rely on a colleague or co-worker who has more experience with those kinds of operations.

One category of activity durations that is often questioned is material procurement durations with long lead times. Because most scheduling provisions require that activity durations be limited to three weeks, activities that represent the procurement of items with long lead times often violate this requirement. It is recommended that if a material procurement activity has a duration that is greater than the maximum activity duration specified in the Contract, and if it appears to be a reasonable amount of time to procure the item, then the reviewer should accept these even though they may violate the activity duration limit.

The reviewer should also be aware that the manipulation of logic relationships and inflation of activity durations are methods that a Contractor can use to bank or sequester float. Doing so may provide the Contractor with a sense of security, but this is false. Such manipulation can just as easily mask an Owner delay as a delay caused by the Contractor. The Contractor should always provide realistic durations, and the reviewer should do her best to ensure that this is the case. It is very difficult to identify instances when the Contractor is attempting to bank or sequester float in activity durations without the knowledge of the Contractor's planned resources, and this has led many Owners to request resource-loaded schedules.

Because the sequencing of work activities and activity durations, in most instances, is determined by the Contractor's means and methods, the reviewer should show restraint in questioning the Contractor's work sequence or durations.

The only instances when the reviewer is justified in asking a Contractor to correct or address questions related to its planned sequence or durations is in instances when the mandatory logic is violated. However, if other logic relationships or activity durations do not appear correct or reasonable, then the reviewer should not hesitate to ask the Contractor to further explain or address the item(s) in question. Opening a line of communication between herself and the Contractor and establishing a consistent understanding of the Contactor's plan and schedule will assist the parties in developing a culture of cooperation, transparency, and trust early in the Project.

5. DO THE CRITICAL PATH AND NEAR CRITICAL PATHS MAKE SENSE?

Finally, since we are reviewing a CPM schedule, it only stands to reason that the critical path, which is "the longest path of work activities in the network," must make sense. With a thorough understanding of the scope of work, the reviewer should be able to recognize if the Project's critical path and near critical paths are reasonable. Typically, the critical path and near critical paths should include major items of work or complicated or complex portions of the Project. For example, if the Project is the construction of a bridge, one would expect the critical path to be the physical construction of the Project starting with the substructure, followed by the superstructure, then the roadway, and finally the striping. In conjunction, one would then expect that the near critical path would consist of the procurement of the steel for the superstructure. However, the reviewer should question the Contractor if a relatively mundane or apparently noncritical work activity, such as landscaping or lighting, is on the critical path or near critical paths. Due to the importance of the critical path to the overall schedule, if there are questions related to the reliability of the critical path, these provide sufficient justification to ask the Contractor for clarification or to revise and resubmit the schedule.

In summary, once the reviewer has performed a thorough review of the submitted schedule and is satisfied that the schedule provides a reasonably accurate depiction of the Contractor's plan to complete the Project based on the Contractual and physical characteristics of the Project, an acceptance or approval should be provided to the Contractor. If the reviewer has any questions or comments, then those should be forwarded to the Contractor within the time period that the Contract documents define for the review of the schedule. The reviewer should also keep in mind that the Contract may also define a time period for the review of a resubmission, if one is necessary.

Many Owners or their agents are reluctant to "approve" a schedule. They fear that an "approval" means they are agreeing with *everything* in the schedule, and if something goes wrong, the Contractor may assert that the Owner shares in the responsibility by virtue of the "approval." To some extent, this is a legal question, but at the minimum, the Owner should "accept" the schedule even if it feels more secure by noting that it is the Contactor's plan and the Owner takes no exception to it. Some Owners believe that it is safer to not respond at all to the schedule submission. However, it may seem logical that if no objection is made by the Owner, then the Owner has given the schedule tacit approval. The authors believe it is far wiser to assign adequate qualified resources and perform a detailed review, provide feedback to the Contractor, jointly resolve any problems or questions, and then ultimately accept the initial schedule and move on, using it as an effective management tool for the construction of the Project.

REVIEWING AND APPROVING SCHEDULE UPDATES

Reviewing and approving schedule updates submitted by the Contractor follows the same process as the review and approval of an as-planned schedule. However, there are two additional items that the reviewer should consider when reviewing a schedule update: verification of the as-built data and the identification and evaluation of the schedule changes made in the new schedule update.

Verifying As-Built Dates and Data

When a schedule is updated, activities that have started, finished, or progressed have their progress recorded through the insertion of actual start dates, actual finish dates, and new remaining durations for activities that have started but not yet finished. The reviewer should find a way to track and verify the actual performance of activities during the update period. The schedule update should contain the most accurate information related to the Project's actual performance. While it is always important to ensure that the as-built information is correct, this is even more important for cost-loaded schedules being used by the Contractor to forecast monthly cash flow projections or by the Owner to pay the Contractor based on the monthly progress reported in the schedule.

Identifying and Evaluating Schedule Changes in the Schedule Update

The other difference between schedule updates and the as-planned schedule is the ability of the Contractor to modify or change the schedule through the insertion or deletion of activities, the insertion or deletion of logic relationships among the activities, and the changing of activity durations not resulting from progress.

Whenever a change is made to the schedule, the change may affect the critical path. In his review of a schedule update, the reviewer should be able to identify all of the changes made to the schedule update and evaluate which of those changes affect the critical path or near critical paths. If the Contractor makes any schedule changes that affect the critical path, near critical paths, or Project milestones, the reviewer should require the Contractor to provide an explanation of why the schedule was altered and explain its effect. Best practice is to require that the Contractor explain all logic changes.

Like the review of the as-planned schedule, once the reviewer has completed his review of the Contractor's schedule update, he should approve or accept the update. The reviewer's goal should be to always have a current version of the Project schedule that provides an accurate representation of the Contractor's current plan to complete the Project. Such updates provide a reliable management tool that can be used to forecast, predict, identify, and resolve problems as they arise.

EARLY COMPLETION SCHEDULES

Most construction contracts specify the maximum time allowed for the Contractor to perform the work. Few, if any, contracts specify a minimum duration. However, a Contractor has both the right and the financial incentive to finish a Project early, if possible. Thus, it is important for Owners and Contractors alike to deal with early completion schedules fairly and effectively.

If early Project completion is intended by the Contractor, the early finish date should be shown on the official Project schedule, and multiple schedules should always be avoided. For the reasons discussed following, Owners and Contractors are sometimes resistant to these practices. Owners typically take issue with early completion schedules because some Contractors will claim for damages if they believe the Owner prevented them from finishing early. It is true that in some cases the Owner may be liable for damages associated with the delay, but this is actually a good argument for always showing the early finish, if intended, on the Project schedule.

The Contractor also benefits from these practices because, typically, it is extremely difficult for a Contractor to convince an Owner or the courts that he planned to finish early when the schedule did not reflect this plan. Nevertheless, some Contractors choose to create an early completion schedule for themselves and their Subcontractors, but give the Owner a schedule reflecting the full Contract duration. The rationale is that if the Contractor or a Subcontractor falls behind schedule, he will not suffer any consequences. This position does not make sense as explained below.

Assume that a Contract's duration is 300 calendar days but that the Contractor plans to complete the work in 200 calendar days. If a Liquidated Damages clause is included in the Contract, the damages will not become effective until day 301. Therefore, even if the Contractor exceeds the 200-day duration through its own actions, it faces no greater exposure to damages from the Owner.

Using two different schedules is counterproductive and can cause more problems. For example, how does a General Contractor respond to a Subcontractor who has been given the 200-day schedule but sees the 300-day schedule posted in the Owner's trailer? Also, if a dispute arises over delays, the Contractor could lose credibility when attempting to explain why there were two schedules and which schedule was the "real" one. He may have to answer questions from an attorney such as "Were you lying when you drafted the 300-day schedule or when you drafted the 200-day schedule?" Obviously, this is a no-win question for the Contractor.

Some Owners will not accept schedules that show a duration less than the Contract duration. Usually, there is no contractual basis for the Owner's insistence on this requirement. In this situation, the Contractor should ask the Owner to show where in the Contract the minimum time is specified. If the Owner still insists on a schedule using all the Contract time even though no minimum time is specified, then the Contractor should submit the same early completion schedule with one additional activity at the end making up the difference between the planned early finish date and the Contract's duration. This activity could be called "Contractor contingency." The schedule should be accompanied by a letter that clearly explains that the Contractor plans to finish early and has added the contingency activity only to get the schedule approved. The letter should also state that the Contractor expects to receive compensation if it is prevented from finishing early because of the Owner's actions.

In summary, an Owner should not panic if it receives an early completion schedule, nor should it presume that a "setup" for a delay claim is in the works. The Owner should instead recognize that effective management of the schedule will be especially important for the particular Project. An Owner should appreciate the early completion schedule because if this is the Contractor's plan, it may have direct ramifications to the Owner. For example, a more rapid construction process may require different funding than the Owner had anticipated. It is better for the Owner to know that up front to prevent any problems once the Project is under way.

The Owner (or Owner's representative) must review an early completion schedule carefully to ensure that it is complete, that the logic is correct, and that the durations are reasonable given the Contractor's resources. Short of requiring resource loading for all schedules, an Owner might consider requiring resource loading for early completion schedules only. At the minimum, when reviewing an early completion schedule, the reviewer must closely review the durations assigned to activities. If necessary, the reviewer should require the Contractor to explain the resources that will be assigned to specific activities so productivity calculations can be made to check the practicability of the plan.

If an early completion schedule is accepted (or at least not rejected), the Owner's representative must closely monitor the work on the Project and note any deviations between planned and actual durations as they occur.

Chapter | two

Types of Construction Delays

WHAT IS A DELAY?

There are a number of definitions for *delay*: to make something happen later than expected; to cause something to be performed later than planned; or to not act timely. Each of these definitions can describe a delay to an activity of work in a schedule. On construction projects, as well as on other projects where a schedule is being used to plan work, it is not uncommon for delays to occur. It is what is being delayed that determines if a Project or some other deadline, such as a milestone, will be completed late. Before any discussion of delay analysis can begin, a clear understanding of the general types of delays is necessary. There are four basic ways to categorize delays:

- Critical or noncritical
- Excusable or nonexcusable
- * Compensable or noncompensable
- * Concurrent or nonconcurrent

The chart shown in Figure 2.1 presents a general overview of how the excusable and nonexcusable categories of delay can be viewed. The discussion that follows elaborates on this simple summary chart.

In the process of determining the effect of a delay on the Project, the analyst must determine whether the delay is critical or noncritical. The analyst must also assess if delays are concurrent. All delays that are identified in the analysis will be either excusable or nonexcusable. Delays can be further categorized into compensable or noncompensable delays. This chapter provides basic definitions of these types of delays.

CRITICAL VERSUS NONCRITICAL DELAYS

In any analysis of delays to a Project, the primary focus is on delays that affect the progress of the entire Project (the Project end date or milestone date) or that are

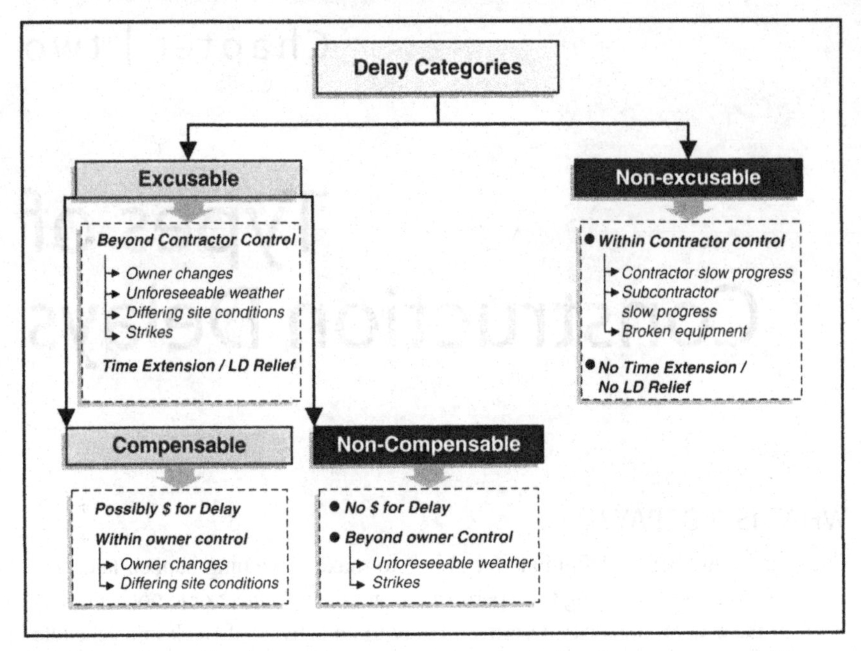

FIGURE 2.1 Note: This figure represents one interpretation. However, excusability and compensability can vary, depending on the contract.

critical to the Project completion. However, many delays occur that do not delay the Project completion date or a milestone date. Delays that affect the Project completion, or in some cases a milestone date, are considered critical delays, and delays that do not affect the Project completion, or a milestone date, are noncritical delays. The concept of "critical" delays emanates from Critical Path Method (CPM) scheduling. While the determination of a critical activity is a major element of CPM scheduling, all projects, regardless of the type of schedule, have "critical" activities. If these activities are delayed, the Project completion date or a milestone date will be delayed. In some contracts, the term "controlling item of work" will be used. Normally, this refers to critical activities or critical work. Regardless of the type of schedule used, all projects have a critical path—the path of activities that if delayed will delay the completion date.

Determining which activities truly control the Project completion date depends on the following:

- The Project itself
- The Contractor's plan and schedule (particularly the critical path)
- The requirements of the Contract for sequence and phasing
- The physical constraints of the Project—how to build the job from a practical perspective

Regardless of how one analyzes a Project and the schedule to find the delays, there is one overriding criterion: The analysis must accurately consider the contemporaneous information when the delays were occurring. "Contemporaneous information" refers to the daily reports, the schedule in effect, and any other job data available that show the circumstances at the time of the delays. Proper research and documentation eliminates the "but-fors" and any other hypotheses contrived to advance predisposed conclusions or desired results.

EXCUSABLE VERSUS NONEXCUSABLE DELAYS

Excusable Delays

All delays are either excusable or nonexcusable. An excusable delay, in general, is a delay that is due to an unforeseeable event beyond the Contractor's or the Subcontractor's control. Normally, based on common general provisions in public agency specifications, delays resulting from the following events would be considered excusable:

- General labor strikes
- Fires
- Floods
- Acts of God
- Owner-directed changes
- Errors and omissions in the plans and specifications
- Differing site conditions or concealed conditions
- Unusually severe weather
- Intervention by outside agencies (such as the EPA)
- Lack of action by government bodies, such as building inspection

These conditions may be reasonably unforeseeable and not within the Contractor's control.

Before the analyst concludes that a delay is excusable based solely on the preceding definitions, he or she must refer to the construction Contract documents. Decisions concerning delays must be made within the context of the specific Contract. The Contract should clearly define the factors that are considered valid delays to the Project that justify time extensions to the Contract completion date. For example, some contracts may not allow for any time extensions caused by weather conditions, regardless of how unusual, unexpected, or severe.

Nonexcusable Delays

Nonexcusable delays are events that are within the Contractor's control or that are foreseeable. These are some examples of nonexcusable delays:

- Late performance of Subcontractors
- Untimely performance by suppliers
- Faulty workmanship by the Contractor or Subcontractors
- A Project-specific labor strike caused by either the Contractor's unwillingness to meet with labor representatives or by unfair labor practices

Again, the Contract is the controlling document that determines if a delay would be considered nonexcusable. For example, some contracts consider supplier delays excusable if the Contractor can prove that the materials were requisitioned or ordered in a timely manner, but the material could not be delivered due to circumstances beyond the control of the Contractor. Other contracts may not allow such delays. The Owner and the Designer or drafter of the Contract specifications must be sure the Contract documents are clear and unambiguous. Similarly, before signing the Contract, the Contractor should fully understand what the Contract defines as excusable and nonexcusable delays.

COMPENSABLE VERSUS NONCOMPENSABLE DELAYS

A compensable delay is a delay where the Contractor is entitled to a time extension and to additional compensation. Relating back to the excusable and nonexcusable delays, only excusable delays can be compensable. A noncompensable delay means that although an excusable delay may have occurred, the Contractor is not entitled to any added compensation resulting from the excusable delay. Thus, the question of whether a delay is compensable must be answered. Additionally, a nonexcusable delay warrants neither additional compensation nor a time extension.

Whether or not a delay is compensable depends primarily on the terms of the Contract. In most cases, a Contract specifically notes the kinds of delays that are noncompensable, for which the Contractor does not receive any additional money but may be allowed a time extension. Contracts distinguish between compensable and noncompensable delays in many ways, some of which are described in the following paragraphs.

Federal Contracts

Federal government contracts normally define strikes, floods, fires, acts of God, and unusually severe weather as excusable but noncompensable delays. Other forms of excusable delays are compensable, such as differing site conditions, Owner-directed changes, and constructive changes that may cause a delay.

No-Damage-for-Delay Clause

Some contracts are more restrictive in defining compensable delays. It is not uncommon for a Contract to use exculpatory language concerning delays. Exculpatory language is language that exculpates, or excuses, a party from some liability. The most general approach used in contracts concerning delays is the broad no-damage-for-delay clause. The wording in this clause can take many forms, but in general the clause states that for any excusable delay the Contractor may be granted a time extension, but no additional compensation will be paid. The time extension is the sole remedy for the Contractor for any type of excusable delay.

Enforcement of these types of clauses, however, has been questioned in the courts. Some courts have been reluctant to strictly enforce these clauses.

If the contractor is delayed in completion of the work under the contract by any act or neglect of the owner or of any other contractor employed by the owner, or by changes in the work, or by any priority or allocation order duly issued by the federal government, or by any unforeseeable cause beyond the control and without the fault or negligence of the contractor, including but not restricted to acts of God or of the public enemy, fires, floods, epidemics, quarantine restrictions, strikes, freight embargoes, and abnormally severe weather, or by delays of subcontractors or suppliers occasioned by any of the causes described above, or by delay authorized by the engineer for any cause which the engineer shall deem justifiable, then:

For each day of delay in completion of the work so caused the contractor shall be allowed one day additional to the time limitation specified in the contract, it being understood and agreed that the allowance of same shall be solely at the discretion and approval of the owner.

No claim for any damages or any claim other than for extensions of time as herein provided shall be made or asserted against the owner by reason of any delays caused by the reasons hereinabove mentioned.

FIGURE 2.2

According to legal views researched, courts have often ruled narrowly on these disclaimers, thus limiting enforceability. Contractors, though, should not assume that these provisions will not be enforced. The paragraph in Figure 2.2 is an example of a broad no-damage-for-delay clause. The wording leaves little doubt that the Contract does not allow compensation for delays, regardless of the cause.

There are many variations in Contract clauses that address every possible situation as to the compensability of delays. However, the broader the clause, the less likely it is to be enforceable. More specific clauses are more readily upheld by the courts. Public contracts at the state and municipal level often contain specific no-damage-for-delay clauses. An example of a no-damage-for-delay clause pertaining to work by utilities is shown in Figure 2.3.

Similarly, the paragraph in Figure 2.4 shows a no-damage-for-delay clause covering work by other Contractors. A no-damage-for-delay clause that specifically covers the review and return of shop drawings is shown in Figure 2.5.

It is understood and agreed that the Contractor has considered in his bid all of the permanent and temporary utility appurtenances in their present or related positions and that additional compensation will not be allowed for delays, inconvenience, or damage sustained by him due to any interference from the said utility appurtenances or the operation of moving them.

FIGURE 2.3

The Contractor shall assume all liability, financial or otherwise, in connection with its Contract, and hereby waives any and all claims against the Department for additional compensation that may arise because of inconvenience, delay, or loss experienced by it because of the presence and operations of other contractors working within the limits of or adjacent to the Project.

FIGURE 2.4

The Contractor should allow thirty calendar days for the review of any shop drawing, samples, catalog cuts, and so on that are required to be submitted in accordance with the Contract. This thirty-day time period will begin on the date the submission is received by the Architect/Engineer (A/E) and terminated on the date it is returned by the A/E. The Contractor should further allow thirty calendar days for each resubmission of any rejected submission. Should any submission not be returned within the thirty calendar days specified, it is understood and agreed that the sole remedy to the Contractor is an extension of the contract time. For each day of delay in completion of the overall project caused by the late return of submissions, the Contractor shall be allowed one day additional to the time limitation specified in the Contract. No claim for any damages or any claim other than for extensions of time as herein provided shall be made or asserted against the Owner by reason of any delays caused by the reasons hereinabove mentioned.

FIGURE 2.5

All parties to a Project should clearly understand the clauses of the Contract concerning delays and time extensions. If a Contractor is considering signing a Contract with such language, it should consult qualified counsel who is familiar with construction litigation and the laws of the jurisdiction in which the clause will be enforced or adjudicated.

When a Contract identifies specific items in a Contract as being noncompensable, it should clearly define each one. For example, if unusually severe weather is a noncompensable delay, the Contract should clearly state the restriction. The Contract may define unusually severe weather as weather not ordinarily expected for the specific time of year and region. The definition in the Contract may further clarify unusual weather as that which exceeds the historical weather data recorded by the National Oceanic and Atmospheric Administration (NOAA) at a specific location. The Corps of Engineers has taken this one step further by specifying in their contracts the exact number of days of rain greater than 0.01 inches that the Contractor can expect during each month of the Project.

While the extent of detail provided by the Corps of Engineers may not be absolutely necessary, the Owner should be sure that the Contract does not have ambiguous wording. Some contracts will list "inclement weather" as an excusable, noncompensable delay, but "inclement" can have many definitions. It is also possible that inclement weather may occur but may not delay the Project. Therefore, the Owner should carefully draft the Contract, and all parties to the Contract must carefully read and clearly understand the compensable and noncompensable delays specified in the Contract.

CONCURRENT DELAYS

The concept of concurrent delay has become a very common presentation as part of some analyses of construction delays. The concurrency argument is not just from the standpoint of determining the Project's critical delays but from the standpoint of assigning responsibility for damages associated with delays to the critical path. Owners will often cite concurrent delays by the Contractor as a reason for issuing a time extension without additional compensation. Contractors will often cite concurrent delays by the Owner as a reason why liquidated damages should not be assessed for its delays. Unfortunately, few Contract specifications include a definition of "concurrent delay" and how concurrent delays affect a Contractor's entitlement to additional compensation for time extensions or responsibility for liquidated damages. To complicate matters further, there is a lack of understanding in the industry concerning the concept of concurrent delay.

Simply stated, concurrent delays are separate delays to the critical path that occur at the same time. While this seems like a simple concept, some presentations of assumed concurrent delays have significantly muddied the waters. The following discussion should clarify the area of concurrent delays.

Concurrent Delays to Separate Critical Paths

The first situation occurs when separate critical paths are concurrently delayed. For example, if shop drawings and bulk excavation are both on the critical path and predecessors to the start of footing excavation and both predecessors are scheduled to finish on the same day, then both predecessors control the start of footing excavation. If the finish of bulk excavation was delayed 30 days, from June 1 to July 1, due to equipment failures and the finish of formwork shop drawings was delayed 30 days, from June 1 to July 1, due to a redesign, then the Project was concurrently delayed by the shop drawings and bulk excavation.

In this situation, typically the delay to bulk excavation is nonexcusable, and the delay to shop drawings is excusable. Most contracts do not specify which delay takes precedence, if any. Assigning 15 days of nonexcusable delay to excavation and 15 days of excusable delay to the shop drawings may be one way to apportion the delay. The existing case law in the jurisdiction of the Project also may offer some explanation as to how this situation has been viewed from a legal perspective. In some jurisdictions, the occurrence of concurrent excusable and nonexcusable delays results in the Contractor receiving a time extension but no additional compensation. Obviously, a carefully drafted Contract should address this potential occurrence.

An example of the Contract addressing this issue can be seen in some Suspension of Work clauses within various Departments of Transportation.

Extract from Suspension of Work Clause

"No adjustment shall be made if the performance by the Contractor would have been prevented by other causes even if the work had not been so suspended, delayed, or interrupted by the Department."

FIGURE 2.6

These clauses provide for an equitable adjustment for suspended work only if there were no other causes of delay to the Project. Applying the suspension of work clause provision shown in Figure 2.6 would result in the Contractor's receiving no time or additional compensation for the redesign because of its concurrent delay to the bulk excavation.

If both 30-day delays were excusable but only one was compensable, then once again you must look to the governing case law. In this instance it has been viewed in some jurisdictions that the noncompensable delay takes precedence, and the Contractor would be issued a 30-day time extension but no additional compensation.

A second situation occurs when there are initially concurrent delays, but one delay ends before the other delay. Back to the previous example, assume the finish of shop drawings was delayed 15 days, from June 1 thru June 16, and the finish of bulk excavation was still delayed 30 days, from June 1 to July 1. In this situation, there were 15 days of concurrent delay and 15 days of delay caused by the late finish of bulk excavation. The resolution of the 15 days of concurrent delay would be approached the same as previously described. The Contractor, however, is not entitled to time or compensation for the 15-day delay caused by the late finish of bulk excavation.

A third situation occurs when there are two or more critical paths and the delay to one path starts before the delay to the other critical path. To evaluate this situation, let's revise the shop drawing and bulk excavation example. Bulk excavation work made no progress from June 1 to July 1 because of a strike. Shop drawings were suspended due to a design change from June 10 to June 25. The concept known as primacy of delay applies in this situation. The Contractor's delay to bulk excavation began before the Owner's delay to the shop drawings. Because of the delay to excavation, the Project had more time to complete the shop drawings, and the delay to the shop drawings never became critical. Bulk excavation was responsible for a 30-day delay, and the Contractor is not entitled to a time extension or additional compensation. The concept of primacy of delay recognizes that critical delays create float on other paths of work, and typically the Contractor and Owner can use this float. Let's take a simple example to explain the concept.

Say we have a Project with a CPM schedule, and in that schedule, there are parallel critical paths of work. Figure 2.7 shows Path A and Path B, each with a duration of 40 days. Both paths of work are scheduled to start the same day. Therefore, an analysis at the start of these paths would show that both paths are concurrently critical.

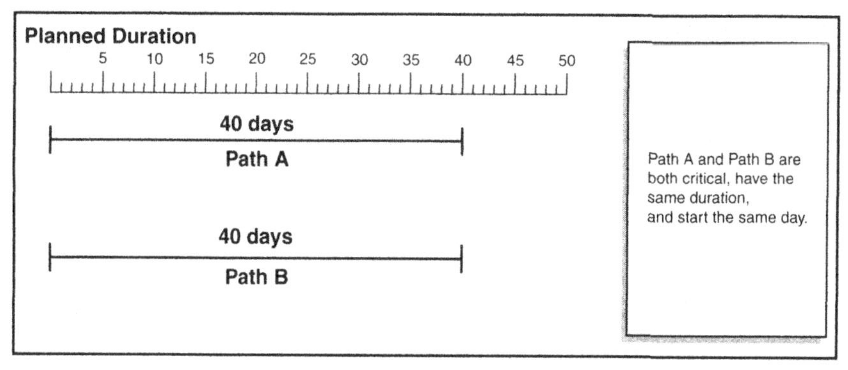

FIGURE 2.7

At the date when these paths are to begin, both Path A and Path B start as planned. Both paths are worked on for the first two days, but Path B stops work because of a design discrepancy. Path A continues to make progress, but Path B does not. If we look at the two paths at the end of 5 days, we should see what is represented in Figure 2.8. In analyzing these two paths of work at the end of day 5, we will recognize that Path A has a remaining duration of 35 days and is still planned to finish on day 40. Path B, however, made 2 days of progress and was not worked on for 3 days. Therefore, at the end of day 5, Path B has a remaining duration of 38 days and is forecast to finish on day 43. Obviously, Path B is the longest path of the two and, for our example, the sole critical path. Once Path A continued to make progress and Path B did not, the critical path shifted from Paths A and B to just Path B.

Let's move out further in time. If we look at the two paths of work at the end of day 20, we see what is shown in Figure 2.9. Path A continued to make progress for all 20 days. Path B made no progress for 10 days (day 3 through

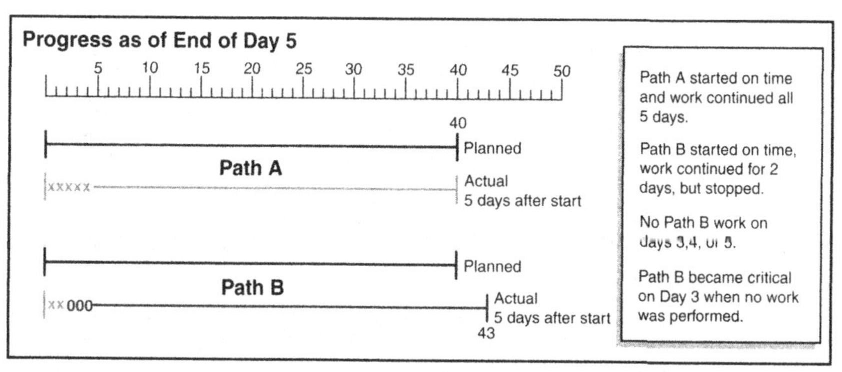

FIGURE 2.8

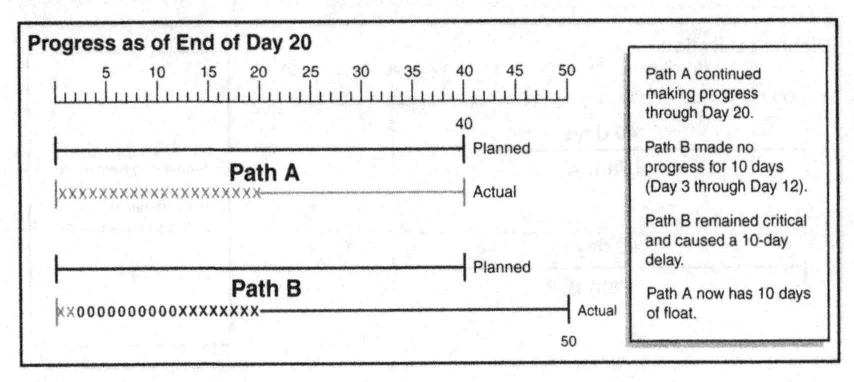

FIGURE 2.9

day 12). So at the end of day 20, Path A is still forecast to complete on day 40, and Path B is now forecast to finish on day 50. Between the earlier analysis at the end of day 5 and the present analysis at the end of day 20, Path B remained critical and Path A has had float beginning on day 3. At this point in time, Path A has 10 days of float.

Moving on out to day 50, as shown in Figure 2.10, we see that Path B continued to make progress and finished as forecast on day 50. Path A, however, had a delay of 10 days from day 31 through day 40. Path A completed on day 50, the same day as Path B. We should recognize that our earlier analysis showed that Path A had picked up 10 days of float. Therefore, even though it experienced 10 days of delay, it caused no overall delay to the Project. Path A was solely critical only on days 1 and 2, it acquired float on day three, and it continued to have float until day 41. Paths A and B were concurrently critical for days 41 through 50.

If an analyst had solely looked at the start and finish dates of these two paths of activities, the analyst might conclude that they both delayed the Project by 10 days, since they both were scheduled to finish on day 40 and both actually

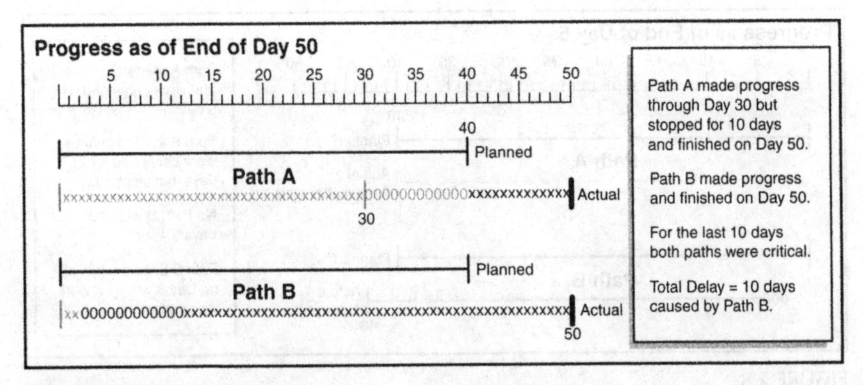

FIGURE 2.10

finished on day 50. The more detailed analysis clearly shows that this is not the case. No delay was caused by Path A, and all 10 days of delay were caused by Path B. In assessing delays during the same time frame, the analyst should perform the analysis on a day-by-day basis to correctly ascertain the exact activities that caused the delay and the correct magnitude of those delays.

Let's go back to our Primacy of Delay example. There are some who advocate a theory of concurrent delay that would grant the Contractor a 15-day, noncompensable time extension for the delay to the shop drawings. This theory of concurrent delay falls within the general category of a "but for" argument. It basically argues that "but for" my delay to the bulk excavation, I would have been delayed by the shop drawings. The major problem with this is that the "but for" assumes that the two delays are unrelated and would have occurred regardless of the other. But the analyst really does not know that. It is entirely possible that the Owner recognized the delay to bulk excavation and recognized that taking additional time on the shop drawing review would have no effect on the critical path of the Project. The "but for" argument is highly subjective and seldom persuasive. It is an argument, not an analysis.

The bottom line is that analyzing concurrent delays to separate critical paths is best done by using a technique to analyze delays that will identify the critical path of the Project every day. The correct analytical approach is discussed in Chapter 3 through 6.

Concurrent Delays to the Same Critical Path

A delay to the Project's critical path can have multiple causes. Assume that the first activity on the critical path of a highway Project is the demolition of the substructure of an existing bridge. As of the schedule update of June 1, demolition was to be finished on June 15. If the start of demolition was delayed from June 1 to June 16 because of a union strike and concurrently the Owner failed to obtain a permit, the two causes of delay are concurrent and run for the same period. The union strike is typically an excusable, noncompensable delay, and the lack of permit is typically an excusable, compensable delay. But recognize that this is not a concurrent delay but an example of concurrent causes for the same delay. Therefore, the analysis of the delay is simple: The critical path was delayed 15 days because of the late start for the demolition of the existing bridge. The only question that remains is the party responsible for the delay. In this case, both the Owner and the Contractor share responsibility. The question of a time extension and of compensability is now a legal one and must be viewed in the context of the Contract and the prevailing case law.

If both causes of the delay are concurrent but not of the same duration, then there may be an apportionment between the two causes. For example, if the causes both started on June 1 but the strike ended on June 10 and the Owner obtained the permit on June 15, then the first 10 days have a shared responsibility. The propriety for a time extension and compensation must be determined from legal precedent for these 10 days. The remaining 5 days of delay, however, appear to be attributable to the lack of a permit, and depending on the specific Contract language, these 5 days may well be both excusable and compensable.

While application of the concept of apportionment appears simple, there are complications. First, there must be a clear definition of what delays are compensable and noncompensable. Second, the analyst must be able to show when the delays started and ended to allow an apportionment. The analyst must review all contemporaneous documents to ensure that the analysis reflects the status of the Project when the delays were occurring. Proper research and documentation of what the parties knew as the delays unfolded will eliminate unsupported "but for" arguments that advance predisposed conclusions or desired results.

Chapter | three

Measuring Delays—The Basics

THE IMPORTANCE OF PERSPECTIVE

> "Reality is a question of perspective; the further you get from the past, the more concrete and plausible it seems."
>
> —Salman Rushdie, *Midnight's Children*

The length of a critical delay is often a question of perspective. Every analyst has a way of illustrating this point, but the classic example is the "ribbon-cutting" story. Consider a Project where in addition to all its other responsibilities, the Contractor must also provide the scissors for the mayor's ribbon cutting at the conclusion of the Project. The Architect rejected the Contractor's original scissor submittal (the Contract specified something larger and grander). The Project Manager shoved the rejected submittal to the bottom of her "to-do" pile, where it languished and was eventually lost. The Project ultimately finished late due to an error in the design of the structural steel. The error necessitated refabrication of steel, delaying the critical structural steel erection work.

At the ribbon-cutting ceremony, it quickly became apparent that the scissors had not been purchased. The Project Manager, at the last minute, ran to the local office supply store and bought the biggest, brightest pair of scissors she could find. She returned to the Project site just as the mayor was about to cut the ribbon. The proceedings were held up only a few seconds as she ran up to the entrance of the new water treatment plant.

After the ribbon-cutting ceremony, the Project Manager met with the Architect to close out the Project. The Contractor sought a time extension due to the steel design error. The Architect rejected the Contractor's request, stating that even without the steel design error, the formal opening of the Project would have been delayed by the lack of scissors to cut the ribbon.

The ribbon-cutting story points out the importance of perspective. Viewed solely from the end of the Project, the lack of a pair of scissors, and the flawed procurement process that caused them to turn up missing, appears to be critical to opening the Project. Given these facts, most of us quickly see the error in the Architect's logic, but what if the scissors are changed to aluminum tank covers? In response to the steel design error, assume that the Project Manager called the fabricator of the aluminum tank covers to let him know that the Project would be a little late, the delivery of the tank covers should be postponed. If the tanks weren't ready when the covers were delivered, they would have to sit before they could be installed and might be damaged. As the Project Manager recommended, the tank covers were delivered later than originally scheduled, but they finally arrived and were installed as the delayed tanks were completed.

In this revised story, the Contractor and the Architect again meet after the ribbon cutting to close out the Project. Again, the Contractor asks for a time extension, and, again, the Architect refuses the request. This time, however, the Architect denies the time extension because the "aluminum tank covers were late." We know all the facts, so we, again, see the error in the Architect's logic. But what if the facts weren't known? What if there was no written record of the Project Manager's conversation with the tank cover fabricator? Absent verifiable facts, is the Architect correct? Is the view from the end of the Project a relevant and valid perspective?

Perhaps it's only the view from the end of the Project that is problematic. What about the view from the beginning? Consider the same Project. As required by the Contract, the Contractor prepared a CPM schedule. The first schedule prepared on a Project is called the initial, baseline, or "as-planned" schedule. It typically depicts only the Contractor's plan, and it doesn't include "as-built" or actual performance information. The critical path of the Project as depicted in the Contractor's as-planned schedule proceeded through the erection of structural steel. During the close-out meeting, the Architect requires the Contractor to prepare an analysis that demonstrates that the steel design error introduced in the first paragraph of this chapter delayed the Project. The Contractor concludes that the best way to evaluate or "measure" the delay associated with the steel design error would be to simply "insert" this delay into its as-planned schedule. This is typically accomplished by developing a minischedule that models the work associated with the problem. This minischedule is called a fragnet (a fragmentary network). According to the Contractor, inserting a fragnet representing the steel design error into the as-planned schedule will show both that the error caused a critical delay and, when the schedule is recalculated and compared to the unaltered as-planned schedule, the magnitude of the delay. If we didn't know anything else, this approach might be acceptable.

But we do know something else. We know that a dispute developed between the Contractor and its steel erector. In fact, the steel erector abandoned the Project. The Contractor was not able to get another erector on site until after the prefabricated steel was delivered to the site. But the Contractor's analysis doesn't consider this problem. The only fragnet inserted into the schedule is the fragnet for the steel error, and this results in a Project delay. Is the Contractor entitled to a time extension for the steel design bust regardless of what else might be going on at the Project site when the delay occurred? Is the view from the beginning of the Project a relevant and valid perspective?

In addition to viewing critical Project delays from the end of the Project or the beginning, another perspective would be to evaluate delays as they occur—in other words, evaluate delays to the Project at the time the delay is experienced. This would avoid the ribbon-cutting error and would force the analyst to consider everything else happening on the Project when the delay occurs. But what if the analyst isn't brought in until long after the Project has been completed? Is the view from the time when the delay actually occurred still relevant and valid, even though the analyst knows what ultimately happens?

The answer to the questions raised so far in this chapter are at the heart of many of the disagreements among analysts regarding the best way to analyze delays on a construction Project. Does the analyst evaluate the delay from the perspective of the beginning of the Project, adding delays to the as-planned schedule, or from the end of the Project, evaluating only those delays that appear to ultimately hold up the Project's completion (the ribbon-cutting example)? Or should the analyst try to put herself in the shoes of the Project Manager at the time the delay occurs? It would be disingenuous to suggest that analysts are united in their answers to these questions. There is, however, an emerging consensus supported not only by many analysts but by case law, as well. First, a little background.

Perspectives—Forward Looking and Backward Looking

Though rarer now, there was a time when delays were sometimes analyzed by "impacting" the as-planned schedule. The as-planned schedule is usually defined as the earliest complete and Owner-approved Project schedule. It represents the Contractor's plan for completion of the Project before any work is actually done. If delays are analyzed using an "impacted as-planned" approach, the delay (or impact) is inserted into the as-planned schedule, and the schedule is then recalculated. The difference between the originally scheduled completion date and the completion date that results from impacting the as-planned schedule is the Project delay attributable to the impact. This type of analysis takes the position that delays should be measured from the perspective of the beginning of the Project, considering only the Project team's original plan and the delay being analyzed. The problems with this analytical approach will be discussed in more detail in another chapter, but here's what a judge had to say about this approach in *Haney v. United States* [30 CCF ¶ 70, 1891], 676 F. 2d 584 (Ct. Cl. 1982).

> We have found that [the contractor's] analysis systematically excluded all delays and disruptions except those allegedly caused by the Government.... We conclude that [his] analysis was inherently biased, and could lead to but one predictable outcome.... To be credible, a contractor's CPM analysis ought to take into account, and give appropriate credit for all of the delays which were alleged to have occurred.

Essentially, the judge's criticism was that the outcome of an impacted as-planned analysis, because it ignores everything other than the as-planned schedule and the delay the analyst is evaluating, was predetermined. It would overstate the delay, if any, associated with the inserted delay. Years of experience analyzing impacted as-planned analyses have confirmed this judgment. They very nearly always overstate the Project delay, predicting Project delays well beyond the actual Project completion date. On this basis, an analysis of delays based solely on the perspective from the beginning of the Project and employing an impacted as-planned analytical technique is flawed and to be avoided.

The logical opposite of an impacted as-planned analysis is the "collapsed as-built." Again, the problems with this analytical approach are discussed in another chapter, but a discussion concerning perspective is appropriate here. Stripped to its essentials, a collapsed as-built analysis is performed by first identifying the "as-built schedule" for the Project. This is essentially a schedule showing how a Project is actually constructed. It is not a schedule that ever existed on the Project, though it is theoretically composed of actual Project events. The analyst creates the "as-built schedule" after the Project is completed. The next step is to identify the delay to be analyzed. Note that this approach is a little like the tail wagging the dog. The delay must first be identified before it can be analyzed. The analysis is performed by removing the delay from the as-built schedule and then rerunning the schedule to see what happens. If the collapsed schedule shows an earlier Project completion date, then the conclusion would be that the delay that was removed was responsible for a Project delay equivalent to the improvement in the Project completion date associated with the collapsed schedule. This analysis presumes that delays are best analyzed from the perspective of the end of the Project.

Setting aside the questions concerning the mechanics of a collapsed as-built analysis, consider what it means. Essentially, the collapsed as-built approach is based on the assumption that all that matters is what *happened,* not what was planned. To understand the problems with this assumption, consider the following example. A Contractor is tasked with excavating a 100-foot rock face and then lining the face with concrete. Excavation began, and the Contractor immediately encountered a problem. It turns out that a fault zone ran through the area of construction. This fault zone was oriented in such a way that as the Contractor removed rock, the rock face that was left tended to slip into the excavated area. This was not only dangerous, but it prevented the Contractor from excavating the planned 100-foot rock face. The Owner and the Contractor met to discuss the problem, and they decided to pin the rock face with rock anchors as the face was excavated in 10-foot lifts. Also, the Owner decided that the concrete lining had to be constructed before the next 10 feet of the rock face could be excavated.

At the conclusion of the Project, the Contractor asked for a time extension to cover the additional time it had expended excavating and lining the rock face in

10-foot lifts as opposed to all at once, as planned. The Owner responded with a collapsed as-built analysis showing that the only delay was the time required to install the rock anchors, which had not been contemplated in the original design. The rock excavation and concrete liner were not "delays," since this work had always been required.

The fallacy of the Owner's analysis was that in addition to the rock-anchor delay, the Contractor was also delayed because the Contractor built the Project in 10-foot lifts rather than all at once, as planned. Because the Owner's delay analysis considered only what happened (the as-built schedule), it could not quantify delays associated with deviations from the Contractor's plan. And this is the essential failure of any analysis based solely on what happened or solely on the perspective from the end of the Project.

If the perspectives from the beginning of the Project and the end of the Project are flawed as the logical basis for analyzing delay, the only perspective remaining is to analyze the Project at the point where the delay actually occurred. An analysis based on this perspective has a name: contemporaneous analysis. Before discussing how such an analysis might be performed, consider this judge's decision.

> Mr. Maurer, appellant's expert, testified about the critical delays to the Project.... The analysis about the critical delays was based on appellant's original schedule, the schedule updates, the daily reports, Project correspondence, and the contract documents. Mr. Maurer described his analysis as a step-by-step process, beginning with the original schedule and proceeding chronologically through the Project, updating the sequence at intervals to see what happens as the Project progressed [(tr. 262) ASBCA No. 34, 645, 90–3 BCA ¶ 12, 173 (1990)].

A second judge's decision is also relevant to this discussion.

> In the absence of compelling evidence of actual errors in the CPMs, we will let the parties "live or die" by the CPM applicable to the relevant time frames [Santa Fe, Inc. VABCA No. 2168, 87–3 BCA ¶ 20677].

Taken together, these decisions have an important message: When analyzing delays, it is important to evaluate Project events relying on the documents in use on the Project at the time the delay occurred. Of particular importance are the original schedule (as-planned schedule) and schedule updates. These contemporaneous schedules form the foundation of a credible analysis of Project delay. To put it in the judge's terms, absent actual errors, the parties will "live or die" by the Project plan and the events as depicted in the schedules that were in place at the time the delay occurred.

In summary, the only valid perspective for the analyst is to adopt a view of the Project contemporaneous to the delay itself—not from the beginning of the Project or the end of the Project. Now, let's see how such an analysis would be performed.

USE THE CONTEMPORANEOUS SCHEDULE TO MEASURE DELAY

In Chapter 1, we defined the contemporaneous schedule, and now we are going to explain why the contemporaneous Project schedule should be used to identify and measure delay. As stated in Chapter 1, a contemporaneous schedule is the Project schedule, which typically consists of the baseline schedule and schedule updates that were used to manage and construct the Project.

It is necessary to use the contemporaneous Project schedules in an analysis of Project delays because they are essentially snapshots of the Project's status at specific moments in time. As snapshots in time, the schedule updates identify what work has been done and the order in which it was completed. Perhaps most important, the contemporaneous Project schedules also capture changes made to the construction plan in reaction to ever-evolving Project conditions.

The contemporaneous Project schedules are the preferred tool to measure Project delay because they were used by the Project participants to manage and construct the Project and provide the most accurate picture of the plan to complete the Project at a moment in time based on the known Project conditions. These attributes provide the analyst with a real-time perspective of the Project and enable the analyst to identify, measure, and assign Project delay using the same information available to the Project participants at that moment in time. By using the contemporaneous schedules and updates, and by tracking delays as they occur throughout the Project, there is no need to attempt to inject information that is known at a later date. Information is incorporated into the analysis in a contemporaneous fashion throughout the analysis. In other words, if the analyst knows that something significant occurs in month ten of the Project, when month ten is being analyzed with the schedule updates from that period, it is then that the information is incorporated.

DO NOT CREATE SCHEDULES AFTER THE FACT TO MEASURE DELAYS

In the absence of contemporaneous schedules, an analyst may feel it would be acceptable to create a schedule after the fact that he believes portrayed the Contractor's intended construction plan. Although the analyst may rely on the Project documentation and his own knowledge of the type of construction being performed, creating a schedule for the sole purpose of measuring and identifying Project delay after the Project is complete negates the objectivity of the analysis. Even though the analyst may do his best to remain objective, the fact remains that he would have complete knowledge of all the facts that pertain to the construction of the Project and all the problems that were encountered during the course of construction. This after-the-fact perspective would influence the after-the-fact schedule and ignore, or at least significantly diminish, the contemporaneous knowledge and thinking of the Project participants before and during the Project.

The analyst may argue that creating an after-the-fact schedule will allow the analysis to be more precise, containing all the facts of the Project. However, it should be noted that schedules created after the fact should not be relied on because there is more than one way to build a Project, and the analyst may choose a different approach than the original planner. And even slight differences in a schedule could affect the results of an analysis. Using a schedule created after the fact to measure and identify Project delay, however, has two weaknesses: The schedule does not depict the original construction plan, and the schedule may include predetermined conclusions concerning delays. There are many ways a construction plan can be represented in a schedule. Preparing one after the fact merely shows the plan the analyst believes was intended. This does not make it correct.

When possible, it is always best to use the contemporaneous Project schedules to measure Project delay. While the analyst may make very minor modifications to the contemporaneous schedule to account for obvious errors, such changes must be made judiciously. This subject is addressed in more detail in Chapter 5.

WHAT TO DO WHEN THERE IS NO SCHEDULE

There are instances when contemporaneous Project schedules cannot be used to measure Project delay. In those cases, the Project schedules either were not developed and maintained or the analyst might determine that the contemporaneous schedules did not accurately depict the plan to construct the Project and would not be a reliable tool to measure Project delay.

When a contemporaneous schedule is not available to measure critical Project delays, the analyst should use an as-built analysis (discussed later in the book) to identify the critical delay, which is based on an as-built diagram. An as-built diagram is prepared using the Project's contemporaneous documents. Those documents may include, but are not limited to, timesheets, inspector daily reports, meeting minutes, Project photos, and so on. When complete, an as-built diagram should depict the order and durations of the Project work activities. The analyst would then proceed as described in the as-built analysis section of this book.

WHAT IS THE AS-PLANNED SCHEDULE?

As its name implies, the as-planned schedule is the schedule created either before construction begins or very early in the first stage of the Project and should represent the Contractor's plan to construct the Project based on the information it had at the time of bid. Most projects have some form of an as-planned schedule. The as-planned schedule is most often the Contractor's original schedule submitted in accordance with the Contract documents.

Forms of As-Planned Schedules

As mentioned in Chapter 1, Project schedules can take many forms. Depending on the construction Project's size and level of complexity, it may be a written narrative of the Contractor's plan, a simple bar chart, or a detailed CPM schedule. Don't dismiss a schedule merely because you believe it lacks detail or because it is a bar chart schedule instead of a detailed CPM schedule. The most important characteristic to remember is that you are trying to identify the earliest and most accurate representation of the Contractor's construction plan.

Identifying the As-Planned Schedule

The analyst must carefully choose the schedule that best represents the Project's as-planned schedule. For example, the Owner may have included a schedule with the bid documents as a guide for the Contractors bidding the work. However, it may be erroneous to use the Owner's version as the as-planned schedule for the Project because the Contractor may plan to construct the Project in a different sequence or manner. Typically, the Contractor's initial schedule submission serves as the Project as-planned schedule. It is common for the Owner's representative to send the initial schedule back to the Contractor for changes or corrections. If the Contractor submits the schedule a second and third time until it is finally accepted by the Owner, chances are that the third schedule submission best represents the as-planned schedule for the Project. In some cases, the Contract may require Owner approval or acceptance of the initial schedule submission as a method of establishing the Project's as-planned schedule.

Reviewing the As-Planned Schedule

After identifying the schedule that most reasonably represents the Contractor's original planned sequence of work, the analyst should review that schedule for sequencing and feasibility. A note of caution: Often the analyst, or the person assessing the Project for delays, reviews the Contractor's schedule and decides that it did not correctly portray (1) the sequencing of the Project or (2) the durations for the activities. The analyst might then change the schedule to reflect his judgment about the errors. The analyst should avoid this practice at all costs. If there are minor errors or inconsistencies in the Contractor's as-planned schedule, they will be accounted for during the analysis of the delays. The decision as to whether or not the Contractor's schedule was practicable is a highly subjective one. Therefore, it is far better to give the Contractor the benefit of the doubt than to disallow, ignore, or even change the schedule.

WHAT IS AS-BUILT INFORMATION?

As-built information is the actual start and finish dates of the Project work activities. One of the best places to find as-built information is in the Project schedule updates, because the periodic updates typically record the dates that specific activities start and finish. Even if the updates contain the Project's as-built information, it is always wise to verify information in the updates, using as many independent sources as possible. For example, the analyst might review the Project daily reports to verify that specific activities started and finished on the dates indicated in the updates.

If the updates do not provide the information required or do not exist, then the analyst has no alternative but to prepare his own as-built diagram, using the contemporaneous Project documents. These documents should be reviewed for possible sources of as-built information:

- Project daily reports
- Project diaries
- Meeting minutes
- Pay requests/estimates
- Inspection reports by the Designer, Owner, lending institution, and so on
- Correspondence
- Memos to the file
- Dated Project photos

Note that in the creation of an as-built diagram, the analyst should document every day that work is recorded for each activity. It is not enough to merely record the start date and then the finish date. While the start and finish dates are extremely important, the determination of whether work was continuous or interrupted may also be significant.

In some instances we have used the term as-built "schedule" with the word *schedule* in quotation marks. This is intentional. An as-built is not truly a schedule but rather a chronicle or history of when specific work is performed.

THE IMPORTANCE OF THE CRITICAL PATH

As stated earlier, the critical path of the Project is the longest path of work activities through the network. Due to the fact that the critical path is solely responsible for determining the date that the Project can finish, it is logical that only delays to the critical path will delay the Project.

In many instances, as the Project progresses and Project conditions change, the critical path can shift to other work paths. This shifting of the critical path will occur when other work paths are either delayed or changed to a point that they now are on the longest path, become critical, and because they are now critical, determine when the Project will finish. If the contemporaneous Project schedules are properly maintained and updated, they will capture the shifting of the critical path as the Project conditions change. Therefore, reliance on the contemporaneous Project schedules to identify the critical path at specific moments in time throughout the Project's duration will enable an analyst performing an analysis of delays after the fact to accurately identify the work activities that were actually critical and correctly assign Project delay to the responsible party.

The analyst should never assume that the critical path is static and remains unchanged throughout the Project. Experience teaches us that the reverse is more often true. Changes in the critical path are normal and should be expected.

UNDERLYING PRINCIPLES FOR ANALYZING A SCHEDULE FOR DELAYS

In most of the methods for analyzing delays that are discussed in this book, the analyst should be following certain general principles during the analysis. Obviously, the specific steps in the analysis will vary, depending on the nature of the available information. In general, the following approach is used during the evaluation of delays.

The first step is a determination of the Contractor's as-planned schedule. For purposes of this discussion, we will use a simple bar chart to demonstrate. Figure 3.1 is the Contractor's as-planned schedule for a Project. To determine what occurred on the Project, the analyst will create an as-built diagram or chart. For this example, Figure 3.2 represents the as-built chart for the actual progress of the work as it occurred on the job.

At this stage of an analysis, there often is the inclination to compare the as-planned schedule with the as-built chart and attempt to reach conclusions concerning what was delayed. The comparison of the as-planned schedule and the as-built chart is shown in Figure 3.3. When we look at Figure 3.3, we might conclude that activity D was delayed by 25 days, from day 35 to day 60. If we know that the Subcontractor performing the work on Activity D showed up the day it started, we might conclude that the delay of 25 days was caused by the late arrival of the Subcontractor—and this conclusion may be totally incorrect.

When we are analyzing delays, we need to start at the beginning of the Project and move through the Project chronologically. As we do this, we should be able to identify each delay as it occurs and update the schedule accordingly. Let's do that with the information we have, using Figure 3.4. When we look at Figure 3.4, we start with the first activity, Activity A, and we compare the

FIGURE 3.1

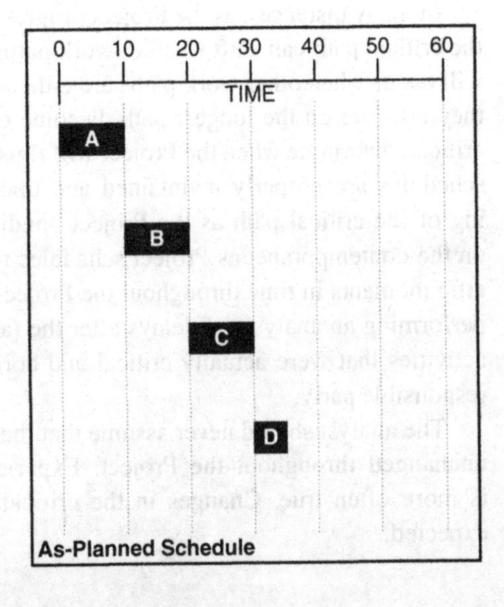

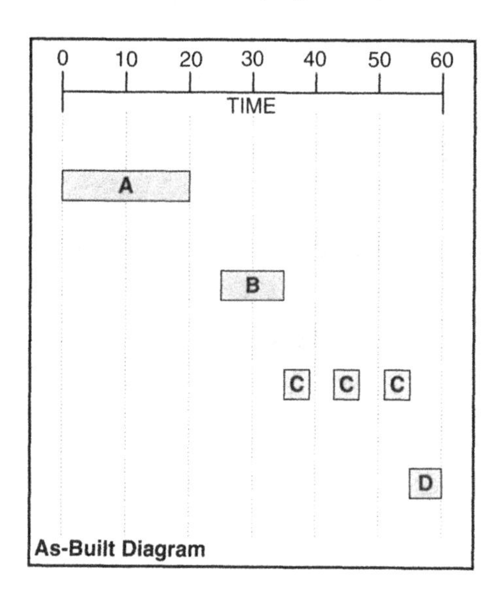

FIGURE 3.2

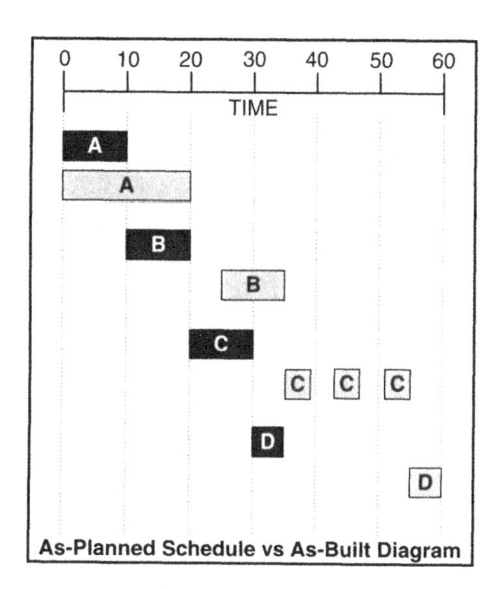

FIGURE 3.3

as-planned schedule for activity A with the as-built information. We can readily see that Activity A started on time but took twice as long to perform as was planned. As opposed to 10 days, Activity A took 20 days. From that we should conclude that Activity A was delayed for 10 days as a result of an extended duration.

Now that we have identified the delay to the first activity, we need to update our schedule for the effect of the delay to Activity A to the remaining planned activities. In Figure 3.5 we show the original planned Activity A and the as-built for Activity A. We then move or update the planned start of Activity B later because the start of Activity B depended on the finish of Activity A. We have

FIGURE 3.4

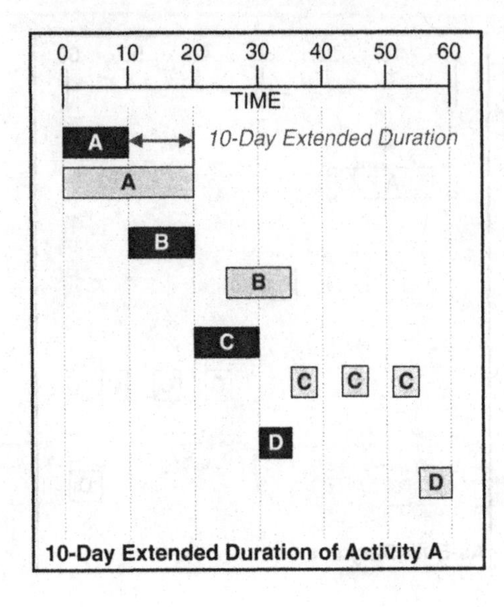

10-Day Extended Duration of Activity A

FIGURE 3.5

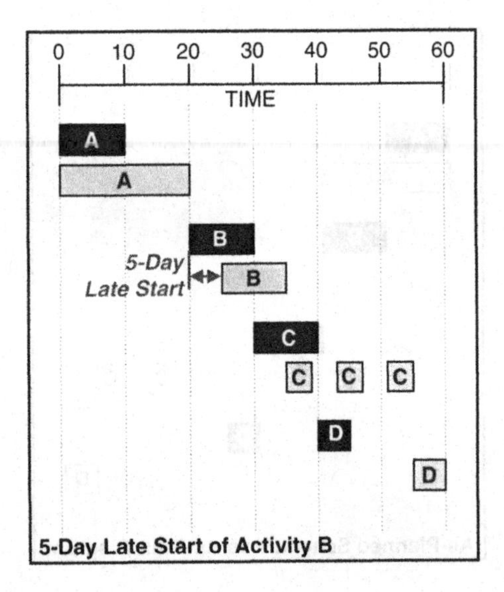

5-Day Late Start of Activity B

updated or "bumped" the start of Activity B because of the delay to Activity A. Now that we have updated the planned schedule for the actual performance of Activity A, we can look at Activity B to see whether it affected the completion of the Project. As we see in Figure 3.5, Activity B started 5 days later than it should have, based on the late finish of Activity A. Therefore, we have identified our second delay: Activity B was delayed 5 days because of a late start.

The process repeats itself for each subsequent activity. In Figure 3.6, we update the schedule. Based on the two preceding delays to Activities A and B, now we can look at Activity C to determine if it experienced any delay. As we can see in Figure 3.6, Activity C was delayed 10 days. Since we plotted our

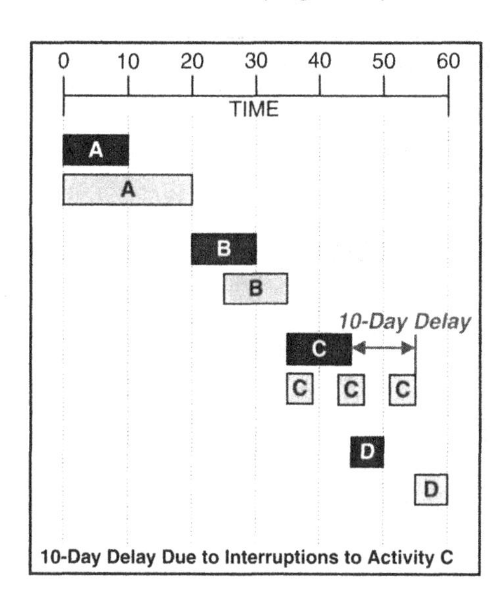

FIGURE 3.6

as-built chart as precisely as we could, we also know that the 10-day delay was the result of two periods where no work was performed on the activity. The last step is identical to the preceding steps. We update the schedule once more and look at Activity D.

Figure 3.7 shows that Activity D caused no delay to the Project. This conclusion is very different from the one we might have initially reached if we had solely looked at a comparison of the as-planned schedule and the as-built chart. Figures 3.8 and 3.9 summarize the results of our analysis.

In one form or another, this stepwise approach starting at the beginning of the Project should be used in almost all analyses of delays. We also note that precise charting of the as-built information is very helpful when the analyst

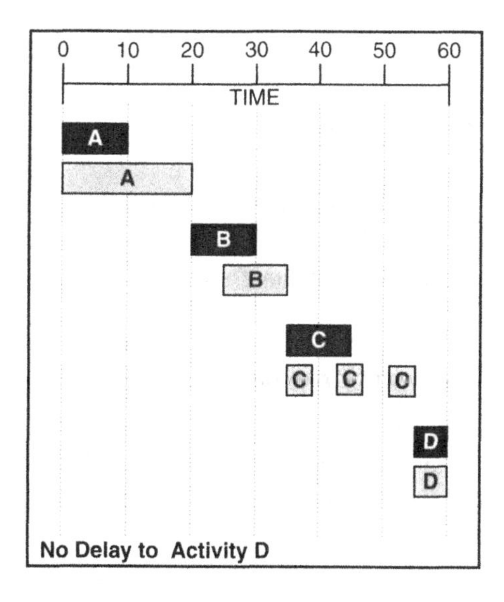

FIGURE 3.7

FIGURE 3.8

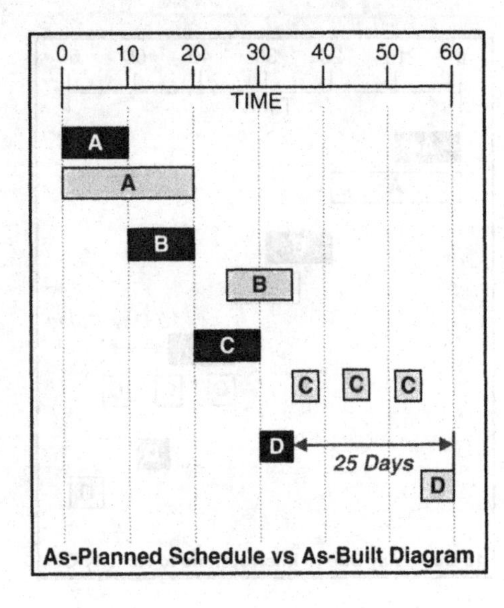

FIGURE 3.9

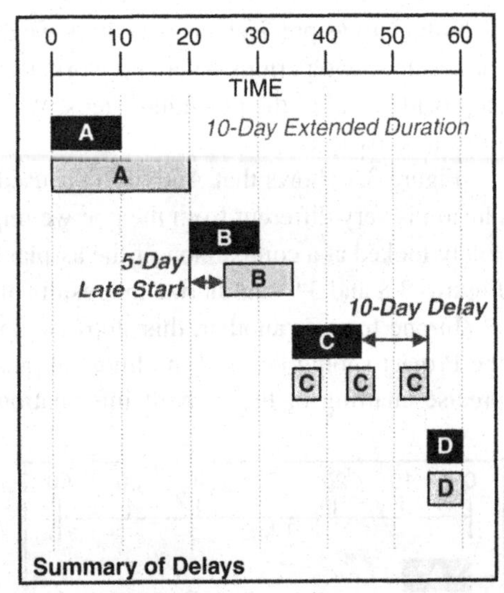

moves to a determination of the cause of the delay or the liability for the delay. For instance, by knowing that Activity C was interrupted, as opposed to just taking longer, you can review the available documentation for reasons why the work would have stopped for those two periods.

THE UNIQUE POSITION OF SUBCONTRACTORS

Because the duration of a Project can only be extended by delays to activities on the Project's critical path, a Contractor's performance period can only be extended when the Project experiences a critical delay. However, this is not necessarily the case for the performance period of a Subcontractor.

While certain Subcontractors may have work to perform during the entire Project period, it is more typical for a trade Subcontractor to have its work become available sometime after the Project work has begun and be required to complete its subcontract work before all the Project work can be or has been completed. As a result, the trade Subcontractor's work may or may not ever show up on the Project's critical path. Still, delays that extend the Subcontractor's work will require the Subcontractor to be on the job longer, thus extending the Subcontractor's performance period.

For example, a masonry Subcontractor may not be able to begin its subcontract brick veneer work until the exterior sheathing has been installed on a building that is expected to take 18 months to construct. The as-planned schedule may show that after the exterior sheathing had been completed on one elevation, the masonry work can begin and will take three months to complete, followed by other exterior and interior finish work. Because the masonry work is planned to follow the expected pace of the exterior sheathing installation, the masonry may never show up on the critical path.

Presuming that the exterior sheathing work experiences delays, the exterior sheathing work is more likely to show up on the critical path than is the masonry work. Yet, the masonry work will be delayed because it will not be able to proceed at the pace planned. As a result, the masonry work takes five months to complete instead of the planned three-month period. The mason claims that its performance period was extended by two months through no fault of its own. It requests additional compensation for extended overhead costs and other delay damages.

In this example, an analysis of delays along the critical path of the Project may not support the mason's request. However, from the facts presented, it is evident that the mason's performance period was extended, and, depending on the provisions of its subcontract, the mason may be able to recover the delay costs caused by others.

If a critical path analysis of the Project does not support the mason's claim, what type of analysis should be performed to determine if the mason's claim has merit? In the preceding simple example, the answer appears straightforward. For most Subcontractors, however, their work is integrated with many aspects of the Project work. Often, the relationships among the various work activities of the Prime Contractor and the various Subcontractors are more complex than the preceding simple example. To complicate matters, when the Subcontractor's work is not on the critical path of the Project, unless constrained in some other way, it will have float. Therefore, any analysis of Subcontractor delays will also involve an examination of the Subcontractor's obligations with respect to activity float.

When investigating potential delays to a Subcontractor's performance, we begin by evaluating the performance requirements of the subcontract. The objective of this evaluation is to determine the period of performance for which the Subcontractor is obligated under the terms of the subcontract. Often, a Subcontractor is required to perform its work according to the Project schedule. Typically, the Prime Contractor reserves its right to modify the schedule as necessary to complete the work in a timely fashion. The subcontract will then require the Subcontractor to perform according to these modifications as well. Because these modifications are not known at the time of the subcontract agreement, there is a certain expectation that the parties have regarding the Subcontractor's performance period. This expectation will usually be a product of the particular negotiation that led to the signing of the subcontract.

While the Subcontractor typically takes on some risk regarding the Prime Contractor's right to modify the schedule, this risk is typically not without limits. The Contract CPM schedule will identify early and late dates for all of the Subcontractor's work activities. When obligated to perform according to the Contract schedule, it is reasonable to conclude that the Subcontractor is obligated to be on site from the projected early start date of its first activity to the late finish of its last work activity. This conclusion recognizes that the work activities do not need to be performed on the early dates for the Project to complete on time. Thus, performance of the work within the early and late date ranges are foreseeable because such performance is, in fact, "according to" the schedule. This remains true, even if such performance affects the continuity of the trade Subcontractor's work activities.

The Prime Contractor, however, may argue that the subcontract allows for modifications to the sequence and duration of the work. Here again, there may be a question as to the degree such modifications are foreseeable. It may be reasonable for the Prime Contractor to argue that because it is responsible to the Owner to complete the Project on time, it must continually assess progress against its plan to complete the work. When the actual progress differs from that planned, it must modify the sequence and duration of future work to ensure an on-time completion. As a result, modifications to the schedule that change the sequence and duration of the Subcontractor's work activities within the original Project performance period may be foreseeable. Much of this argument, however, will depend on the nature and extent of these changes. Unlimited modifications to the sequence and duration of the work are generally not anticipated by the parties.

Through careful evaluation of the subcontract and the understandings and circumstances leading to the subcontract agreement, the Subcontractor's planned performance period can be determined. Unlike the Prime Contractor's contract performance period, which is generally expressed in the Contract, the parties may be unable to agree on the subcontract period of performance. In such cases, the parties will prepare their respective arguments based on the subcontract performance period they believe to be correct.

Once the subcontract period of performance has been established, a comparison to the Subcontractor's actual performance period provides a measure of the total delay experienced by the Subcontractor. But this is only the beginning of the story. Next, it is necessary to determine the causal link between the actions of the parties and the delays incurred in order to determine if the Subcontractor's delays were caused by others.

In order to determine the cause of any delays to the subcontract period of performance, it is necessary to determine the critical path of the Subcontractor's work. The critical path of the Subcontractor's work is the longest path of activities leading from the first work activity to be performed by the Subcontractor to the last. This path may consist of some or all of the Subcontractor's work activities, as well as work activities performed by others. Because these activities are integrated within the entire schedule network, the analyst cannot simply isolate the Subcontractor's work activities and evaluate the paths among these in a vacuum. Many of the Subcontractor's work activities will be driven by activities being performed by others, and all of these relationships must be considered in the analysis.

As a result of these complexities, it may not be possible to determine the longest path between the Subcontractor's start and end points through electronic analysis of the schedules. As an alternative, it may be necessary to determine the Subcontractor's critical path through a detailed evaluation of the Subcontractor's daily work progress. This evaluation is similar to the As-Built Delay Analysis discussed in Chapter 6.

This process begins with the preparation of a detailed as-built diagram that tracks all of the Subcontractor's actual performance. This performance is then compared to all of the available planned performance information. To begin with, the analyst determines if the Subcontractor was able to meet planned durations for its work and, if not, why not. Did the Subcontractor provide sufficient resources to accomplish the work? Was the Subcontractor given access to the work as anticipated or was it required to perform its work under conditions that differed from those it expected to encounter? Was the Subcontractor in control of the pace of the work, or was something else controlling the pace?

We also look at the sequence of the Subcontractor's work to see if it differed from that planned and, if so, why. As the work progresses, we also consider all of the subcontract work remaining and the precedent requirements of that work. For example, if the Subcontractor was delayed in one area of the Project, was there other available work for it to perform?

By evaluating the Subcontractor's as-built work performance moving forward through the Project and considering the work that remains, we can determine the critical path of the Subcontractor's work and those factors that extended the work along this path.

When a Project is managed by a well-thought-out and periodically updated CPM schedule, the analyst has many tools at his disposal to help determine the delays to the Prime Contractor's performance period. To begin with, the Contract will usually state the Contract performance period, and the longest path through the Project can easily be determined by the scheduling program. However, in the case of delays to the Subcontractor's performance period, these tools are less effective. As a result, we must apply a more in-depth knowledge of both the subcontracting process and of the Project management process in order to determine the most appropriate way to resolve disputes related to Subcontractor delays.

Chapter | four

Delay Analysis Using Bar Chart Schedules

Later in this book, we will address how to perform a delay analysis in the ideal circumstances, when a detailed CPM schedule was created as the original as-planned schedule for the Project. However, many Projects are scheduled using a bar chart. For Projects with many interrelated activities, a bar chart is not as desirable as a CPM schedule, but one can still perform a meaningful and accurate analysis using a bar chart. However, as the level of existing detail and the quality of information decrease, the delay analysis becomes more subjective. This chapter describes how one goes about performing a delay analysis when the Project schedule is a bar chart. There is nothing inherently wrong with scheduling a Project with a bar chart. Bar charts were in use long before the Critical Path Method was ever created. As some professionals are quick to point out, the Empire State Building was scheduled with a bar chart and not a CPM. In fact, a detailed bar chart can provide almost as much information as a CPM schedule.

Figure 4.1 is a simple bar chart for the construction of a bridge. Though it does not contain a significant number of activities, it does show the general sequence of work for the construction of the bridge. Based on this simple bar chart as a starting point, the Project Manager could easily define each activity in more detail.

Figure 4.2 is a more detailed bar chart for the one shown in Figure 4.1. This more detailed bar chart more clearly defines the proposed work plan of the Contractor. In this bar chart, each major activity is broken down into the work on the respective piers and spans, which provides the analyst and the Contractor and Owner with a detailed picture of the plan for construction.

With very little effort, the Project Manager or Project Scheduler can modify the bar chart in Figure 4.2 to show the interrelationships among the activities, as shown in Figure 4.3. Study of this schedule shows that with minimal effort, one could produce a CPM schedule for the Project. In fact, most bar charts for

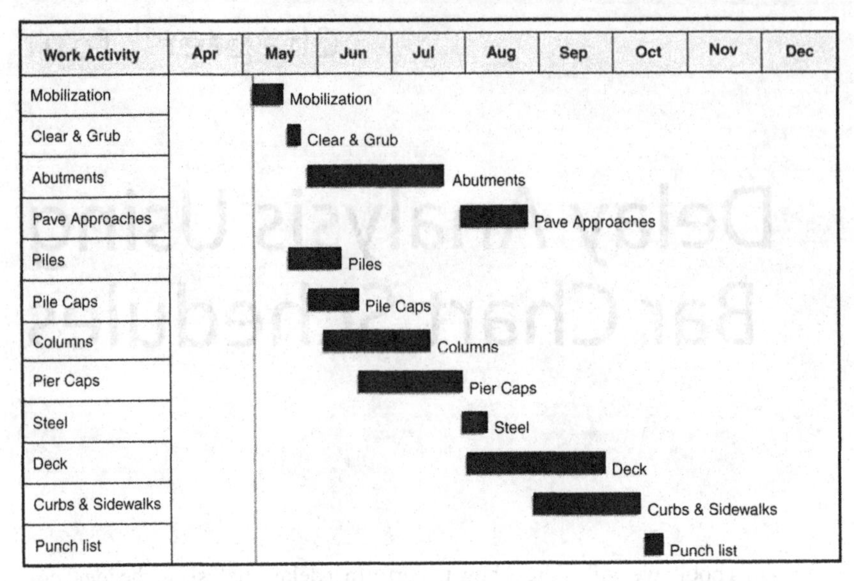

FIGURE 4.1

Projects do not contain as much detail as that in Figure 4.3, and often not even as much as the bar chart in Figure 4.2. In general, most bar charts suffer from the following major shortcomings that diminish their usefulness as a management tool and their effectiveness in measuring delays:

- Lack of detail—too few activities for the amount and complexity of the work
- No indication of the interrelationships among the activities
- No definition of the critical path of the Project

Obviously, these weaknesses hamper the ability of the analyst to perform a delay analysis, but they do not make it impossible. If nothing else, the bar chart is helpful in that it defines the plan for constructing the Project, and it can be used as the basis for an analysis of delays.

DEFINING THE CRITICAL PATH

The first step in analyzing a bar chart is to define the critical path. Every Project has a critical path. Even if a Project does not have a CPM schedule, it does not mean that it does not have a critical path. The following definitions illustrate this point.

Basic CPM

In CPM scheduling, the drafter of the schedule prepares a logic or network diagram. Once durations are assigned to the activities in the diagram, the critical path is calculated. This is a purely arithmetic process. The definition of the

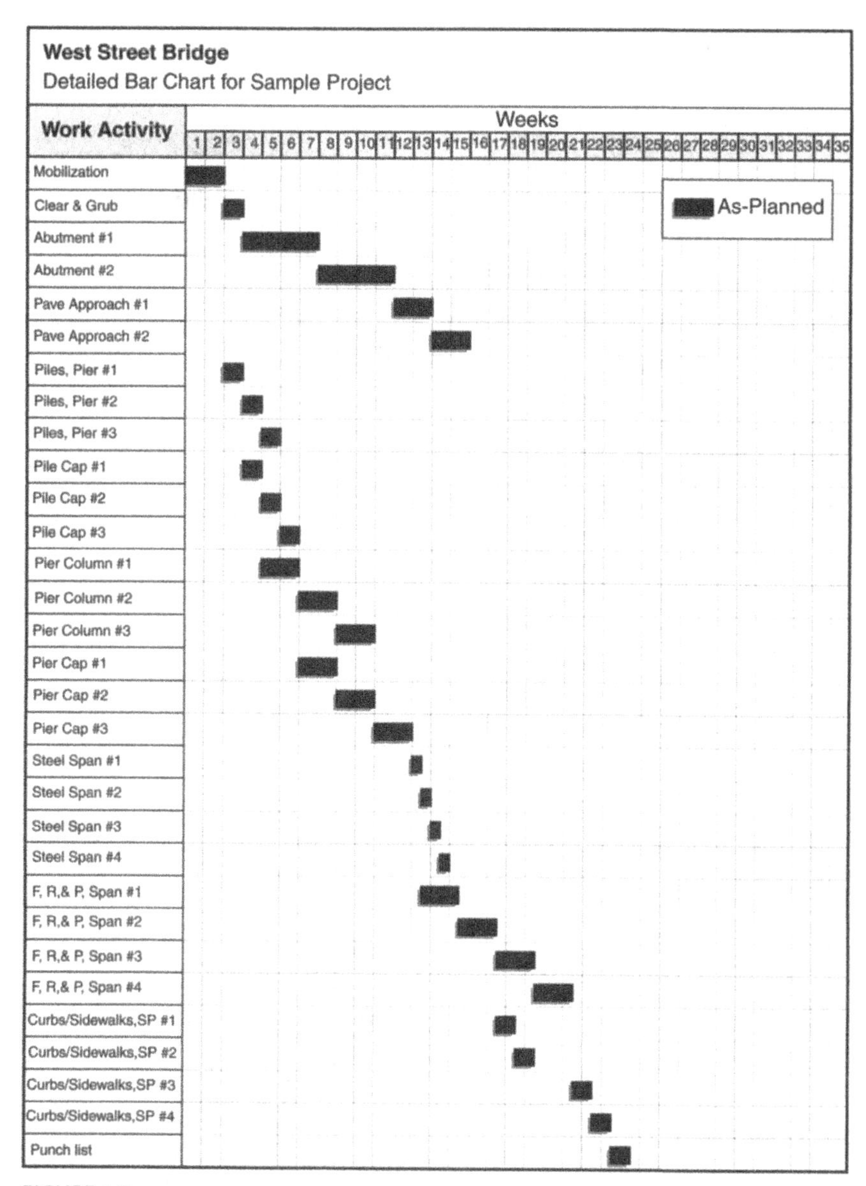

FIGURE 4.2

critical path is *the longest path of work activities through the network diagram.* Thus, the path with the longest duration defines the shortest possible duration for the Project. The Project cannot finish until every path has been traversed. A delay to an activity on the critical path will delay the completion date of the Project. Whether the critical path is defined in a CPM schedule or a bar chart, every Project has a series of interrelated activities that will control the Project completion time.

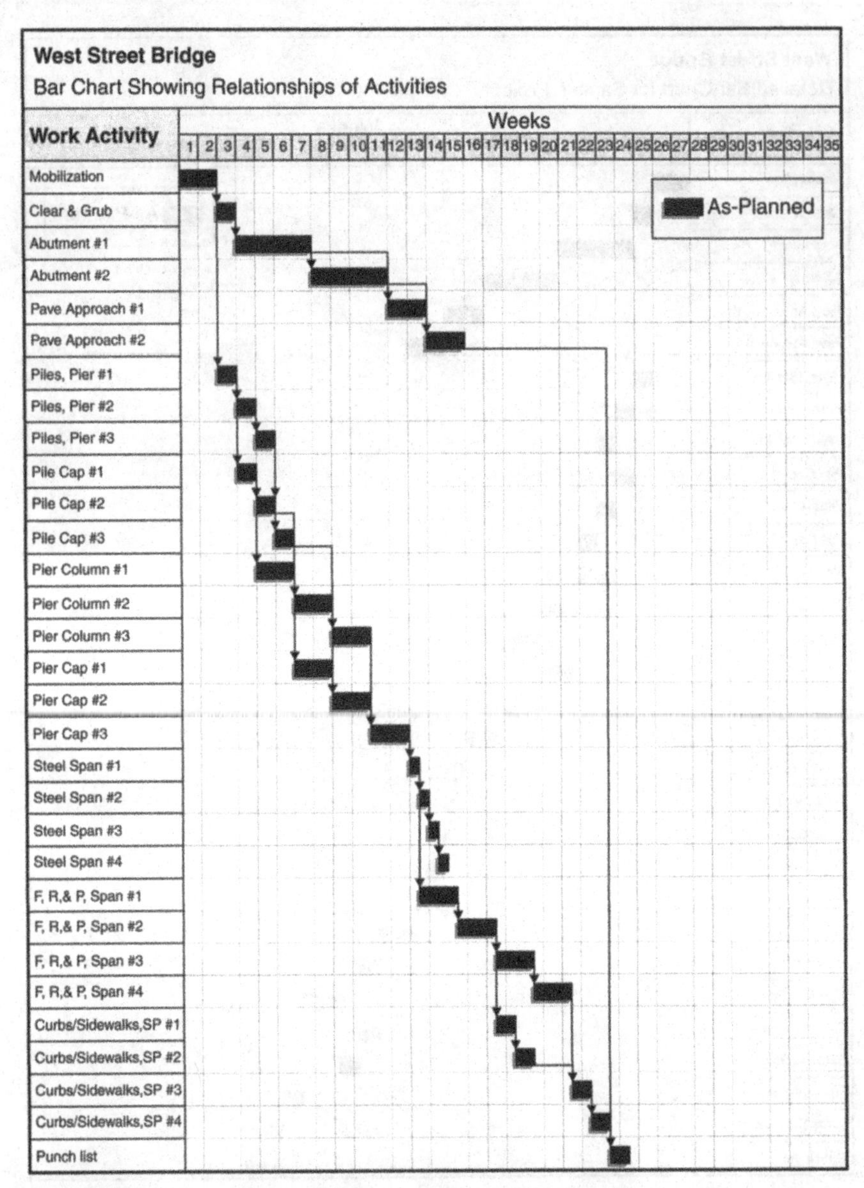

FIGURE 4.3

Identifying the Critical Path on a Bar Chart

Having established the fact that a critical path exists in a bar chart schedule, the delay analyst should identify this critical path. The analyst must review the bar chart in detail for obvious conclusions about the sequence of work. These conclusions may be based on Project documentation that might clarify the thought process that went into creating the bar chart or defining the planned work sequence. Documentation that can be helpful includes the Contract documents (which may dictate staging or phasing), preconstruction meeting minutes, internal Contractor documentation, or Project correspondence.

Practical knowledge of the type of Project and the physical construction requirements is also necessary to reach a reasonable conclusion about the Project's critical path. For example, to analyze a bar chart of a high-rise structure, the analyst must know that interior finishes usually cannot start until the building is "dried-in," that the normal sequence of the progression of trades is from the bottom up, and that it is common for trades to follow behind one another as the building progresses upward, instead of waiting until the preceding trade completely finishes its work.

A note of caution: The analyst should resist the temptation to interpret the bar chart schedule based solely on his or her own experience. Merely because one has performed work in a particular sequence in the past does not mean that the Contractor on the Project being analyzed has planned it the same way. Unless the bar chart is extremely brief, the analyst should be able to glean some indication of the overall plan and sequence of activities to determine the critical path.

Referring to Figure 4.3, we will define the critical path for the sample bridge Project. The critical path starts with the mobilization activity, with a duration of two weeks. This is obvious, since no other activity is scheduled to occur during this period. We show this first critical activity in Figure 4.4. The next two activities on the schedule are the clear and grub activity and the piles at pier #1. In reviewing the sequence of activities, the clear and grub activity is related to the abutment and approach work.

The abutment and approach work is scheduled to finish well before the end of the Project and does not appear to be related to the schedule of activities for bridge construction. The analyst can determine from the Contract documents that the approaches are not concrete as is the bridge deck; consequently, there is no physical reason to coordinate the concrete placement for the bridge with the approach construction. The only possible relationship could be the ability to move the concrete placing equipment onto the bridge superstructure. However, since the schedule reflects that the deck work is to start before workers complete either of the approaches, the analyst concludes that the equipment can be located independently of the approach work. Therefore, it appears that the abutment/approach path is not on the critical path for the Project.

The critical path must move through the piles and piers. When viewing this bar chart, the analyst sees that the work is "stair-stepped" through the specific activities for each pier. Thus, the piles at pier #1 are complete and the piles at pier #2 are starting, while the pile cap at pier #1 is concurrently constructed. Based on the graphic representation, the critical path appears to follow these activities (shown in Figure 4.5):

==> Piles—Pier #1
==> Pile Cap #1
==> Pier Column #1
==> Pier Cap #1

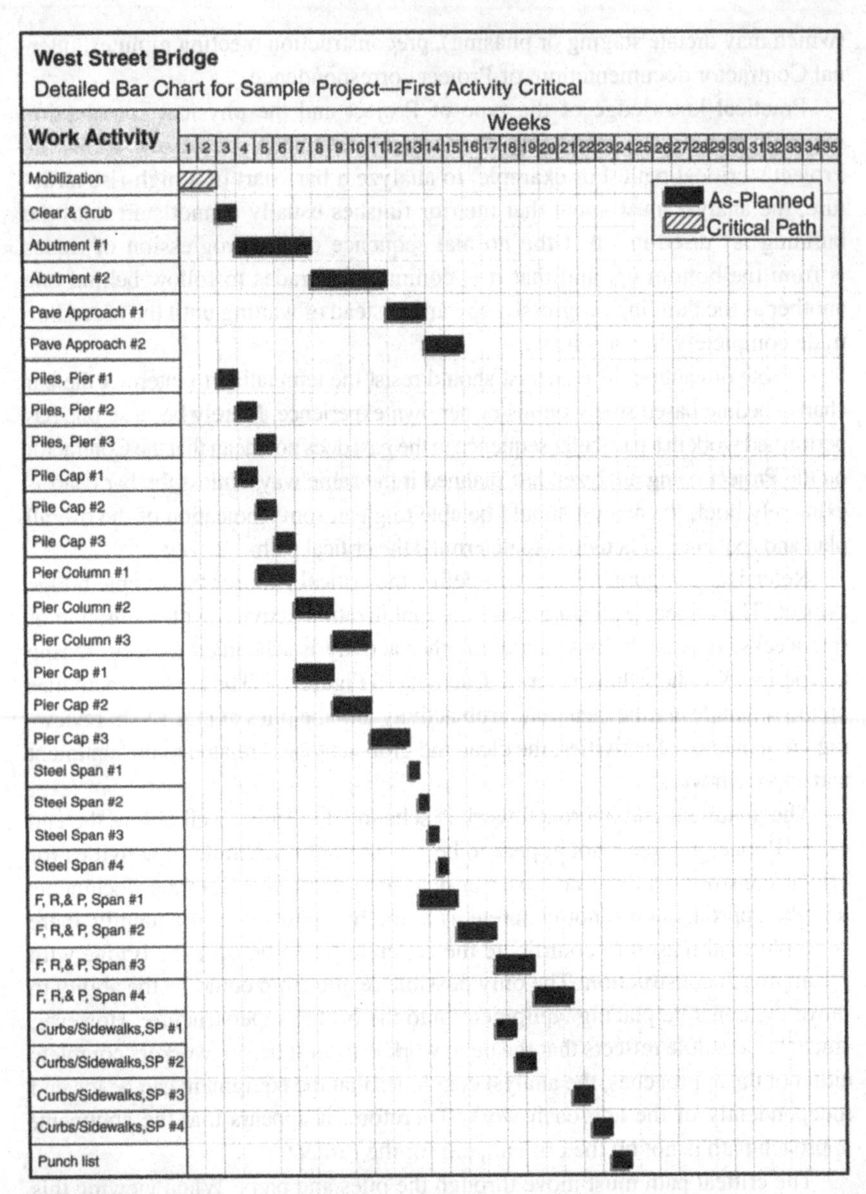

FIGURE 4.4

At this point, the analyst recognizes that the pier columns and the pier caps each have two-week durations, and the next activity—steel erection—does not begin until all pier caps are completed. Therefore, all pier column and pier cap activities are most likely on the critical path, not just the first piles, piers, columns, and caps. Using similar reasoning, the steel for span #1 is critical, and then the path continues through deck placement for all spans. Next are the curbs and sidewalks for spans #3 and #4 and, finally, the punch list work.

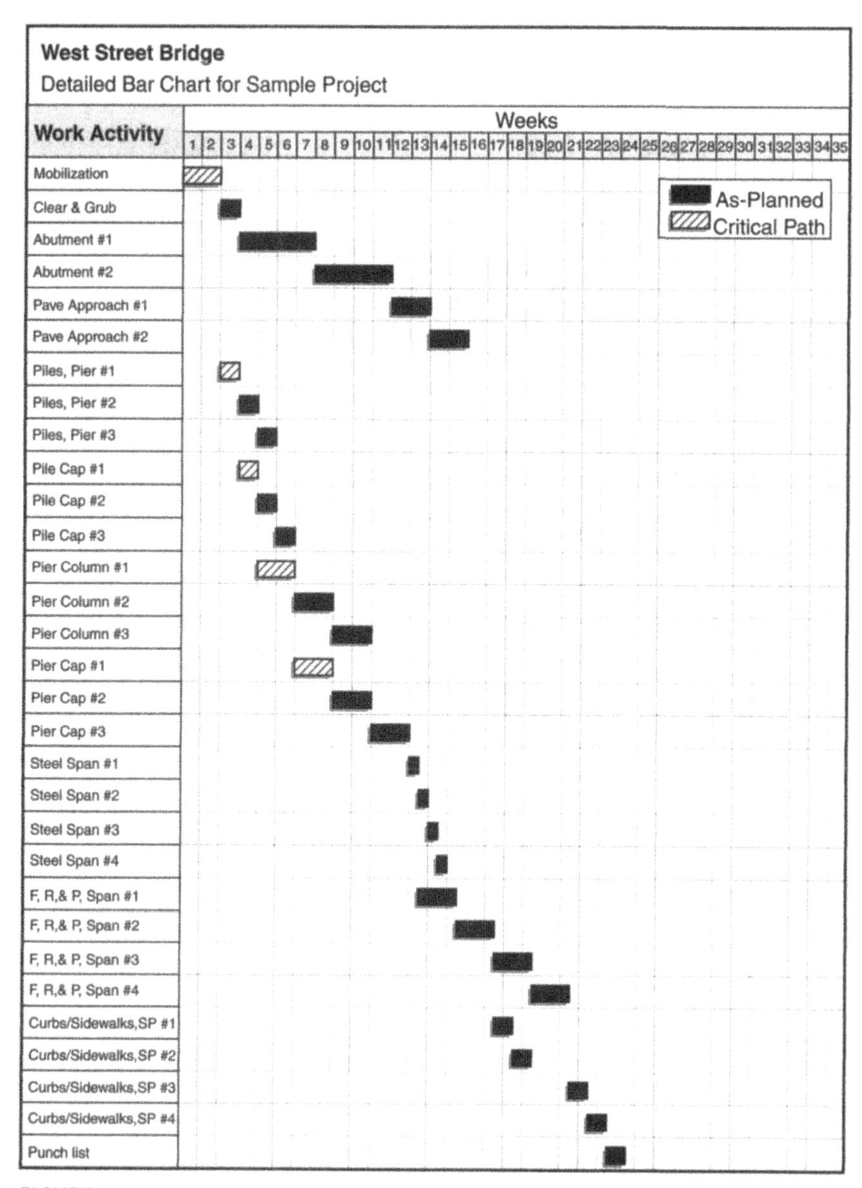

FIGURE 4.5

Thus, the overall critical path from the bar chart (Figure 4.6) is:

==> Mobilization
==> Piles—Pier #1
==> Pile Cap #1
==> Pier Column #1
==> Pier Column #2

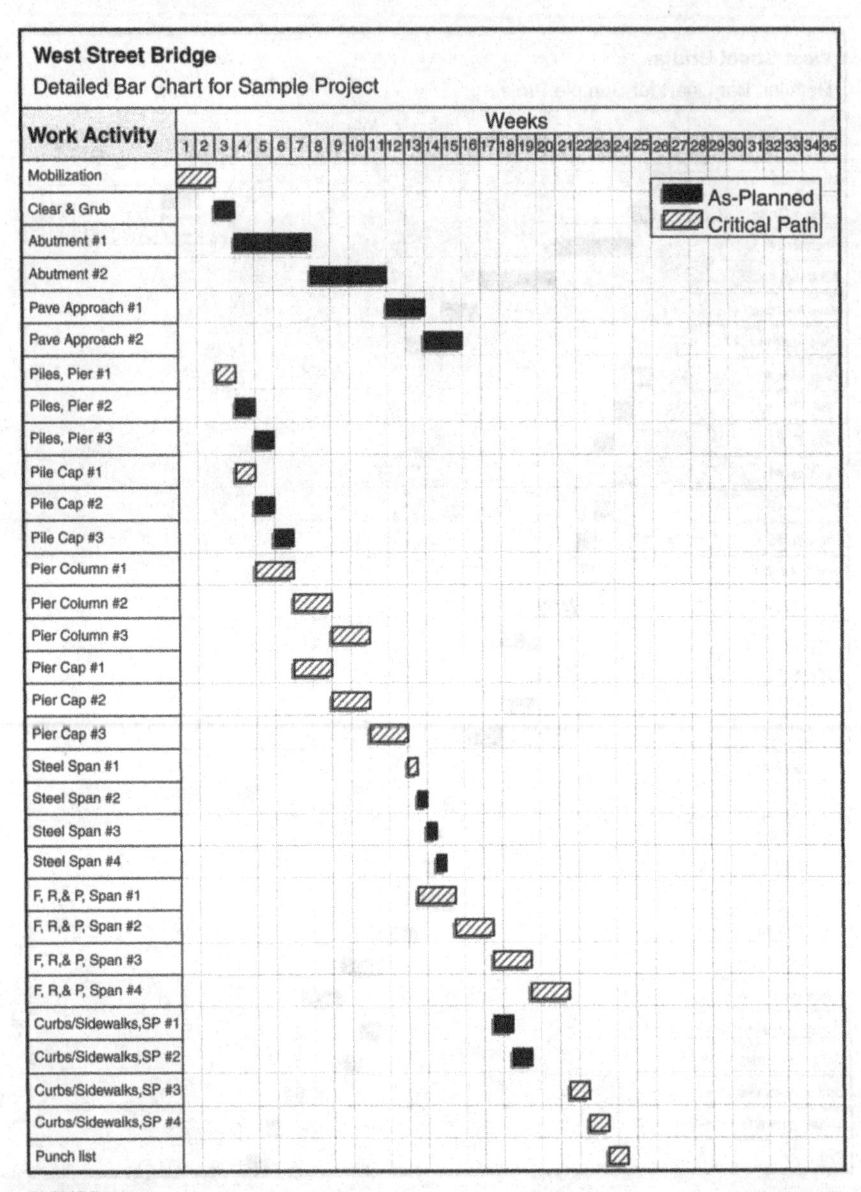

FIGURE 4.6

==> Pier Cap #1
==> Pier Column #3
==> Pier Cap #2
==> Pier Cap #3
==> Steel Span #1
==> F, R, & P Span #1

==> F, R, & P Span #2
==> F, R, & P Span #3
==> F, R, & P Span #4
==> Curbs & Sidewalks Span #3
==> Curbs & Sidewalks Span #4
==> Punch list

The analyst could reach a similar conclusion working with the less detailed bar chart alone (see Figure 4.1). This would, however, require that the analyst make more assumptions about the work on the separate piers. As was noted in the preceding discussion, contemporaneous documentation can help the analyst define the Contractor's planned sequence in more detail. The less detailed the bar chart, the more assumptions required by the analyst to determine the Project's critical path.

QUANTIFYING DELAYS USING BAR CHART SCHEDULES

The process of quantifying the delays using a bar chart is similar to the process that will be described later in this book when a CPM schedule is available. To start the process, the analyst must prepare a detailed as-built diagram that shows as specifically as possible when the work was actually performed. Figure 4.7 is the as-built diagram for this Project. Once the as-built has been prepared, the analysis can proceed. As the as-built diagram (Figure 4.7) shows, the mobilization activity started on schedule (the first day of week one) and finished on schedule (by the end of week two). The remaining activities, however, did not proceed in the same manner as the as-planned schedule had predicted. For this example, rather than analyze the entire Project, the first three delays will be analyzed to demonstrate the delay analysis methodology that can be applied to the rest of the Project.

As the as-built diagram (Figure 4.7) shows, the pile driving at piers #1, #2, and #3; the pile caps at piers #1 and #2; and pier column #1 were accomplished as-planned in the three weeks immediately following the mobilization activity. The clear and grub activity, however, did not proceed as planned but started two weeks late and finished in a duration of one week (the planned duration). If the previous conclusions concerning the critical path were correct, the delay to the start of clearing and grubbing should not have resulted in a delay to the Project. To check this conclusion, the analyst can "update" the bar chart as of the end of week five, as shown in Figure 4.8.

As can be seen in Figure 4.8, the Project is still on schedule, but the abutment and approach work has been moved over, delayed, or "bumped" in time because of the delay to the clear and grub activity. As expected, there is no delay to the critical path. The adjusted schedule (Figure 4.8) shows the as-built condition for the first five weeks of the Project and the adjusted as-planned activities for the remainder of the job.

Based on the as-built information, the analyst decides to "update" the schedule as of the end of week 11. The as-built diagram (Figure 4.7) shows that

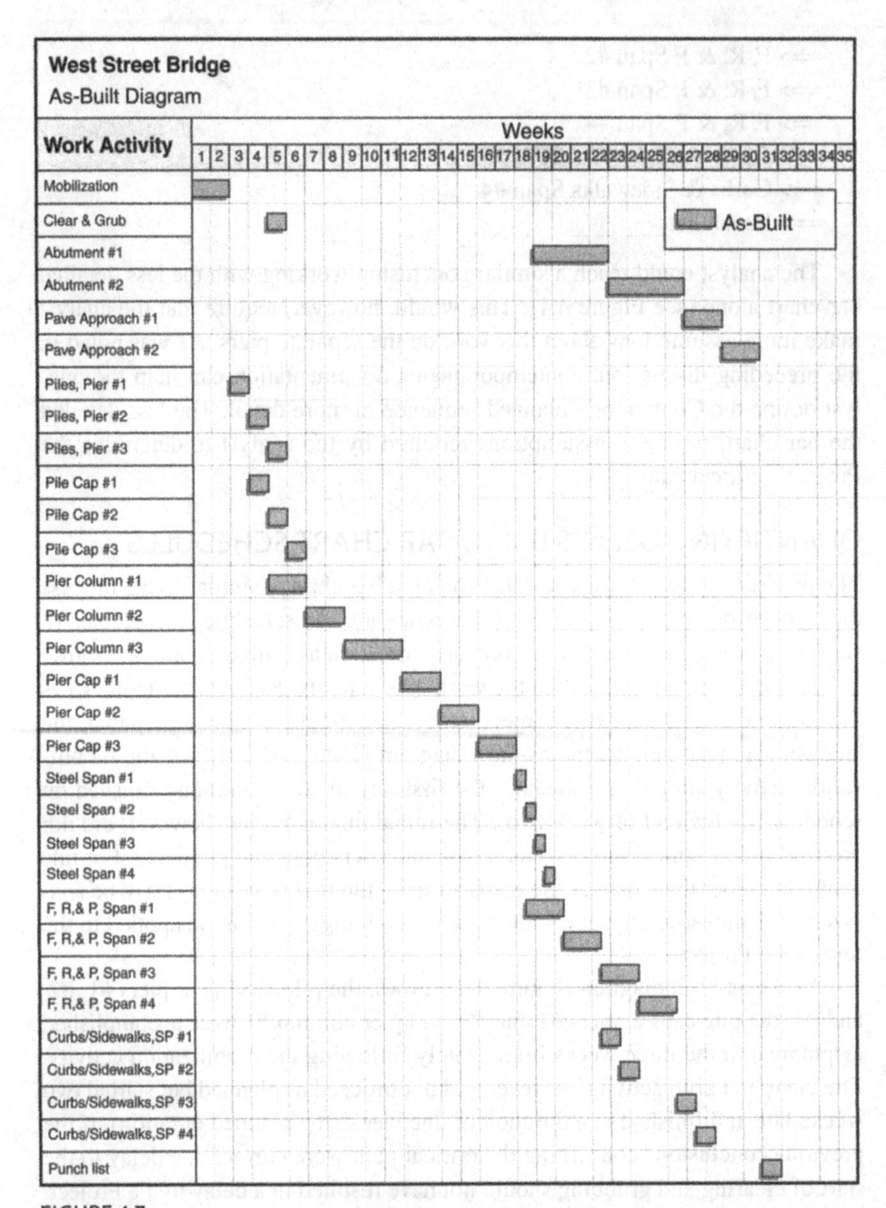

FIGURE 4.7

the abutment/approach work has not yet begun and that the pier cap work also has not yet begun. Pier columns #1 and #2 were completed on schedule. Pier column #3, however, took one week longer to complete than planned. The adjusted schedule for week 11 is shown in Figure 4.9. Based on the updated and adjusted schedule presented in Figure 4.9, the analyst concludes that the Project is now five weeks behind schedule. The delay was caused by the late start of pier

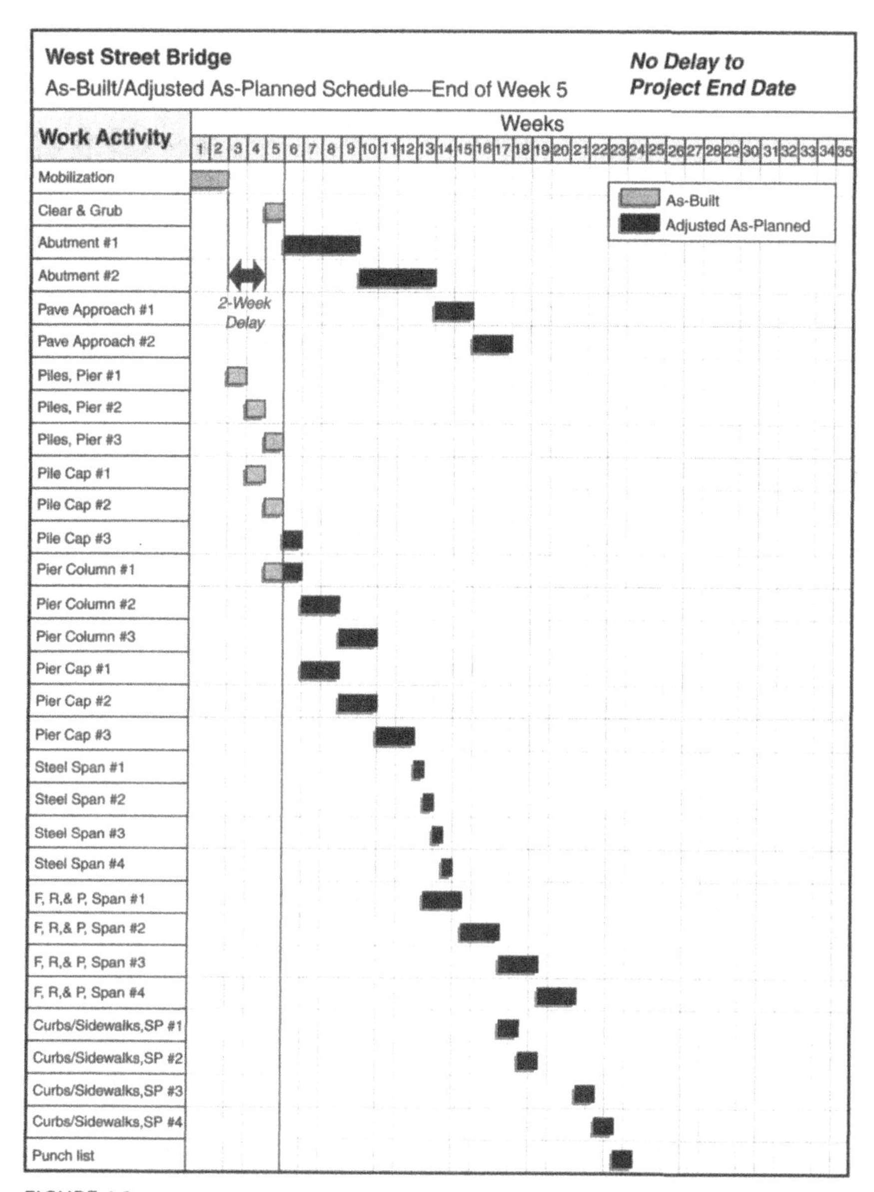

FIGURE 4.8

cap #1 work. Although pier column #3 was late in finishing and was on the original critical path, once the pier cap #1 activity did not start on time, the critical path shifted to depend solely on the pier cap work. The pier column #3 activity was effectively given float by virtue of the delay to the pier cap activity.

Next, the analyst decides to "update" the schedule at the end of week 18. This point is chosen because the activities along the pier/deck path continued

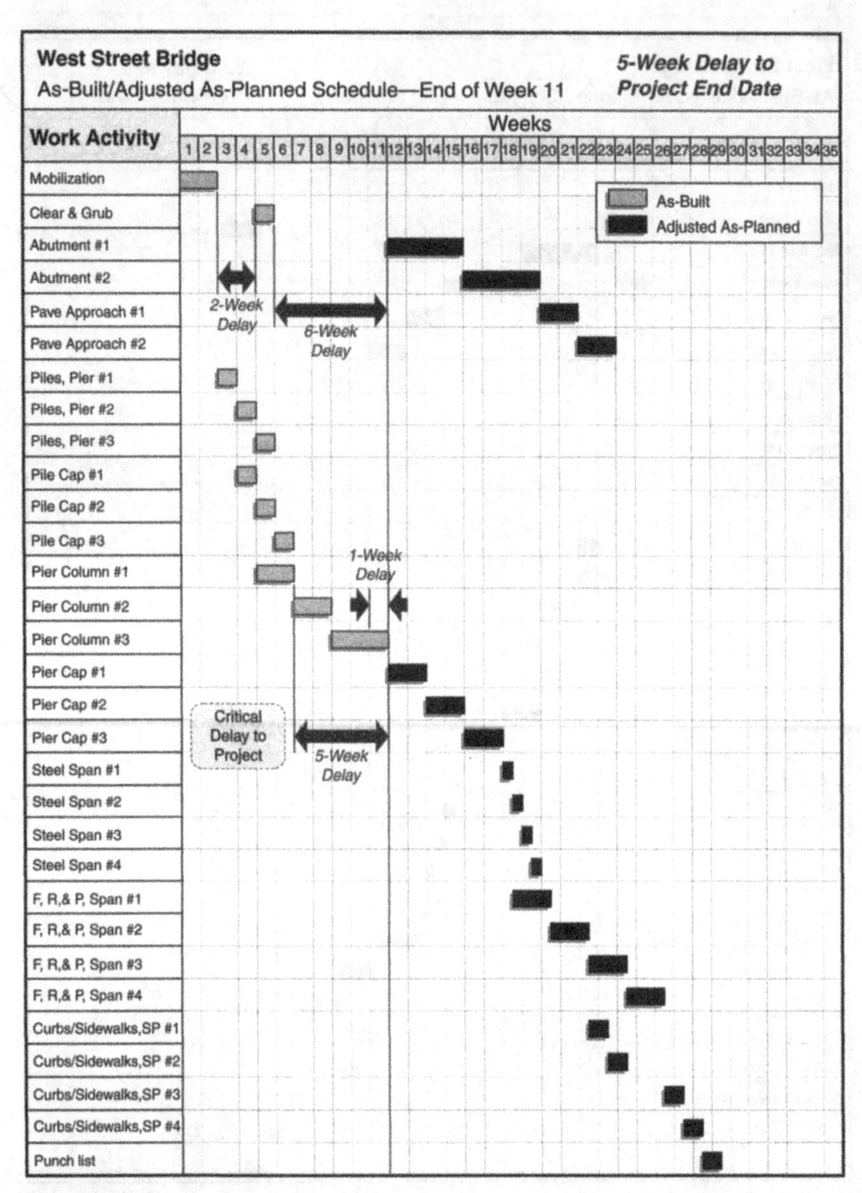

FIGURE 4.9

in accordance with the adjusted schedule, but the abutment/approach work did not. It is in the middle of week 15 that the critical path shifts. The as-built and updated as-planned schedule at the end of week 18 is shown in Figure 4.10. Since the Project had been delayed five weeks as of the last update, the additional delay since that update is two and one-half weeks. The activities on the pier/deck path were not delayed any further since the last update. Therefore, the additional delay appears to be the result of a lack of

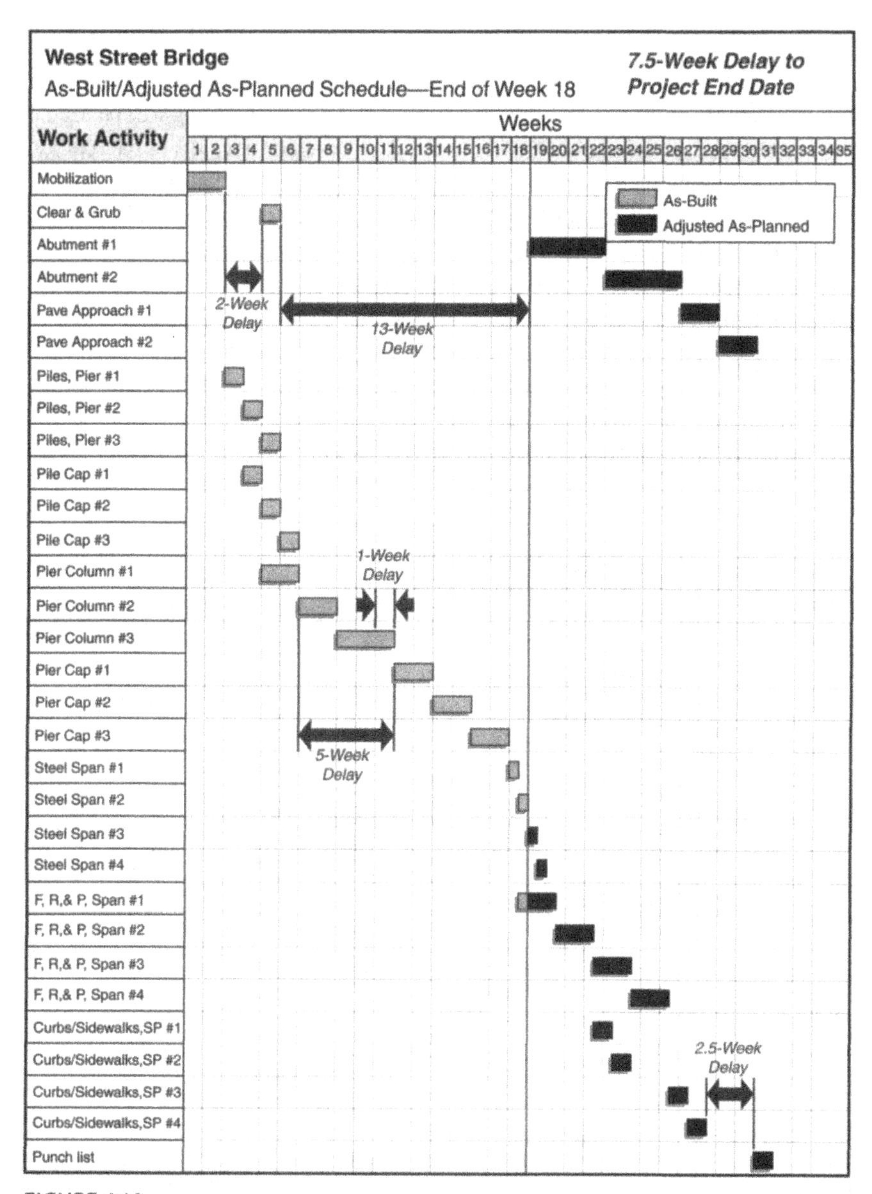

FIGURE 4.10

progress on the abutment/approach path. Since this is an additional two and one-half weeks, the critical path shifted two and one-half weeks before this update, in the middle of week 15.

The example just given uses the same methodology described in Chapter 3. The analysis was made step-by-step at various points in time. The exact method used to perform the analysis and the accuracy of the results depend on the detail of the as-planned schedule and the available as-built information.

Example Delay Analysis of Potential Changes with Bar Charts

To further illustrate the process of determining delays with a bar chart, the following example of the construction of a simple, three-story building will be used. Figure 4.11 is a bar chart schedule for the Project. Based on a review of the schedule, the critical path has been shown with cross-hatched bars. During the course of the Project, there were four changes that occurred that might have affected the Project completion date. The analyst has been asked to analyze each of these and determine what delay, if any, they caused to the Project completion date. In order to perform the analysis, the analyst prepared an as-built diagram based on the Project daily reports and other contemporaneous information. The as-built diagram with the changed work highlighted in grey is shown in Figure 4.12.

The analyst, having read this book, understands that he cannot just look at a comparison of the as-planned schedule and the as-built diagram and draw

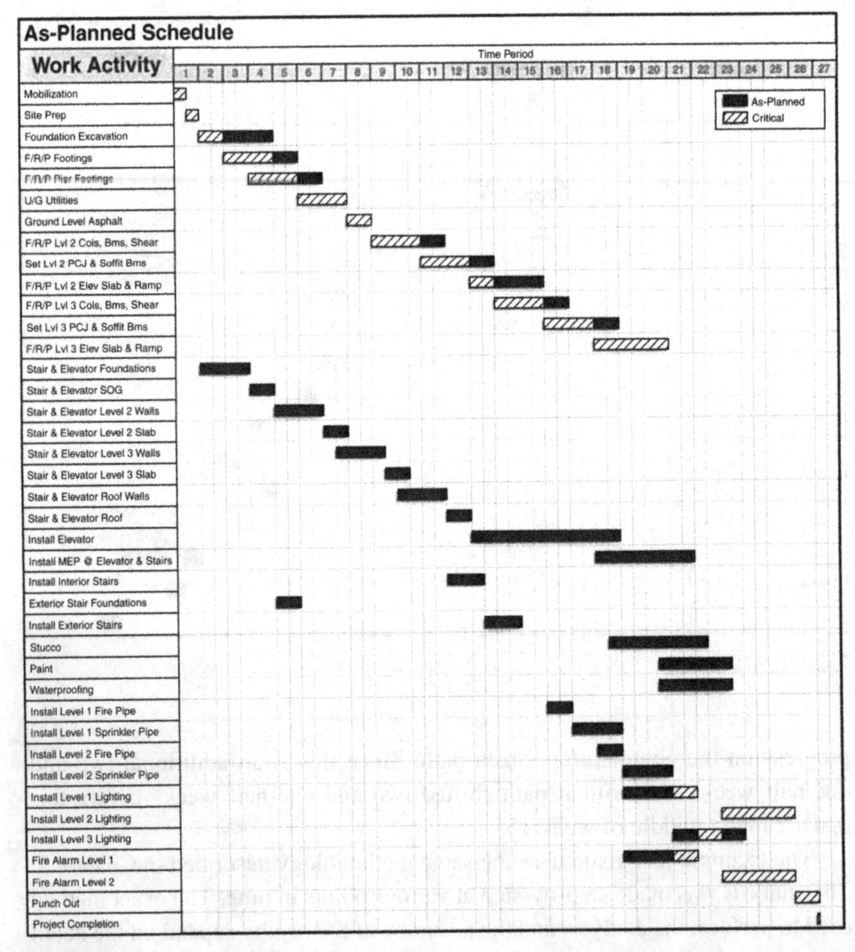

FIGURE 4.11

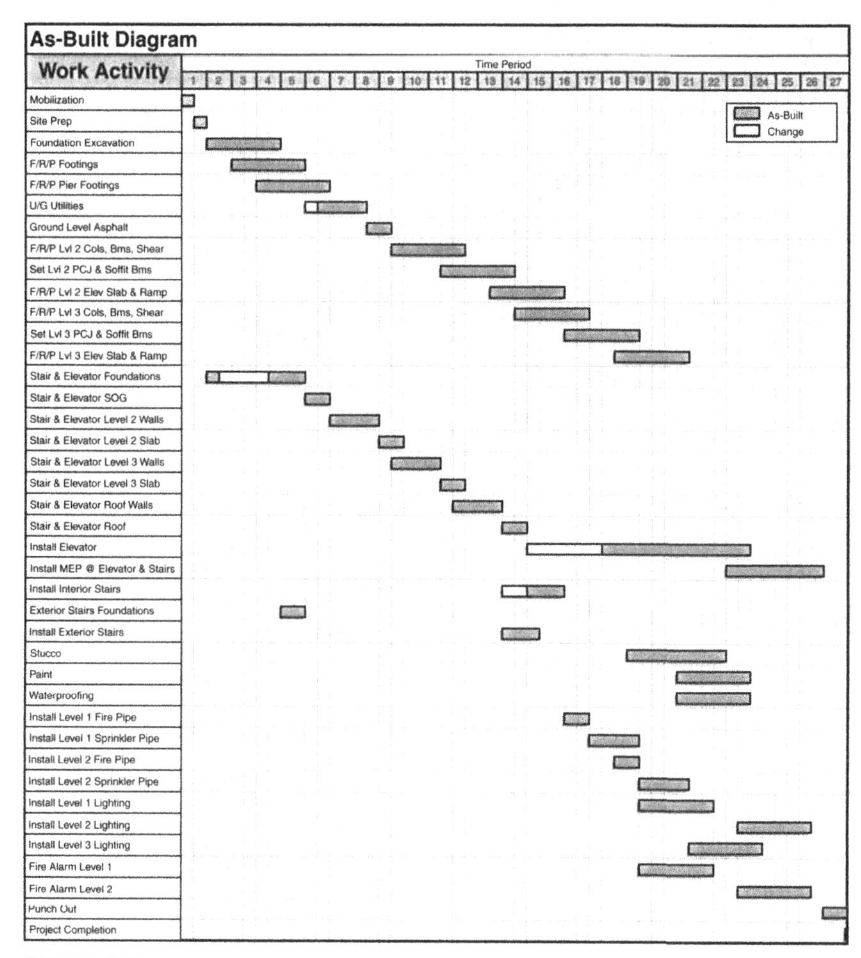

FIGURE 4.12

conclusions. Instead, the analysis must apply the changes as they occur and measure any delay, as it would have occurred. Similarly, the analyst could be performing the analysis contemporaneously as the Project progresses. In this case, the as-built diagram would be prepared up to the date of the change and any delays determined at that time.

Figure 4.13 shows the Project as of time period 3.5. The actual progress is plotted in grey, the change in white, and the remaining work in cross-hatched and black bars, similar to the as-planned schedule. As can be seen from a comparison between the as-planned schedule, Figure 4.11, and the updated schedule with the changed work, Figure 4.13, the Project is still scheduled to finish at the end of time period 26. Therefore, no delay was caused by the first change.

The next change to the Project occurs during time period 6. The analyst has updated the bar chart to include the changed work, shown in Figure 4.14, with

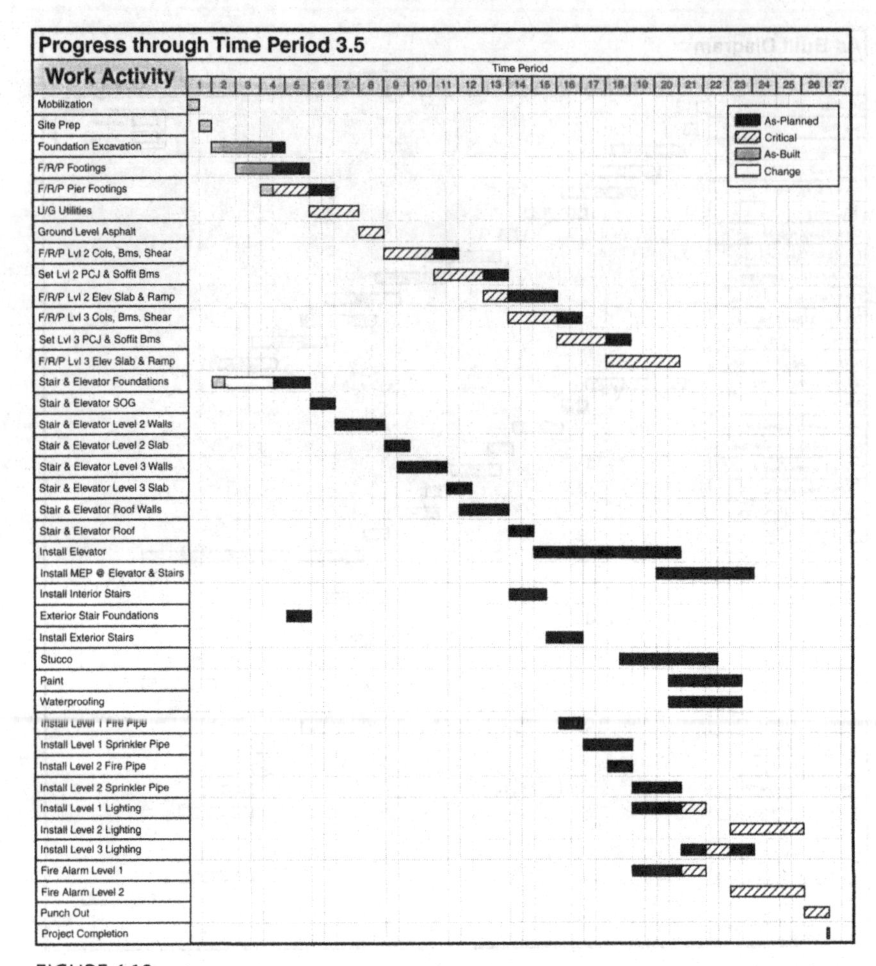

FIGURE 4.13

the actual progress in grey, the changed work in white, and the future work in cross-hatched and black bars. As can be seen from a comparison between the first update of the schedule, Figure 4.13, and the present update of the schedule, Figure 4.14, the Project has been delayed one-half time period and will now finish in the middle of time period 27. One can also see that change #2 affected the critical work of the underground utilities. So the analyst has determined that a delay has occurred because of change #2 and the delay is one-half time period in duration.

The next change to the Project, change #3, occurred during time period 14. The analyst has updated the bar chart to include the changed work. This is shown in Figure 4.15, with the actual progress in grey, the changed work in white, and the future work in cross-hatched and black bars. As can be seen from a comparison between the preceding update and this update, the end date of the Project has not changed; therefore, the changed work did not affect the critical path, and no delay resulted.

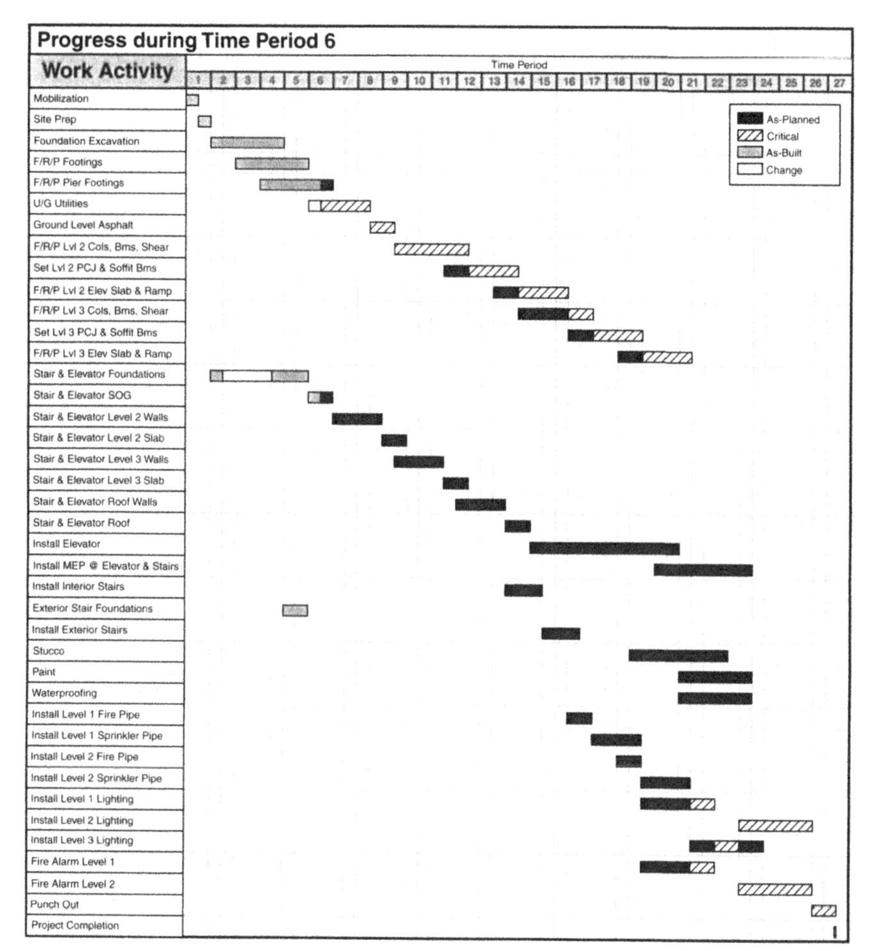

FIGURE 4.14

The next change, change #4, occurred during time periods 15 through 17. The analyst has updated the bar chart schedule to include the changed work. This is shown in Figure 4.16, with the actual progress in grey, the changed work in white, and the future work in cross-hatched and black bars. In reviewing this update and comparing it with the preceding update, it is obvious that the Project end date has moved to a later date. The Project will now complete at the end of time period 27 and has been delayed half a time period. Note that the critical path has changed. Because of the duration of the changed work, a path that previously had float is now the critical path and caused the delay that was measured.

This simple example demonstrates how one can update a bar chart and determine delays contemporaneously while the Project progresses and also identify changes in the critical path.

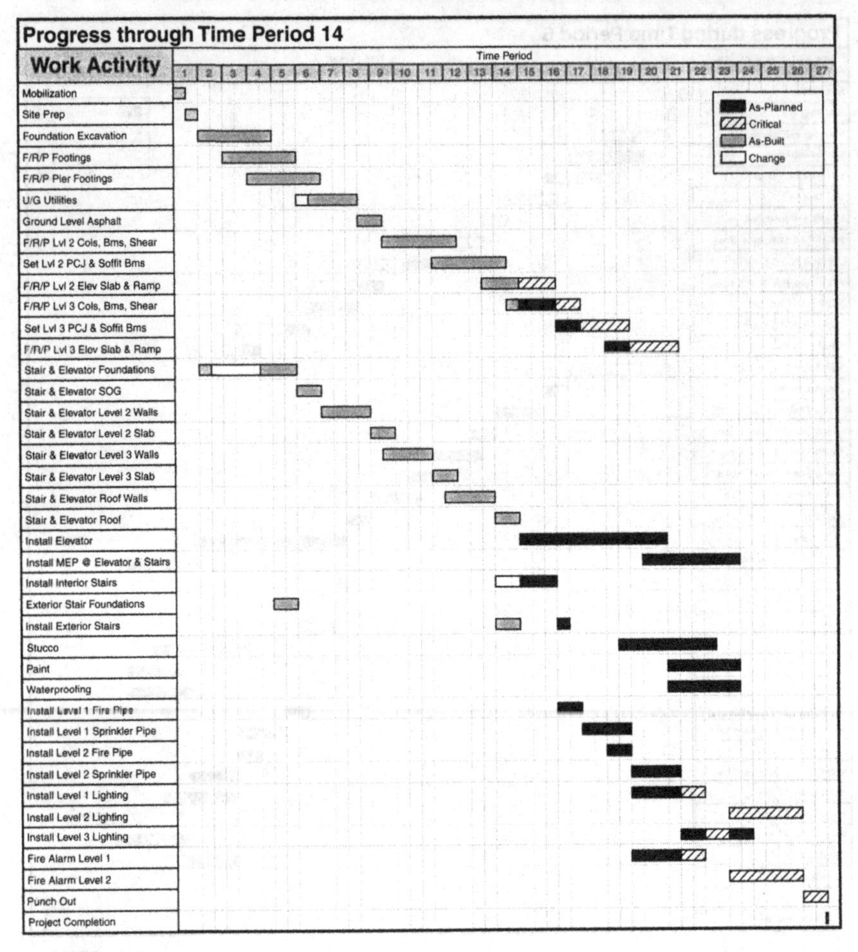

Progress through Time Period 14
Work Activity
Time Period
1 2 3 4 5 6 7 8 9 10 11 12 13 14 15 16 17 18 19 20 21 22 23 24 25 26 27
As-Planned
Critical
As-Built
Change
Mobilization
Site Prep
Foundation Excavation
F/R/P Footings
F/R/P Pier Footings
U/G Utilities
Ground Level Asphalt
F/R/P Lvl 2 Cols, Bms, Shear
Set Lvl 2 PCJ & Soffit Bms
F/R/P Lvl 2 Elev Slab & Ramp
F/R/P Lvl 3 Cols, Bms, Shear
Set Lvl 3 PCJ & Soffit Bms
F/R/P Lvl 3 Elev Slab & Ramp
Stair & Elevator Foundations
Stair & Elevator SOG
Stair & Elevator Level 2 Walls
Stair & Elevator Level 2 Slab
Stair & Elevator Level 3 Walls
Stair & Elevator Level 3 Slab
Stair & Elevator Roof Walls
Stair & Elevator Roof
Install Elevator
Install MEP @ Elevator & Stairs
Install Interior Stairs
Exterior Stair Foundations
Install Exterior Stairs
Stucco
Paint
Waterproofing
Install Level 1 Fire Pipe
Install Level 1 Sprinkler Pipe
Install Level 2 Fire Pipe
Install Level 2 Sprinkler Pipe
Install Level 1 Lighting
Install Level 2 Lighting
Install Level 3 Lighting
Fire Alarm Level 1
Fire Alarm Level 2
Punch Out
Project Completion

FIGURE 4.15

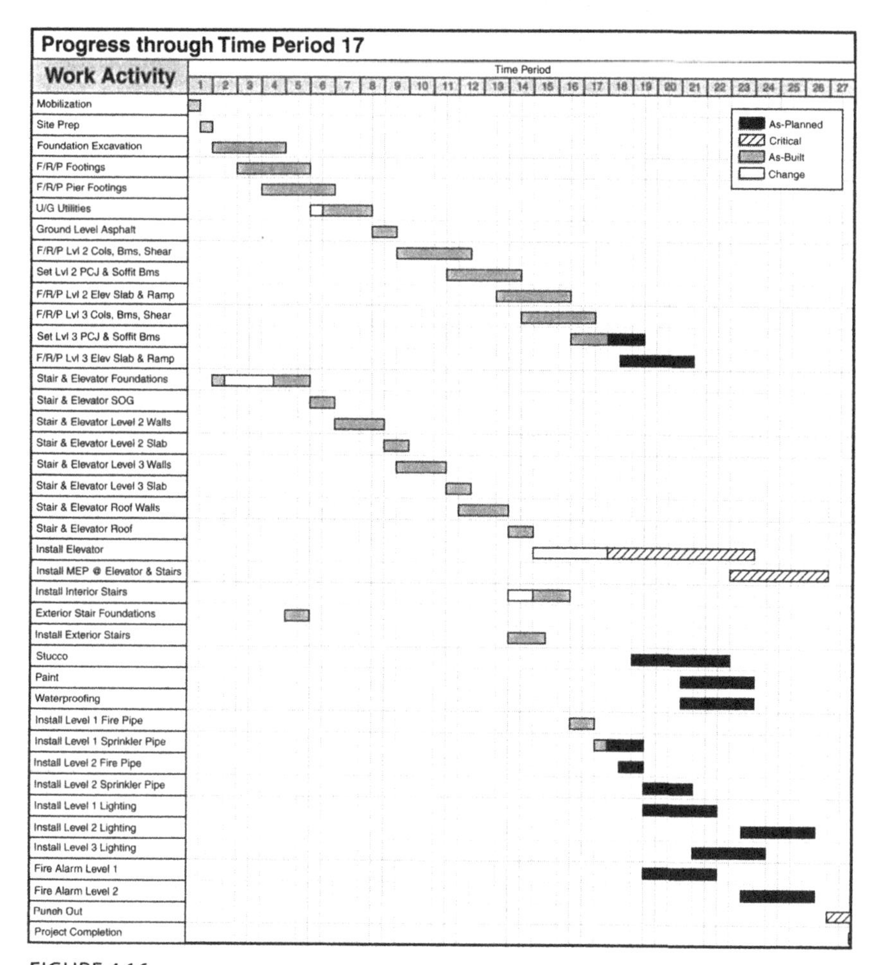
Progress through Time Period 17
Work Activity
Time Period
1 2 3 4 5 6 7 8 9 10 11 12 13 14 15 16 17 18 19 20 21 22 23 24 25 26 27
As-Planned
Critical
As-Built
Change
Mobilization
Site Prep
Foundation Excavation
F/R/P Footings
F/R/P Pier Footings
U/G Utilities
Ground Level Asphalt
F/R/P Lvl 2 Cols, Bms, Shear
Set Lvl 2 PCJ & Soffit Bms
F/R/P Lvl 2 Elev Slab & Ramp
F/R/P Lvl 3 Cols, Bms, Shear
Set Lvl 3 PCJ & Soffit Bms
F/R/P Lvl 3 Elev Slab & Ramp
Stair & Elevator Foundations
Stair & Elevator SOG
Stair & Elevator Level 2 Walls
Stair & Elevator Level 2 Slab
Stair & Elevator Level 3 Walls
Stair & Elevator Level 3 Slab
Stair & Elevator Roof Walls
Stair & Elevator Roof
Install Elevator
Install MEP @ Elevator & Stairs
Install Interior Stairs
Exterior Stair Foundations
Install Exterior Stairs
Stucco
Paint
Waterproofing
Install Level 1 Fire Pipe
Install Level 1 Sprinkler Pipe
Install Level 2 Fire Pipe
Install Level 2 Sprinkler Pipe
Install Level 1 Lighting
Install Level 2 Lighting
Install Level 3 Lighting
Fire Alarm Level 1
Fire Alarm Level 2
Punch Out
Project Completion

FIGURE 4.16

Chapter | five

Delay Analysis Using CPM Schedules

USING CPM SCHEDULES TO MEASURE DELAYS

This chapter explains how to perform a delay analysis with a Critical Path Method (CPM) schedule. Unfortunately, a detailed explanation of every nuance of a delay analysis for a CPM schedule is beyond the scope of this book, but this chapter covers the basic principles.

The theory behind CPM scheduling is that the network of activities is designed to model the way in which the Project will be constructed. If the network closely models the Project's plan, the predictions calculated from the schedule will be reliable. With this objective in mind, CPM software developers have worked to improve this modeling capability. The earliest such innovations involved the calendar and enabled the scheduler to define the anticipated Project workdays. This was followed by the ability to define multiple calendars and assign each activity to an appropriate calendar. Now, the schedule could consider in its calculating process the fact that concrete would be placed on a weekday and cure over the weekend; this better model resulted in a more accurate forecast date for stripping the forms.

Current scheduling software allows the scheduler to define a variety of relationship types between activities. The relationship between two activities can be defined so one activity starts after the completion of the other, both start together, both finish together, or one finishes when the other starts. All of these relationships can be defined with a lag so some positive or negative amount of time is factored into the relationship.

Also, the dates, durations, and float of activities can be constrained in a variety of ways. For example, activity early and late start and finish dates can be constrained to ensure that an activity will not be forecast to start or finish earlier or later than a specified date. An activity can also be constrained to start on a particular day. An expected finish date can also be imposed on an activity, causing the program to calculate the activity's remaining duration. Mandatory dates may also be imposed, overriding the network logic leading to and from the constrained activity. Additionally, an activity's float can be constrained so its late dates are forced to equal its early dates or so it will start as late as possible.

In addition to activity constraints, the way the schedule considers progress can also be controlled by the scheduler. For example, the schedule can be calculated such that progress made on an activity out of sequence with the flow of the network will override the schedule logic preceding it or the logic of the network can be retained despite such out-of-sequence progress. All of these features have served to enhance the scheduler's ability to model the plan to construct the Project. Each of these constraints supplements the logic of the schedule—and each has its place.

IDENTIFYING THE AS-PLANNED SCHEDULE

As noted in the preceding chapter, the as-planned schedule is defined as the schedule submitted by the Contractor at the beginning of the Project. Normally, the Owner requires some review, and possibly approval, of this schedule. Since the schedule submittal process can produce many different versions of the schedule, when possible and appropriate, the final approved schedule is used for analysis.

CORRECTING VERSUS LEAVING ERRORS

A delay analysis should rely on the contemporaneous Project schedules as the basis of analysis to the maximum extent possible. By using the management tools employed by the parties during the Project, the analyst is able to adopt the perspective of the Project Manager at the time of a particular delay and avoid applying the "wisdom" of hindsight. Reliance on the contemporaneous Project schedules helps keep the analysis objective and it guards against the analyst's drawing erroneous conclusions. For example, was the steel delivered late allowing more time to construct the foundations or was the steel delivery pushed back because the foundations were late? Analyses that stray far from the contemporaneous schedules, relying on after-the-fact creations, are usually biased and unpersuasive.

Assessing the reliability of the CPM schedules to form the basis of an analysis of delays may be one of the most important decisions that the analyst has to make. While it is true that there may be sufficient cause to abandon the contemporaneous schedules as an analytical tool, this decision should be made carefully and with good reason. No schedule is perfect. Most contain or omit logic that could be characterized as errors. But many of these are minor and have no effect on the Project's critical path. And most errors are self-correcting, meaning that if the schedules are properly analyzed, the actual progress of the work will establish the correct answer.

To better understand this principle of self-correction, consider the following example. Presume that in a sequence of critical path work involving the installation of wall studs, rough-in wiring, and drywall, the scheduler omits the activity for rough-in wiring. The schedule presumes that drywall installation will begin at the completion of the stud installation; no time has been allotted for roughing in wire. One might argue that the schedule is flawed because the Contractor cannot really complete all of the Contract work on time. But what really happens? When the stud installation is actually completed, the drywall work fails to start. Instead, the Contractor first roughs in the wire. The Contractor is assigned a critical Project delay equal to the late start of the drywall work. And so the error self-corrects. Had the Contractor included rough-in wiring in the original schedule, it would have done so in a way that did not increase the Contract time. The same choice is available when the omission is discovered. Perhaps the Contractor can perform the rough-in wiring concurrent with some of the wall stud and drywall work such that no delay occurs. Either way, the error in the schedule is self-correcting.

So when is the schedule no longer good enough? The contemporaneous CPM schedule should only be abandoned as an analytical tool when there is evidence that the schedule is not a reliable model of the Contractor's plan or when the schedule contains such gross errors as to render its predictions useless. Often, a schedule that has been significantly revised to mitigate a large number of days of delay will begin to contain gross errors of the type that cause the schedule to model an unexecutable plan. In such cases, the plan being modeled by the schedule and the actual execution of the Project depart, and the schedule is no longer a reliable tool to measure Project delay.

Here, a word of caution is appropriate. The proper use of advanced scheduling features should not be mistaken as scheduling errors. The mere existence of an additional calendar, a lag, or an activity constraint is not in and of itself an error. While the use of advanced scheduling features may make the task of analyzing the schedule to identify and measure delays more challenging, understanding how these features affect the calculation of the schedule allows the analyst to preserve the contemporaneous schedule as an objective analytical tool.

Still, the analyst must resist the temptation to alter the schedule by adding or deleting activities. Similarly, the analyst must not change the logic or durations to produce a schedule that seems more representative of the schedule that should have been used on the Project. This practice can produce a completely erroneous analysis.

If the analyst notes serious errors in the logic of the schedule, he or she should consider not accepting the Contractor's schedule as a valid tool to measure the delays. The validity of the schedule is subjective; therefore, the analyst should always seek help from a qualified scheduling consultant before making this determination. If, indeed, the schedule does not reflect the reality of the job progress, or does not reasonably represent the Contractor's plan for performing the work, then it may be wiser to abandon the schedule and perform a delay analysis using the as-built approach described in Chapter 6.

Upon reviewing the CPM schedule, the analyst may question the validity of the durations assigned to specific activities based on his or her own knowledge of the Project, estimating skills, and experience. However, if the reviewer does not know the specific resources that the Contractor planned to apply to the work, the durations in question should not be dismissed as erroneous. After all, an experienced and creative Contractor can devise the most expedient method to build the Project, and this may well require less time than one would normally estimate. In the same vein, the Contractor may decide to apply fewer resources to particular activities and have durations longer than one might normally estimate. Neither of these decisions on the part of the Contractor makes the schedule incorrect. Without specific Contract language constraining the Contractor's sequence or imposing milestone dates, the execution of the Project is the Contractor's responsibility.

One last thought about modifying schedules: While changes to schedule logic should be avoided, some analysis techniques require the addition of "dummy" logic to allow the schedule software to assist in the analysis of delays. Dummy logic is logic that has no net effect on the calculation of the schedule but is required to allow the program to identify certain key aspects of, or milestones within, the schedule. For example, in order to identify the critical path on a complex schedule, it may be necessary to run the schedule through the longest path filter. In some scheduling software applications, this filter will determine the longest path from the first activity in the schedule to the last activity in the schedule. If the last activity is not the Project completion milestone being analyzed, the filter may not identify the longest path to the completion milestone. In such cases, it may be necessary to add dummy logic to focus the filter on the path being analyzed. Such analysis techniques should not be confused with the types of logic changes that "correct" or alter the contemporaneous schedules in a way that reduces or limits the reliability of the results. Again, understanding how the advanced features of modern scheduling software affect the calculation of the schedule is essential in order to properly apply analytical techniques while preserving the contemporaneous schedule as an objective analytical tool. The use of scheduling software and other software tools in the schedule analysis process is discussed later in this chapter.

CPM SCHEDULES AND THE CRITICAL PATH

There is no question that the enhanced modeling capabilities of current CPM software have made CPM schedules more complex. In fact, in reaction to these advanced scheduling capabilities, some analysts have started a "back-to-basics" movement, attempting to convince Owners to write scheduling specifications that prohibit the use of many of these advanced features. Such fears, however, are misplaced. When properly applied, the superior modeling features of CPM schedules can benefit Owners and Contractors alike. But to guard against misuse, Contractors, Owners, and schedule analysts must all become proficient in the application of these scheduling tools.

Too often, scheduling analysts are inclined to conclude that the schedules cannot be used because "the critical path makes no sense." Such an important decision, however, must be closely scrutinized. And while it may sound too simplistic, the first question that should be asked is "How have you identified the critical path?" In simpler times, as was discussed in an earlier chapter of this book, the critical path was simply the path with zero float. This "foundation" of the scheduling world was an intuitive conclusion, drawn from an understanding of the way the schedule calculates dates through a network of finish-to-start relationships with no other logic constraints. But this foundation has been shaken by the power of the modern CPM schedule. In a current CPM schedule, an activity may have zero float for a variety of reasons, including something as simple as a zero total float constraint.

With the additional logic that can be added through the use of the advanced features now available in most CPM scheduling programs, the longest path may have zero float along its entire path, a portion of its path, or nowhere on the path. This means that it is possible, even likely, that certain activities on the critical path will actually have float. Again, this is okay. If these features have been used properly, a close look at the schedule logic will reveal why the float on these activities makes sense. Often, this float is a function of a constraint that is modeling a Contract requirement or other known construction restriction.

A properly implemented CPM schedule is a dynamic tool that lives throughout the duration of the Project. The critical path is equally dynamic. As the Project progresses, the schedule should be updated to reflect both the actual Project progress and any logic revisions that are necessary to better model the plan to complete the Project. Some logic revisions may be necessary to reflect a change in the Contractor's plan. Other revisions to the logic may be necessary even though the plan hasn't changed. This is because, as the actual progress is entered and the schedule is recalculated, certain logic needed to improve the model may become apparent. For example, the previous update may have forecast a start date for work in an environmentally sensitive area within a time period allowed in the permit. However, in the current update, the schedule is now forecasting the work to start during a restricted period, highlighting the need to constrain the start to occur after the restricted period has passed. This does not mean that the original schedule was flawed. Rather, such revisions reflect the normal dynamics of the Project management process.

The critical path may be so dynamic that it changes from day to day. While such a degree of change is unusual, critical path shifts between updates are quite common. This results from the fact that as the Project progresses, the lengths of the paths relative to one another change. For example, as the steel erection work on the longest path makes progress, the remaining duration of that path becomes less. Conversely, as the masonry work on a shorter path fails to make progress, the remaining duration of that path remains the same. If this condition continues, the day will come when the remaining duration of the masonry work path will equal the remaining duration of the steel erection path. On that day, both steel erection and masonry are concurrently critical. On the following day, the lack of progress of the masonry work begins to critically delay the Project.

Understanding how to identify shifts in the critical path is essential to properly allocate critical Project delays. In the preceding example, the lack of progress of masonry work does not delay the Project until its path of work becomes the longest path. The analysis techniques employed by the analyst should be such that the critical path of the Project is known for every day of the Project. The advanced features of current CPM software have given us powerful tools that provide a high level of precision in the analysis process. While these precise calculations still need to be interpreted in the context of the specific facts of each Project, the broad-brush techniques that used to be commonplace are things of the past.

While the activities along the critical path may have varying float values, many schedules still have critical paths with zero float. For simplicity purposes, most of the examples used in this chapter will refer to zero float critical paths.

IDENTIFYING SCHEDULE UPDATES FOR THE PURPOSE OF MEASURING DELAYS

Hopefully, the Contractor will have updated the CPM schedule at intervals throughout the Project. When CPM updates are available, the analyst can readily perform the delay analysis for the entire Project. The bridge example introduced earlier will be used to demonstrate the delay analysis process using CPM updates.

The Project is a simple four-span bridge with two reinforced concrete abutments and three piers. The piers have pile foundations, concrete pile caps, concrete pier columns, and concrete pier caps. The bridge has a steel superstructure, SIP metal deck forms, a reinforced concrete deck, curbs, and sidewalks.

Figure 5.1 is the network diagram submitted by the Contractor. A total float sort for this schedule is shown in Figure 5.2. The submitted schedule has one calendar and no constraints. As a result, we can rely on float as an indication of the critical path. As explained previously, this may not always be the case. As can be seen in the schedule, the Contractor did not include any activities for procurement or for shop drawing preparation, submission, and approval. Other than this oversight, the logic and durations for the remaining activities appear reasonable based on the information available to the analyst at this time.

First Update

The first update of the CPM schedule for the example Project is shown in Figure 5.3. This schedule report is the total float sort—the easiest type of analysis report to use for determining what happened to the schedule since the last update (in this case, the original schedule). As the schedule update shows, the Project has fallen behind schedule by seven workdays. One can recognize this by noting that the greatest negative float is seven days. Since the schedule is calculated in workdays (in this case), that means the Project is seven workdays behind schedule.

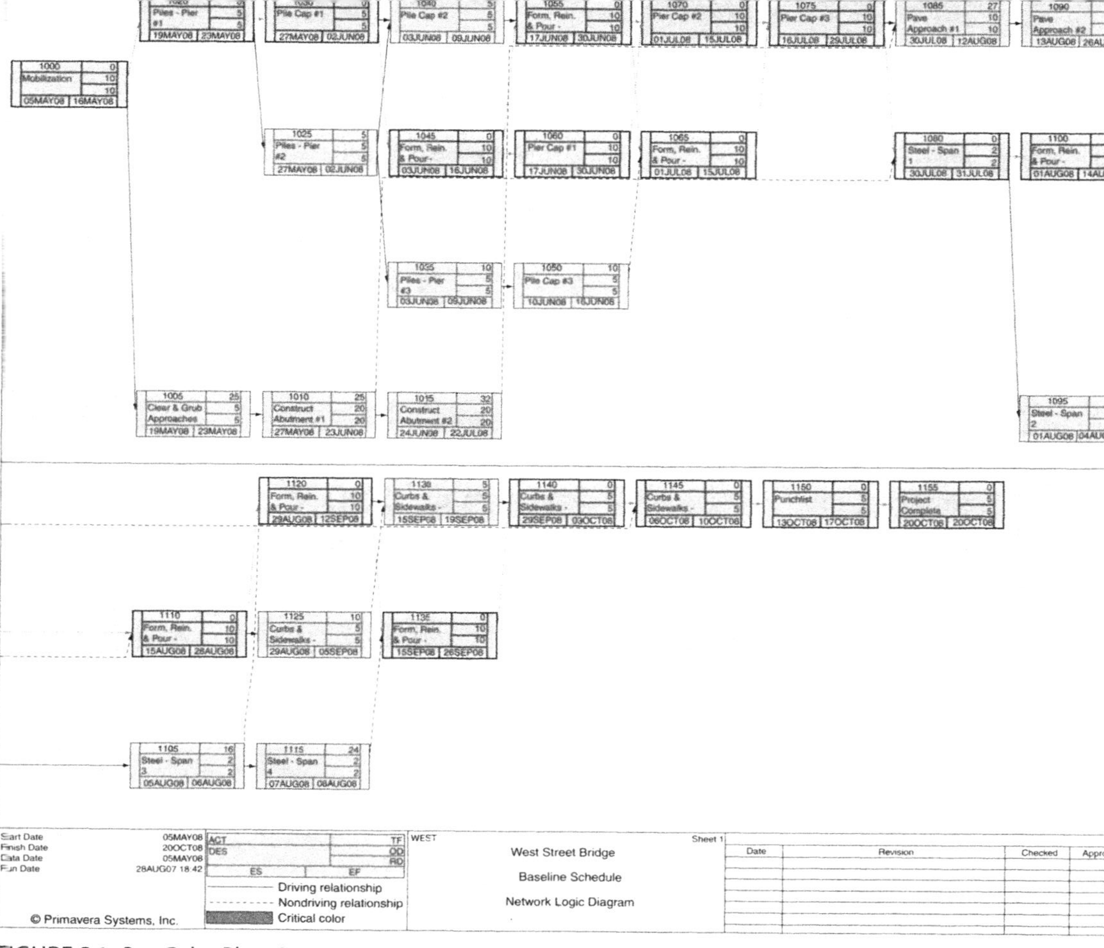

FIGURE 5.1 See Color Plate 2.

PRIMAVERA PROJECT PLANNER

West Street Bridge

REPORT DATE 22FEB08 RUN NO. 10
11:40

START DATE 5MAY08 FIN DATE 20OCT08

Schedule Report - Sort by TF, ES

DATA DATE 5MAY08 PAGE NO. 1

ACTIVITY ID	ORIG DUR	REM DUR	%	CODE	ACTIVITY DESCRIPTION	EARLY START	EARLY FINISH	LATE START	LATE FINISH	TOTAL FLOAT
1000	10	10	0	MOB	Mobilization	5MAY08	16MAY08	5MAY08	16MAY08	0
1020	5	5	0	PILE	Piles - Pier #1	19MAY08	23MAY08	19MAY08	23MAY08	0
1030	5	5	0	PILC	Pile Cap #1	27MAY08	2JUN08	27MAY08	2JUN08	0
1045	10	10	0	COLS	Form, Rein. & Pour - Column #1	3JUN08	16JUN08	3JUN08	16JUN08	0
1055	10	10	0	COLS	Form, Rein. & Pour - Column #2	17JUN08	30JUN08	17JUN08	30JUN08	0
1060	10	10	0	PRCP	Pier Cap #1	17JUN08	30JUN08	17JUN08	30JUN08	0
1065	10	10	0	COLS	Form, Rein. & Pour - Column #3	1JUL08	15JUL08	1JUL08	15JUL08	0
1070	10	10	0	PRCP	Pier Cap #2	1JUL08	15JUL08	1JUL08	15JUL08	0
1075	10	10	0	PRCP	Pier Cap #3	16JUL08	29JUL08	16JUL08	29JUL08	0
1080	2	2	0	STEL	Steel - Span 1	30JUL08	31JUL08	30JUL08	31JUL08	0
1100	10	10	0	DECK	Form, Rein. & Pour - Span 1	1AUG08	14AUG08	1AUG08	14AUG08	0
1110	10	10	0	DECK	Form, Rein. & Pour - Span 2	15AUG08	28AUG08	15AUG08	28AUG08	0
1120	10	10	0	DECK	Form, Rein. & Pour - Span 3	29AUG08	12SEP08	29AUG08	12SEP08	0
1135	10	10	0	DECK	Form, Rein. & Pour - Span 4	15SEP08	26SEP08	15SEP08	26SEP08	0
1140	5	5	0	CURB	Curbs & Sidewalks - Span 3	29SEP08	3OCT08	29SEP08	3OCT08	0
1145	5	5	0	CURB	Curbs & Sidewalks - Span 4	6OCT08	10OCT08	6OCT08	10OCT08	0
1150	5	5	0	PNCH	Punchlist	13OCT08	17OCT08	13OCT08	17OCT08	0
1155	1	1	0	PNCH	Project Complete	20OCT08	20OCT08	20OCT08	20OCT08	0
1025	5	5	0	PILE	Piles - Pier #2	27MAY08	2JUN08	3JUN08	9JUN08	5
1040	5	5	0	PILC	Pile Cap #2	3JUN08	9JUN08	10JUN08	16JUN08	5
1130	5	5	0	CURB	Curbs & Sidewalks - Span 2	15SEP08	19SEP08	22SEP08	26SEP08	5
1095	2	2	0	STEL	Steel - Span 2	1AUG08	4AUG08	13AUG08	14AUG08	8
1035	5	5	0	PILE	Piles - Pier #3	3JUN08	9JUN08	17JUN08	23JUN08	10
1050	5	5	0	PILC	Pile Cap #3	10JUN08	16JUN08	24JUN08	30JUN08	10
1125	5	5	0	CURB	Curbs & Sidewalks - Span 1	29AUG08	5SEP08	15SEP08	19SEP08	10
1105	2	2	0	STEL	Steel - Span 3	5AUG08	6AUG08	27AUG08	28AUG08	16
1115	2	2	0	STEL	Steel - Span 4	7AUG08	8AUG08	11SEP08	12SEP08	24
1005	5	5	0	CG	Clear & Grub Approaches	19MAY08	23MAY08	24JUN08	30JUN08	25
1010	20	20	0	ABUT	Construct Abutment #1	27MAY08	23JUN08	1JUL08	29JUL08	25
1085	10	10	0	PVAP	Pave Approach #1	30JUL08	12AUG08	8SEP08	19SEP08	27
1090	10	10	0	PVAP	Pave Approach #2	13AUG08	26AUG08	22SEP08	3OCT08	27
1015	20	20	0	ABUT	Construct Abutment #2	24JUN08	22JUL08	8AUG08	5SEP08	32

FIGURE 5.2

PRIMAVERA PROJECT PLANNER — West Street Bridge

REPORT DATE 22FEB08 RUN NO. 18 — START DATE 5MAY08 FIN DATE 20OCT08*
11:38
Schedule Report - Sort by TF, ES — DATA DATE 5JUN08 PAGE NO. 1

ACTIVITY ID	ORIG DUR	REM DUR	%	CODE	ACTIVITY DESCRIPTION	EARLY START	EARLY FINISH	LATE START	LATE FINISH	TOTAL FLOAT
1000	10	0	100	MOB	Mobilization	5MAY08A	16MAY08A			
1020	5	0	100	PILE	Piles - Pier #1	19MAY08A	23MAY08A			
1030	5	5	0	PILC	Pile Cap #1	5JUN08	11JUN08	27MAY08	2JUN08	-7
1045	10	10	0	COLS	Form, Rein. & Pour - Column #1	12JUN08	25JUN08	3JUN08	16JUN08	-7
1055	10	10	0	COLS	Form, Rein. & Pour - Column #2	26JUN08	10JUL08	17JUN08	30JUN08	-7
1060	10	10	0	PRCP	Pier Cap #1	26JUN08	10JUL08	17JUN08	30JUN08	-7
1065	10	10	0	COLS	Form, Rein. & Pour - Column #3	11JUL08	24JUL08	1JUL08	15JUL08	-7
1070	10	10	0	PRCP	Pier Cap #2	11JUL08	24JUL08	1JUL08	15JUL08	-7
1075	10	10	0	PRCP	Pier Cap #3	25JUL08	7AUG08	16JUL08	29JUL08	-7
1080	2	2	0	STEL	Steel - Span 1	8AUG08	11AUG08	30JUL08	31JUL08	-7
1100	10	10	0	DECK	Form, Rein. & Pour - Span 1	12AUG08	25AUG08	1AUG08	14AUG08	-7
1110	10	10	0	DECK	Form, Rein. & Pour - Span 2	26AUG08	9SEP08	15AUG08	28AUG08	-7
1120	10	10	0	DECK	Form, Rein. & Pour - Span 3	10SEP08	23SEP08	29AUG08	12SEP08	-7
1135	10	10	0	DECK	Form, Rein. & Pour - Span 4	24SEP08	7OCT08	15SEP08	26SEP08	-7
1140	5	5	0	CURB	Curbs & Sidewalks - Span 3	8OCT08	14OCT08	29SEP08	3OCT08	-7
1145	5	5	0	CURB	Curbs & Sidewalks - Span 4	15OCT08	21OCT08	6OCT08	10OCT08	-7
1150	5	5	0	PNCH	Punchlist	22OCT08	28OCT08	13OCT08	17OCT08	-7
1155	1	1	0	PNCH	Project Complete	29OCT08	29OCT08	20OCT08	20OCT08	-7
1040	5	5	0	PILC	Pile Cap #2	12JUN08	18JUN08	10JUN08	16JUN08	-2
1130	5	5	0	CURB	Curbs & Sidewalks - Span 2	24SEP08	30SEP08	22SEP08	26SEP08	-2
1025	5	3	40	PILE	Piles - Pier #2	29MAY08A	9JUN08		9JUN08	0
1095	2	2	0	STEL	Steel - Span 2	12AUG08	13AUG08	13AUG08	14AUG08	1
1125	5	5	0	CURB	Curbs & Sidewalks - Span 1	10SEP08	16SEP08	15SEP08	19SEP08	3
1035	5	5	0	PILE	Piles - Pier #3	10JUN08	16JUN08	17JUN08	23JUN08	5
1050	5	5	0	PILC	Pile Cap #3	17JUN08	23JUN08	24JUN08	30JUN08	5
1105	2	2	0	STEL	Steel - Span 3	14AUG08	15AUG08	27AUG08	28AUG08	9
1005	5	5	0	CG	Clear & Grub Approaches	5JUN08	11JUN08	24JUN08	30JUN08	13
1010	20	20	0	ABUT	Construct Abutment #1	12JUN08	10JUL08	1JUL08	29JUL08	13
1115	2	2	0	STEL	Steel - Span 4	18AUG08	19AUG08	11SEP08	12SEP08	17
1015	20	20	0	ABUT	Construct Abutment #2	11JUL08	7AUG08	8AUG08	5SEP08	20
1085	10	10	0	PVAP	Pave Approach #1	8AUG08	21AUG08	8SEP08	19SEP08	20
1090	10	10	0	PVAP	Pave Approach #2	22AUG08	5SEP08	22SEP08	3OCT08	20

FIGURE 5.3

Earlier, float was defined as the difference between when an activity could start or finish and when the activity must start or finish. In the original schedule (Figures 5.1 and 5.2), the critical path of the Project has zero float. Thus, the activities with zero float can start/finish and must start/finish on the same day. These activities cannot experience delays without delaying the scheduled completion or end date of the Project. A note of caution: The use of float to determine the critical path is correct in this specific example. This may not always hold true. This was discussed in Chapter 1 and will also be discussed later in this chapter.

Negative float indicates that the activity is delayed and will start or finish on a date that will delay the overall Project end date. When negative float appears in a schedule update, the critical path is the path of activities with the greatest negative float. Therefore, in Figure 5.3, the activities with negative seven workdays of float make up the critical path at the time of the update.

While the update clearly shows that the Project is behind schedule by seven workdays, it also shows the delay in calendar days. Refer to Activity 1155, *Project Complete*, the last activity on the critical path. The scheduled late finish date for this activity is October 20, 2008. The update shows, however, that the early finish date is October 29, 2008. This represents the current forecast completion date and, in this case, the delayed Project completion date. The Project is now nine calendar days behind schedule. It should be noted that the negative float results from the decision to constrain the late finish of the completion activity to October 20, 2008. Had this constraint not been set, the completion activity would have the same early and late finish date of October 29, 2008 and would have zero float.

While the schedule update shows that the Project has a seven-workday or nine-calendar-day delay, the analyst must also determine which specific activity or activities caused the delay. (Note that the two-day difference is due to workdays versus calendar days.)

By reviewing the completion dates of the finished activities and the status of each activity, and comparing the updated schedule to the original schedule, the analyst determines that Activity 1030, *Pile Cap #1*, has caused the delay. From the schedule in Figure 5.3, the analyst notes that all preceding activities on the critical path were finished either on or ahead of the original schedule. However, Activity 1030 has not yet started. Therefore, this is the critical delay for the Project for the period of this update.

Second Update

The second update for the Project appears in Figure 5.4. This schedule report is also a total float sort and shows the present critical path with the greatest negative float. In this update, the Project is now an additional four workdays, or six calendar days, behind schedule, for a total Project delay of 11 workdays, or 15 calendar days—seven workdays + four workdays = 11 workdays; nine calendar days + six calendar days = 15 calendar days.

PRIMAVERA PROJECT PLANNER

West Street Bridge

REPORT DATE 22FEB08 RUN NO. 27
11:37

Schedule Report - Sort by TF, ES

START DATE 5MAY08 FIN DATE 20OCT08*

DATA DATE 7JUL08 PAGE NO. 1

ACTIVITY ID	ORIG DUR	REM DUR	%	CODE	ACTIVITY DESCRIPTION	EARLY START	EARLY FINISH	LATE START	LATE FINISH	TOTAL FLOAT
1000	10	0	100	MOB	Mobilization	5MAY08A	16MAY08A			
1020	5	0	100	PILE	Piles - Pier #1	19MAY08A	23MAY08A			
1025	5	0	100	PILE	Piles - Pier #2	29MAY08A	9JUN08A			
1030	5	0	100	PILC	Pile Cap #1	5JUN08A	11JUN08A			
1040	5	0	100	PILC	Pile Cap #2	12JUN08A	19JUN08A			
1045	10	0	100	COLS	Form, Rein. & Pour - Column #1	12JUN08A	30JUN08A			
1035	5	0	100	PILE	Piles - Pier #3	20JUN08A	27JUN08A			
1055	10	8	20	COLS	Form, Rein. & Pour - Column #2	1JUL08A	16JUL08		30JUN08	-11
1065	10	10	0	COLS	Form, Rein. & Pour - Column #3	17JUL08	30JUL08	1JUL08	15JUL08	-11
1070	10	10	0	PRCP	Pier Cap #2	17JUL08	30JUL08	1JUL08	15JUL08	-11
1075	10	10	0	PRCP	Pier Cap #3	31JUL08	13AUG08	16JUL08	29JUL08	-11
1080	2	2	0	STEL	Steel - Span 1	14AUG08	15AUG08	30JUL08	31JUL08	-11
1100	10	10	0	DECK	Form, Rein. & Pour - Span 1	18AUG08	29AUG08	1AUG08	14AUG08	-11
1110	10	10	0	DECK	Form, Rein. & Pour - Span 2	2SEP08	15SEP08	15AUG08	28AUG08	-11
1120	10	10	0	DECK	Form, Rein. & Pour - Span 3	16SEP08	29SEP08	29AUG08	12SEP08	-11
1135	10	10	0	DECK	Form, Rein. & Pour - Span 4	30SEP08	13OCT08	15SEP08	26SEP08	-11
1140	5	5	0	CURB	Curbs & Sidewalks - Span 3	14OCT08	20OCT08	29SEP08	3OCT08	-11
1145	5	5	0	CURB	Curbs & Sidewalks - Span 4	21OCT08	27OCT08	6OCT08	10OCT08	-11
1150	5	5	0	PNCH	Punchlist	28OCT08	3NOV08	13OCT08	17OCT08	-11
1155	1	1	0	PNCH	Project Complete	4NOV08	4NOV08	20OCT08	20OCT08	-11
1060	10	7	30	PRCP	Pier Cap #1	30JUN08A	15JUL08		30JUN08	-10
1005	5	5	0	CG	Clear & Grub Approaches	7JUL08	11JUL08	24JUN08	30JUN08	-8
1050	5	5	0	PILC	Pile Cap #3	7JUL08	11JUL08	24JUN08	30JUN08	-8
1010	20	20	0	ABUT	Construct Abutment #1	14JUL08	8AUG08	1JUL08	29JUL08	-8
1130	5	5	0	CURB	Curbs & Sidewalks - Span 2	30SEP08	6OCT08	22SEP08	26SEP08	-6
1095	2	2	0	STEL	Steel - Span 2	18AUG08	19AUG08	13AUG08	14AUG08	-3
1015	20	20	0	ABUT	Construct Abutment #2	11AUG08	8SEP08	8AUG08	5SEP08	-1
1085	10	10	0	PVAP	Pave Approach #1	9SEP08	22SEP08	8SEP08	19SEP08	-1
1125	5	5	0	CURB	Curbs & Sidewalks - Span 1	16SEP08	22SEP08	15SEP08	19SEP08	-1
1090	10	10	0	PVAP	Pave Approach #2	23SEP08	6OCT08	22SEP08	3OCT08	-1
1105	2	2	0	STEL	Steel - Span 3	20AUG08	21AUG08	27AUG08	28AUG08	5
1115	2	2	0	STEL	Steel - Span 4	22AUG08	25AUG08	11SEP08	12SEP08	13

FIGURE 5.4

The analyst determines this delay by noting the 11 days of negative float shown on the critical path, beginning with Activity 1055, *Form, Rein., and Pour, Column #2*, and on the early and late finish dates of Activity 1155, *Project Complete.* The delay in this case had two causes: (1) The first three workdays of delay were due to the late finish of Activity 1045, *Form, Rein., and Pour, Column #1*; and (2) the last one workday of delay was caused by the lack of progress on Activity 1055, *Form, Rein., and Pour, Column #2*.

The analyst reached these conclusions by referring to the projected finish date for Activity 1045 on the preceding update (Figure 5.3) of June 25, 2008. From Figure 5.4, the analyst sees that the actual finish date was June 30, 2008, three workdays later than scheduled on the preceding update. Likewise, an analysis of Activity 1055 shows that it actually started on July 1, 2008, but the progress recorded is only two workdays, which is one workday behind the scheduled progress. Three workdays (late finish) + one workday (slow progress) = four workdays delay.

Third Update

Figure 5.5 is the Total Float schedule report for the third update for the Project. As of this update, the Project is now an additional 11 workdays, or 15 calendar days, behind schedule, for a total Project delay of 22 workdays or 30 calendar days. The analyst reaches this conclusion by reviewing Activities 1080, *Steel–Span #1*, and 1155, *Project Complete*, as was done for the preceding updates. However, in this case, Activity 1080, *Steel–Span #1*, has not yet begun, despite the fact that the scheduled start date in the previous update was August 14, 2008.

A careful review of the documents shows a delay for this activity because the shop drawings for the material were not submitted and approved. This conclusion exemplifies how the contemporaneous analysis approach compensates for minor errors in the schedule. The original schedule did not include activities for procurement or shop drawings. While an analyst may be tempted to consider adding these items to the schedule, their omission becomes obvious without having to tamper with the original schedule. The delay analysis for the remainder of the Project would proceed using the same methodology used for the first three updates.

USE OF SCHEDULING SOFTWARE AND OTHER SOFTWARE TOOLS IN THE QUANTIFICATION OF DELAYS

Advances in computer technology and software have changed the capabilities of construction scheduling software over the years. Today's scheduling software runs faster, is more powerful, and contains numerous options that allow the Project Manager and scheduler to organize their specific plan for resource allocation, cost forecasts, and work sequence to complete the Project.

With the changing needs of Project Managers and the variance in capabilities and cost, software companies have diversified their products in order to

PRIMAVERA PROJECT PLANNER

West Street Bridge

REPORT DATE 22FEB08 RUN NO. 25
11:35
Schedule Report - Sort by TF, ES

START DATE 5MAY08 FIN DATE 20OCT08*

DATA DATE 29AUG08 PAGE NO. 1

ACTIVITY ID	ORIG DUR	REM DUR	%	CODE	ACTIVITY DESCRIPTION	EARLY START	EARLY FINISH	LATE START	LATE FINISH	TOTAL FLOAT
1000	10	0	100	MOB	Mobilization	5MAY08A	16MAY08A			
1020	5	0	100	PILE	Piles - Pier #1	19MAY08A	23MAY08A			
1025	5	0	100	PILE	Piles - Pier #2	29MAY08A	9JUN08A			
1030	5	0	100	PILC	Pile Cap #1	5JUN08A	11JUN08A			
1040	5	0	100	PILC	Pile Cap #2	12JUN08A	19JUN08A			
1045	10	0	100	COLS	Form, Rein. & Pour - Column #1	12JUN08A	30JUN08A			
1035	5	0	100	PILE	Piles - Pier #3	20JUN08A	27JUN08A			
1060	10	0	100	PRCP	Pier Cap #1	30JUN08A	14JUL08A			
1055	10	0	100	COLS	Form, Rein. & Pour - Column #2	1JUL08A	15JUL08A			
1005	5	0	100	CG	Clear & Grub Approaches	7JUL08A	11JUL08A			
1050	5	0	100	PILC	Pile Cap #3	7JUL08A	10JUL08A			
1010	20	0	100	ABUT	Construct Abutment #1	14JUL08A	11AUG08A			
1065	10	0	100	COLS	Form, Rein. & Pour - Column #3	17JUL08A	31JUL08A			
1070	10	0	100	PRCP	Pier Cap #2	17JUL08A	30JUL08A			
1075	10	0	100	PRCP	Pier Cap #3	31JUL08A	13AUG08A			
1015	20	0	100	ABUT	Construct Abutment #2	12AUG08A	28AUG08A			
1080	2	2	0	STEL	Steel - Span 1	29AUG08	2SEP08	30JUL08	31JUL08	-22
1100	10	10	0	DECK	Form, Rein. & Pour - Span 1	3SEP08	16SEP08	1AUG08	14AUG08	-22
1110	10	10	0	DECK	Form, Rein. & Pour - Span 2	17SEP08	30SEP08	15AUG08	28AUG08	-22
1120	10	10	0	DECK	Form, Rein. & Pour - Span 3	1OCT08	14OCT08	29AUG08	12SEP08	-22
1135	10	10	0	DECK	Form, Rein. & Pour - Span 4	15OCT08	28OCT08	15SEP08	26SEP08	-22
1140	5	5	0	CURB	Curbs & Sidewalks - Span 3	29OCT08	4NOV08	29SEP08	3OCT08	-22
1145	5	5	0	CURB	Curbs & Sidewalks - Span 4	5NOV08	11NOV08	6OCT08	10OCT08	-22
1150	5	5	0	PNCH	Punchlist	12NOV08	18NOV08	13OCT08	17OCT08	-22
1155	1	1	0	PNCH	Project Complete	19NOV08	19NOV08	20OCT08	20OCT08	-22
1130	5	5	0	CURB	Curbs & Sidewalks - Span 2	15OCT08	21OCT08	22SEP08	26SEP08	-17
1095	2	2	0	STEL	Steel - Span 2	3SEP08	4SEP08	13AUG08	14AUG08	-14
1125	5	5	0	CURB	Curbs & Sidewalks - Span 1	1OCT08	7OCT08	15SEP08	19SEP08	-12
1105	2	2	0	STEL	Steel - Span 3	5SEP08	8SEP08	27AUG08	28AUG08	-6
1115	2	2	0	STEL	Steel - Span 4	9SEP08	10SEP08	11SEP08	12SEP08	2
1085	10	10	0	PVAP	Pave Approach #1	29AUG08	12SEP08	8SEP08	19SEP08	5
1090	10	10	0	PVAP	Pave Approach #2	15SEP08	26SEP08	22SEP08	3OCT08	5

FIGURE 5.5

provide viable and cost-effective software for each type of Project. Some of the more popular construction software applications on the market today are produced by Primavera Systems, Inc., and Microsoft Corporation. Other competitive software is available, but the majority of the current construction Projects utilize these software applications.

No matter how powerful the software becomes, the capabilities of the user are still the most important ingredient in using scheduling software as an effective management tool. As a result, no matter what software is chosen, the Project Manager *must* be aware of the different scheduling capabilities and options for each software because selecting or unselecting particular options can mean a world of difference in how the software mathematically forecasts the plan to complete the Project.

As with creating and updating a schedule, an analyst must have a familiarity with scheduling terminology and be able to accurately interpret the data and results displayed by the schedule. It is also necessary that the analyst be familiar with the specific software used to create and update the schedules, given the different scheduling options available in each software package. However, no matter what software was used to manage the Project schedule, the basic principles of analyzing a Project for delays and improvements remain the same.

Once an analyst has familiarized herself with the software used to manage the Project schedule, the analyst should gather all of the Contractor's schedules throughout the duration of the Project—the as-planned schedule and all subsequent schedule updates. If possible, the analyst should get "electronic copies," a copy of the computer file, for each of the schedules. Electronic copies allow the analyst to access all of the activity and Project data contained within the schedules, whereas "hard copies" or paper copies only allow the analyst to view the information that is available on the printout. Hard copy printouts can be easily manipulated to show only the information the hard copy provider wants the analyst to see, and they often lack information (logic ties, relationship lags, scheduling options) that is vital to determine what the contractor's contemporaneous plan was for completing the Project.

For the remainder of this chapter, the explanations will be based on the assumption that the analyst has electronic copies of all of the Project schedules. Because Primavera software is the most widely used software in the industry, terminology from Primavera scheduling software is used.

As previously stated, an analyst should strive to use all of the Project schedules that were used to manage the Project to quantify Project delays and improvements. Project delays and improvements between schedule updates should be separated into two classifications: (1) delays and improvements due to work progress or lack of work progress, and (2) delays and improvements due to revisions to the Project schedule. Quite often, these two types of delays and improvements are merged and, as a result, provide an inaccurate analysis of what is delaying or improving the Project. Despite common belief, work progress delays and improvements and schedule revision delays and improvements are easy to separate, especially with the software applications on the market.

Work Progress Delays and Improvements

These are the steps required to analyze work progress delays and improvements between two schedules:

1. Use the plan (activities, durations, logic relationships, resources, etc.) shown on the schedule with the earlier data date (Schedule 1).
2. Define the critical path and near-critical paths in Schedule 1. Near-critical paths are network paths where the total duration of the path is not as large as the critical path but could become the critical path if progress is not obtained on its activities during the update period.
3. Update the progress made to the Schedule 1 plan, using the actual start, actual finish dates, and remaining durations from the schedule with the later data date (Schedule 2).
4. Assess how the progress, or lack of progress, made to the Schedule 1 plan affected the critical path of the Project on a daily basis between Schedule 1 and Schedule 2. Remember that the critical path is dynamic and can change between Schedule 1 and Schedule 2, based on the progress or lack of progress.
5. Determine how the progress or lack of progress on the critical path between Schedule 1 and Schedule 2 affected the forecast completion date of the Project.

Primavera software allows the user to have Schedule 1 open while also viewing the progress from Schedule 2. This is done by making Schedule 2 a "Target Schedule" of Schedule 1. Primavera defines a Target Schedule as *a Project plan that can be compared to the current schedule to measure progress.* In Primavera 6, Targets are called Baselines and serve similar functions.

When Schedule 2 is made a Target Schedule of Schedule 1, the user can organize the columns to show a comparison of the two schedules' progress data, including remaining duration, early start and actual start dates, and early finish and actual finish dates. Organizing schedule printouts in this manner provides the user an easy side-by-side comparison of the expected progress and the actual progress obtained during the time period being analyzed. Figure 5.6 is a screen shot of the critical path, or longest path, of Schedule 1, with the progress from the Target Schedule, Schedule 2. From Figure 5.6, the user can identify the following progress on the critical path identified in Schedule 1, between Schedule 1 and Schedule 2:

- Activity 44, *Erect Zone 4.3:* 14 days of progress (SCH1 Rem Dur–SCH2 Rem Dur) with a new Actual Finish date of February 21, 2007.
- Activity 72, *Erect Zone 3.2:* four days of progress with a new Actual Start date of February 22, 2007.

Primavera Project Planner - [EX99]

Activity ID	Activity Description	Orig Dur	SCH 1 Rem Dur	SCH 2 Rem Dur	SCH 1 % Comp	SCH 1 Early Start	SCH 2 Actual Start	SCH 1 Early Finish	SCH 2 Actual Finish
32	Erect Zone 4.2	10	0	0	100	27NOV06A	27NOV06	08DEC06A	08DEC06
44	Erect Zone 4.3	32	14	0	56	11DEC06A	11DEC06	30JAN07	21FEB07
72	Erect Zone 3.2	25	25	21	0	31JAN07	22FEB07	06MAR07	
73	Erect Zone 5	20	20	20	0	07MAR07		03APR07	
74	Erect Zone 6	20	20	20	0	04APR07		01MAY07	
75	Erect Zone 7	25	25	20	0	02MAY07		05JUN07	
448	Roofing Complete West	15	15	15	0	06JUN07		26JUN07	
169	Framing 12th Floor West	8	8	8	0	20JUN07		29JUN07	
147	Framing 11th Floor West	8	8	8	0	02JUL07		11JUL07	
106	Framing 10th Floor West	8	8	8	0	12JUL07		23JUL07	
388	Framing 9th Floor West	8	8	8	0	24JUL07		02AUG07	
345	Framing 8th Floor West	8	8	8	0	03AUG07		14AUG07	
304	Framing 7th Floor West	8	8	8	0	15AUG07		24AUG07	
257	Framing 6th Floor West	8	8	8	0	27AUG07		05SEP07	
210	Framing 5th Floor West	8	8	8	0	06SEP07		17SEP07	
211	HVAC Rough-In	8	8	8	0	18SEP07		27SEP07	
212	Plumbing Rough-In	8	8	8	0	28SEP07		09OCT07	
213	Fire Protection	5	5	5	0	28SEP07		04OCT07	
214	Electrical Rough-In	5	5	5	0	05OCT07		11OCT07	
215	MEP Rough-In Inspections	1	1	1	0	12OCT07		12OCT07	
216	Insulation	2	2	2	0	15OCT07		16OCT07	
217	Framing & Insulation Inspection	1	1	1	0	17OCT07		17OCT07	
218	Rock Walls	5	5	5	0	18OCT07		24OCT07	
219	Finish Walls	5	5	5	0	25OCT07		31OCT07	

12JAN07 Fri

2007

JAN FEB MAR APR MAY JUN JUL AUG SEP OCT

ne 4.2

Erect Zone 4.3

Erect Zone 3.2

Erect Zone 5

Erect Zone 6

Erect Zone 7

Roofing Complete West

Framing 12th Floor West

Framing 11th Floor West

Framing 10th Floor West

Framing 9th Floor West

Framing 8th Floor Wes

Framing 7th Floor W

Framing 6th Flo

Framing 5th

HVAC Rc

Plum

Fire Pr

Elect

MEP

Ins

Fra

R

Progress View

Longest Path

FIGURE 5.6

Due to the progress on Activities 44 and 72 and the finish-to-start relationship in Schedule 1 between Activities 44 and 72, there was a total of 18 days of progress on the critical path identified in Schedule 1. However, Figure 5.6 did not detail the progress, or lack of progress, made to near-critical paths from Schedule 1. Did the critical path shift to another work path between Schedule 1 and Schedule 2? If the critical path did shift, when did that occur, and what progress, or lack of progress, was made on that path while it was critical? These questions can easily be answered using the delay analysis techniques already discussed in previous chapters, in conjunction with the same progress comparison shown in Figure 5.6 between Schedule 1 and its Target Schedule, Schedule 2. The only difference is that instead of viewing only progress to the longest path, as in Figure 5.6, the user would view progress to all activities and organize activities first by Total Float and second by Early Start. Although Total Float is not always an accurate way to determine the critical path, it is a helpful tool to identify near-critical paths.

Once the progress comparison is created between Schedule 1 and Schedule 2 in Primavera, sorted by Total Float and Early Start, the analyst is then ready to compare the progress of the critical path between Schedule 1 and Schedule 2 to the progress of the near-critical paths between Schedule 1 and Schedule 2. Again, the critical path is dynamic in nature and can change on a daily basis depending on the progress obtained by each path. Therefore, the analyst must compare the duration of the critical path to the duration of near-critical paths on a daily basis in order to ensure that the critical path did not shift and that the analysis has accounted for all delays and improvements.

Schedule Revision Delays and Improvements

Following the analysis of progress delays and improvements, the analyst should determine if the critical path was delayed, improved, or shifted, due to schedule revisions. Schedule revisions are changes made to a schedule that are not the result of work progress on the Project. Some examples of schedule revisions include added and deleted logic relationships; changed logic relationships; increased durations; changed activity descriptions; added and deleted activities; changes in the work calendar; and changed, added, and deleted constraints. At a minimum, these schedule revisions should be analyzed to determine whether they affect the critical path of the Project.

To simplify the analysis of schedule revisions, there are software applications on the market that compare two schedules, identify all differences between the schedules, and then provide a detailed report of the differences. Claim Digger, a software application now available from Primavera, is one software application that makes analyzing schedule revisions much easier.

When critical Project delay is mitigated by logic revisions, it is possible that one party may be responsible for the delay, while the other is credited with the savings. Typically, unless directed by the Owner or derived from a waiver of a Contract requirement, the savings are credited to the Contractor. This is because the Contractor is generally entitled to determine the means and methods used to execute the work. When it is necessary to alter the means and methods to mitigate delay, the Contractor takes on the risk of successfully executing the work in accordance with the new plan.

While crediting the Contractor with this projected savings may seem unfair because the Contractor has not actually accomplished the savings, if all of the critical Project delays and savings are identified throughout the Project, the correct answer will result. This is the case because the Contractor is expected to complete the work according to its revised plan. For example, if the Contractor mitigates a delay by shortening the duration of a critical path activity from 20 days to ten days, the Contractor is credited with a ten-day savings. If, when performing the work, the Contractor takes 20 days to perform the work and it is still critical, the Contractor will be assigned a ten-day delay. Thus, if the Contractor cannot actually accomplish the savings, no net savings is credited.

Although software applications may identify all changes made between two schedules, it is still the responsibility of the analyst to determine whether schedule revisions are truly schedule revisions. For example, if a schedule revision report identifies that an activity duration has been reduced, should it be considered a schedule revision or work progress on the activity? Typically, if the activity has already received an actual start and the duration has been reduced, the reduced duration is most likely the result of progress. However, if a change has occurred, such as a reduction in the scope of work for that activity, then a reduced duration may represent a schedule revision instead of work progress. A review of Project correspondence or as-built information near the time of the schedule revisions should assist the analyst in determining whether the reduced duration was a result of work progress or a schedule revision.

Changed actual start dates and actual finish dates is another area that may require interpretation by the analyst. If an earlier schedule identified that Activity Z started on December 1, 2006, and a schedule revision report identified that the actual start for Activity Z had been changed to February 1, 2007, which actual start date is accurate for Activity Z? A quick method of determining the accuracy of the actual start date is to reference as-built information and Project correspondence. If the as-built information identifies that Activity Z did not actually start until February 1, 2007, and its progress was assessed according to the December 1, 2006, actual start date identified in the earlier schedule, then the analyst should consider revising its analysis based on the new actual start date of February 1, 2007. If the analyst does not have access to as-built information, or the as-built information does not yield enough information to identify when Activity Z actually started, then the analyst might consider the most recent schedule information, the actual start of February 1, 2007, to be the most up-to-date, accurate as-built information for the Project. As with any change made to schedules being analyzed, corrections to as-built dates should not be automatic. The analyst must consider all of the factors related to making such a change and how these factors will affect the reliability of the results. In some cases it may be necessary to analyze and present the findings based on both dates.

Other Helpful Software Tools

AS-BUILT DIAGRAMS WITH MICROSOFT EXCEL

As-built information can be depicted in a number of ways, depending on the level of detail required by the analyst. As-built diagrams can be plotted manually on graph paper, but it is often more efficient to create as-built diagrams using Microsoft Excel (Excel). With Excel, the analyst can organize the information in the same manner as graph paper, but in Excel, the information can easily be reorganized (using the copy and paste functions), can be color coded (to add emphasis, separate areas, phasing, Subcontractors, etc.), and the analyst can add notes from the as-built information into the individual cell or as a comment. Excel also allows the analyst to have a larger amount of information than graph paper, customize the look of the printouts, and send as-built diagrams as a file.

Figure 5.7 is an as-built diagram created in Excel that details as-built information for concrete work between May 1, 2006, and May 24, 2006. Figure 5.7 depicts a late start and slower-than-expected progress to the reinforcement of slab on the 4th floor. The shaded highlights denote nonwork days according to the Contractor's schedule. While reading through the as-built information, the analyst noted that the Owner's daily log on May 3, 2006, stated that reinforcement of the 4th floor slab could not start because the Contractor was waiting on the rebar fabricator. The Owner's daily logs also stated that on May 7, 2006, the Contractor received the rebar and began reinforcement of the 4th floor on May 8, 2006. Including this level of detail in the as-built diagram will provide the analyst a clearer understanding of exactly when the delays are occurring to specific activities and will assist the analyst in determining the cause of delays to the critical path of the Project.

SCHEDULE ANALYSIS WITH CASE SOFTWARE

The analysis of Project schedules can take a long time, depending on many factors. The analysis of a three-year Project with monthly updates could take the analyst weeks to complete, no matter the magnitude of the Project. If it is a large Project and has several thousand activities, the analysis would likely take additional time due to the increased number of near-critical paths that the analyst would need to evaluate against the critical path. The mathematical comparisons between the near-critical path duration and the critical path duration become increasingly difficult for schedules containing many leads/lags, constraints,

ACTIVITY	MAY 2006																							
	1	2	3	4	5	6	7	8	9	10	11	12	13	14	15	16	17	18	19	20	21	22	23	24
Fly Forms - 4th Floor	x	x																						
Reinforce Slab - 4th Floor								x	x	x														
Pour Slab - 4th Floor											x													
Form Columns - 4th to 5th Floor													x		x									
Pour Columns - 4th to 5th Floor																x								
Strip Slab and Columns- 4th Floor																							x	
Fly Forms - 5th Floor																	x	x						
Reinforce Slab - 5th Floor																			x	x		x		
Pour Slab - 5th Floor																							x	
Form Columns - 5th to 6th Floor																								
Pour Columns - 5th to 6th Floor																								
Strip Slab and Columns- 5th Floor																								

FIGURE 5.7

Waiting for rebar fabrication
Received rebar
Only 10 ironworkers on 9th

multiple calendars, and/or lack actual start and actual finish dates. Although it is certainly possible to perform the mathematics of a schedule analysis manually, it is often more efficient and cost effective to evaluate numerous, complex schedules using software applications. Computer-Aided Schedule Evaluation (CASE) software, currently owned by Trauner Consulting Services, Inc., is one software application that automates the contemporaneous analysis and often decreases the time it takes to perform a schedule analysis by at least 50 percent. Why only 50 percent? Just as in the use of any other analysis software application (Primavera, Microsoft Project, Excel, Claim Digger, CASE, etc.), the analyst must still correctly interpret the results provided by the software and verify the results of the schedule analysis using as-built information and Project correspondence.

Quantifying delays can be a very arduous process for an analyst to complete. However, with the software tools available, analysts can greatly improve efficiency without sacrificing the quality of their analysis. Proper training in the use of any software is essential because, again, the software is only as effective as the analyst using it.

CHAPTER 5 EXAMPLES

Example 5-1: Sample Church Construction

Consider the construction of a church by XYZ Contracting, Inc. The activities have been simplified and the duration of the Project shortened in order to fit the example appropriately. In its Contract with the Owner, the Contractor received its notice to proceed in May 2007 and was required to complete the Project by October 2, 2007. Before starting construction, the Contractor created a baseline schedule reflecting its original plan for the work. After construction began, the Contractor created updates of its schedule on the first day of each month.

Two months into construction, the Project was almost a month behind schedule. Using the baseline schedule and the first two monthly updates, this example will describe the process of determining the critical path of the Project and identifying the activities that caused the critical Project delays.

BASELINE SCHEDULE

As shown in Figure 5.8, the Contractor submitted its baseline schedule showing a planned start of construction on June 1, 2007, and a forecast Project completion of October 1, 2007, one day earlier than the required completion date. Note that the Project completion activity had one day of total float.

The baseline schedule shows that the Project consisted of two separate sequences of activities, or paths, for the Contractor to successfully complete the Project. The building construction path shows a planned construction period from June 1, 2007, to October 1, 2007. The site improvements path shows a planned construction period from June 1, 2007, to September 17, 2007.

The building construction was the longest path to complete the Project. Any delay to the building construction would delay the completion of the Project.

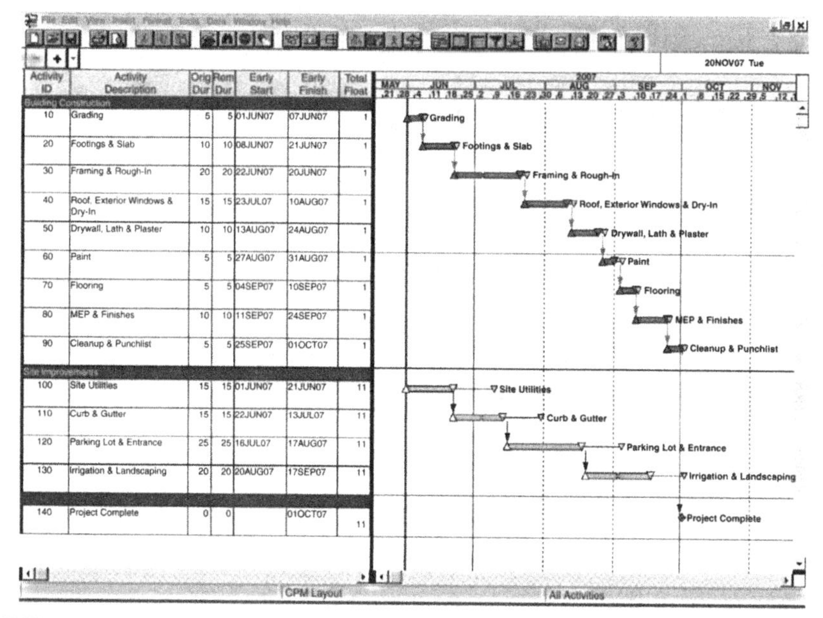

Activity ID	Activity Description	Orig Dur	Rem Dur	Early Start	Early Finish	Total Float
Building Construction						
10	Grading	5	5	01JUN07	07JUN07	1
20	Footings & Slab	10	10	08JUN07	21JUN07	1
30	Framing & Rough-In	20	20	22JUN07	20JUL07	1
40	Roof, Exterior Windows & Dry-In	15	15	23JUL07	10AUG07	1
50	Drywall, Lath & Plaster	10	10	13AUG07	24AUG07	1
60	Paint	5	5	27AUG07	31AUG07	1
70	Flooring	5	5	04SEP07	10SEP07	1
80	MEP & Finishes	10	10	11SEP07	24SEP07	1
90	Cleanup & Punchlist	5	5	25SEP07	01OCT07	1
Site Improvements						
100	Site Utilities	15	15	01JUN07	21JUN07	11
110	Curb & Gutter	15	15	22JUN07	13JUL07	11
120	Parking Lot & Entrance	25	25	16JUL07	17AUG07	11
130	Irrigation & Landscaping	20	20	20AUG07	17SEP07	11
140	Project Complete	0	0		01OCT07	11

FIGURE 5.8

Given the October 2, 2007, Contract completion date, however, the Contractor's October 1, 2007, planned completion for the Project had one day of float relative to the Contract-required completion date. Even though the schedule shows one day of float, as of June 1, 2007, the critical path of the Project ran through the construction of the building.

The site improvements path had 11 days of total float in the baseline schedule. The 11 days of float indicate that the site improvements could be delayed by as much as 11 workdays without causing a delay to the completion date. Relative to the critical building construction path, the site improvements had ten days of float (11 – 1), or could be delayed as much as ten days without becoming critical. For example, if additional landscaping work was added to the Contractor's scope of work resulting in a September 25, 2007, planned completion of the site improvements, the October 1, 2007, Project completion date would not be delayed.

JULY 1, 2007, UPDATE

On July 1, 2007, the Contractor submitted its first monthly update to the Owner, as shown in Figure 5.9. The schedule had a data date of July 1, 2007, represented by the vertical line on the bar chart portion of the schedule. In other words, all the information in the schedule before July 1, 2007, was based on actual progress, and all the information after July 1, 2007, represented the Contractor's updated plan to complete the work as of July 1, 2007.

As of July 1, 2007, the longest path of the Project remained through the building construction. As shown in Figure 5.9, this path had negative 14 days of

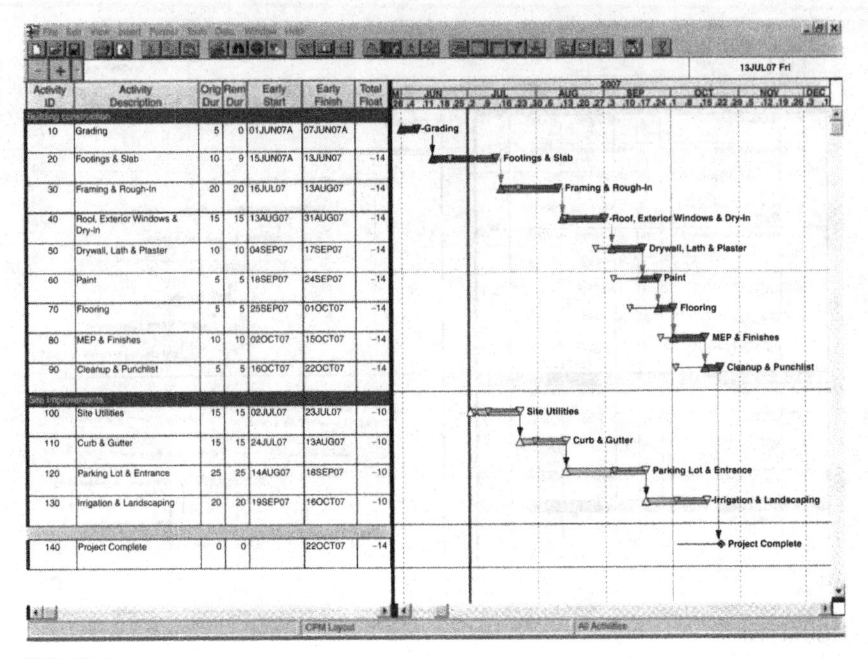

Activity ID	Activity Description	Orig Dur	Rem Dur	Early Start	Early Finish	Total Float
Building construction						
10	Grading	5	0	01JUN07A	07JUN07A	
20	Footings & Slab	10	9	15JUN07A	13JUN07	-14
30	Framing & Rough-In	20	20	16JUL07	13AUG07	-14
40	Roof, Exterior Windows & Dry-In	15	15	13AUG07	31AUG07	-14
50	Drywall, Lath & Plaster	10	10	04SEP07	17SEP07	-14
60	Paint	5	5	18SEP07	24SEP07	-14
70	Flooring	5	5	25SEP07	01OCT07	-14
80	MEP & Finishes	10	10	02OCT07	15OCT07	-14
90	Cleanup & Punchlist	5	5	16OCT07	22OCT07	-14
Site Improvements						
100	Site Utilities	15	15	02JUL07	23JUL07	-10
110	Curb & Gutter	15	15	24JUL07	13AUG07	-10
120	Parking Lot & Entrance	25	25	14AUG07	18SEP07	-10
130	Irrigation & Landscaping	20	20	19SEP07	16OCT07	-10
140	Project Complete	0	0		22OCT07	-14

FIGURE 5.9

float. In other words, as of July 1, 2007, the completion of building construction was 14 workdays behind schedule relative to the Contract completion date. The site improvements had negative ten days of float. While the site improvements were not planned to complete until ten workdays after the Contract completion date, this work was not driving the completion date of the Project. The building construction, with the greater amount of negative float, was critical to the completion of the Project.

The forecast Project completion date in the July 1, 2007, schedule was October 22, 2007—21 days later than planned in the Contractor's baseline schedule. To determine the cause of the 21-day delay, the baseline schedule must first be reviewed for comparison to this first update. After determining the critical path of the baseline schedule, the planned dates for the critical path activities are compared to the actual, or as-built, dates included in the first update. This analysis should be performed on a day-by-day basis for each workday between the June 1, 2007, baseline schedule and July 1, 2007, update.

According to the baseline schedule, the critical path started through the grading work. This work was planned to start on June 1, 2007, and complete on June 7, 2007. According to the progress reported in the July 1, 2007, schedule, the grading work started and completed as planned. Given the June 7, 2007, completion of the grading, the footings and slab work should have started on June 8, 2007, but it did not start until June 15, 2007. To determine how many workdays of delay occurred, the Project calendar must be observed. This is shown in Figure 5.10.

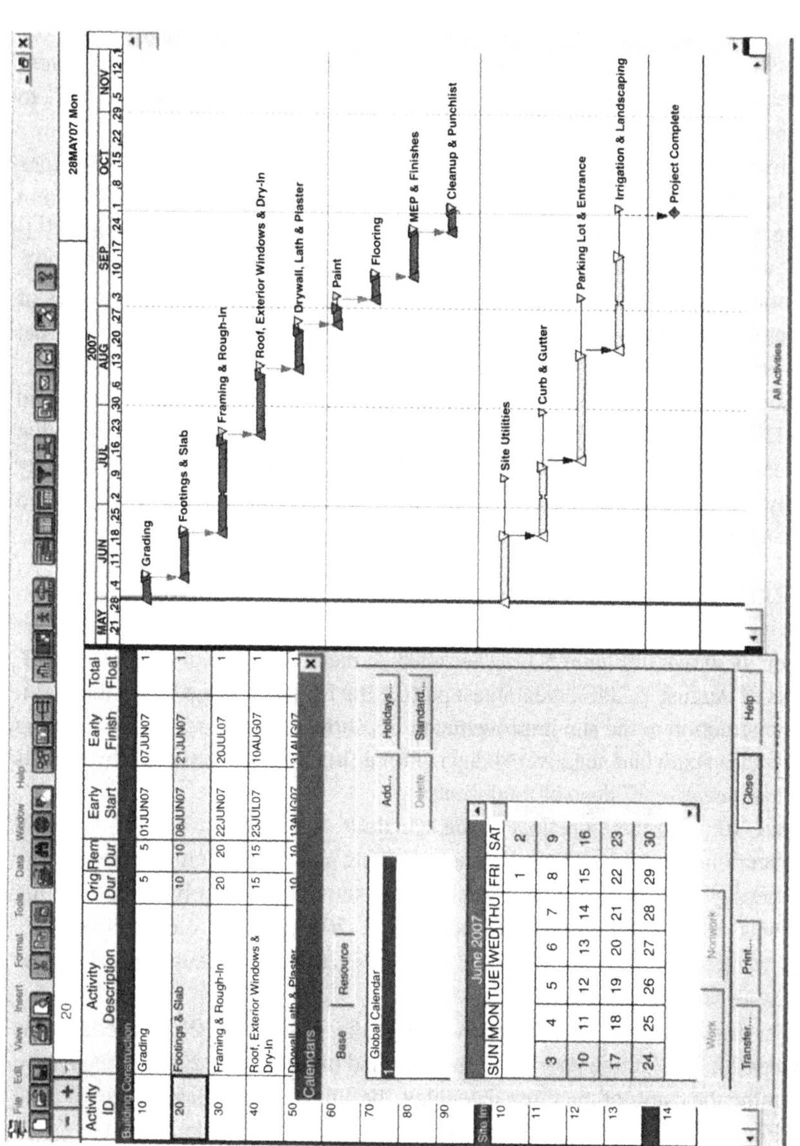
28MAY07 Mon
Activity ID
Activity Description
Orig Dur
Rem Dur
Early Start
Early Finish
Total Float
Building Construction
Grading
Footings & Slab
Framing & Rough-In
Roof, Exterior Windows & Dry-In
Drywall, Lath & Plaster
Paint
Flooring
MEP & Finishes
Cleanup & Punchlist
Site Utilities
Curb & Gutter
Parking Lot & Entrance
Irrigation & Landscaping
Project Complete
Calendars
Base
Resource
Global Calendar
June 2007
SUN MON TUE WED THU FRI SAT
Add...
Delete
Holidays...
Standard...
Work
Nonwork
Transfer...
Print...
Close
Help
All Activities

FIGURE 5.10

According to the preceding Project calendar, the Contractor planned its operations based on a five-day workweek. Note that the weekend days are shaded in the calendar, representing the days the Contractor did not plan to perform any work on the Project. As shown in the Project calendar, the footings and slab work started five workdays later than planned. To determine the delay to the Project completion date, the Project calendar must again be observed (Figure 5.11). The five-workday late start of the footings and slabs caused a five-workday, or seven-calendar-day delay to the Project completion date—from October 1, 2007, to October 8, 2007.

Given the June 15, 2007, start and ten-workday planned duration, the footings and slabs would have been expected to complete by June 28, 2007. As shown in Figure 5.9, however, this work only made one workday of progress as of the July 1, 2007, update. After accounting for the one workday of progress on June 15, 2007, the footings and slab work made no progress for the remaining ten workdays of the period. This would have delayed the Project completion date by an additional ten workdays, or 14 calendar days—from October 8, 2007, to October 22, 2007.

The total delay to the Project completion date between June 1, 2007, and July 1, 2007, was 21 calendar days (7 + 14), or 15 workdays (5 + 10). Note that between the June 1, 2007, and July 1, 2007, schedules, the total float changed by 15 days—from one day to negative 14 days. This change in total float reflected the delay to the Project in workdays.

AUGUST 1, 2007, UPDATE

On August 1, 2007, the Contractor submitted its second monthly update to the Owner, as shown in Figure 5.12. The schedule had a data date of August 1, 2007.

As of August 1, 2007, the longest path of the Project changed from the building construction to the site improvements. As shown in Figure 5.12, the building construction path had negative 14 days of total float, while the site improvements path had negative 17 days of total float.

According to the total float in the schedule, as of August 1, 2007, building construction was 14 workdays behind schedule, and the site improvements were 17 workdays behind schedule. Checking the activities themselves, the building path was planned to complete on October 22, 2007, but the site improvements were not planned to complete until October 25, 2007. As of August 1, 2007, the site improvements were critical.

The forecast Project completion date in the August 1, 2007, schedule was October 25, 2007, three days later than planned in the July 1, 2007, schedule. To determine the cause of the three-day delay, the July 1, 2007, planned critical path (see Figure 5.9) was compared to the progress reported in the August 1, 2007, schedule.

As of the July 1, 2007, schedule, the critical path ran through the footings and slab work. This work was ongoing as of July 1, 2007, and was planned to complete on July 13, 2007. According to the August 1, 2007, update, the footings and slab work completed as planned. Following the completion of the footings and slab work, Figure 5.12 also shows that the framing and rough-in

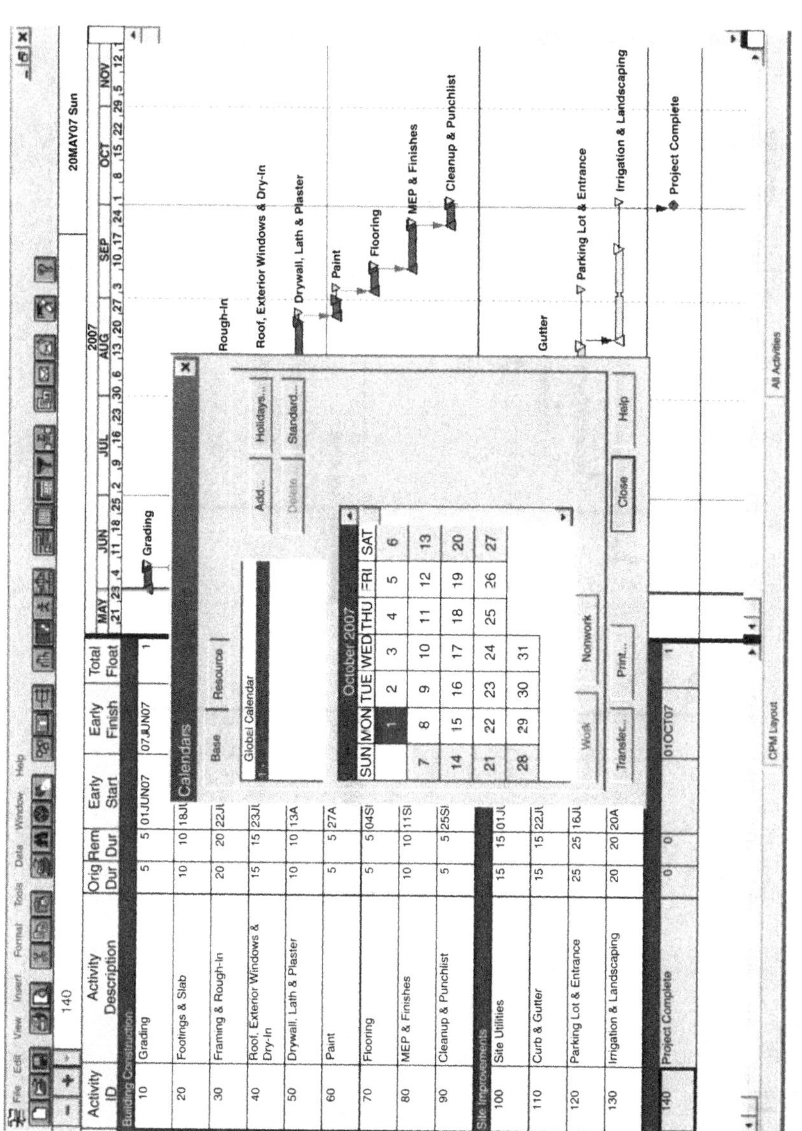

20MAY07 Sun
Calendars
Global Calendar
October 2007
Grading
Rough-In
Roof, Exterior Windows & Dry-In
Drywall, Lath & Plaster
Paint
Flooring
MEP & Finishes
Cleanup & Punchlist
Gutter
Parking Lot & Entrance
Irrigation & Landscaping
Project Complete

FIGURE 5.11

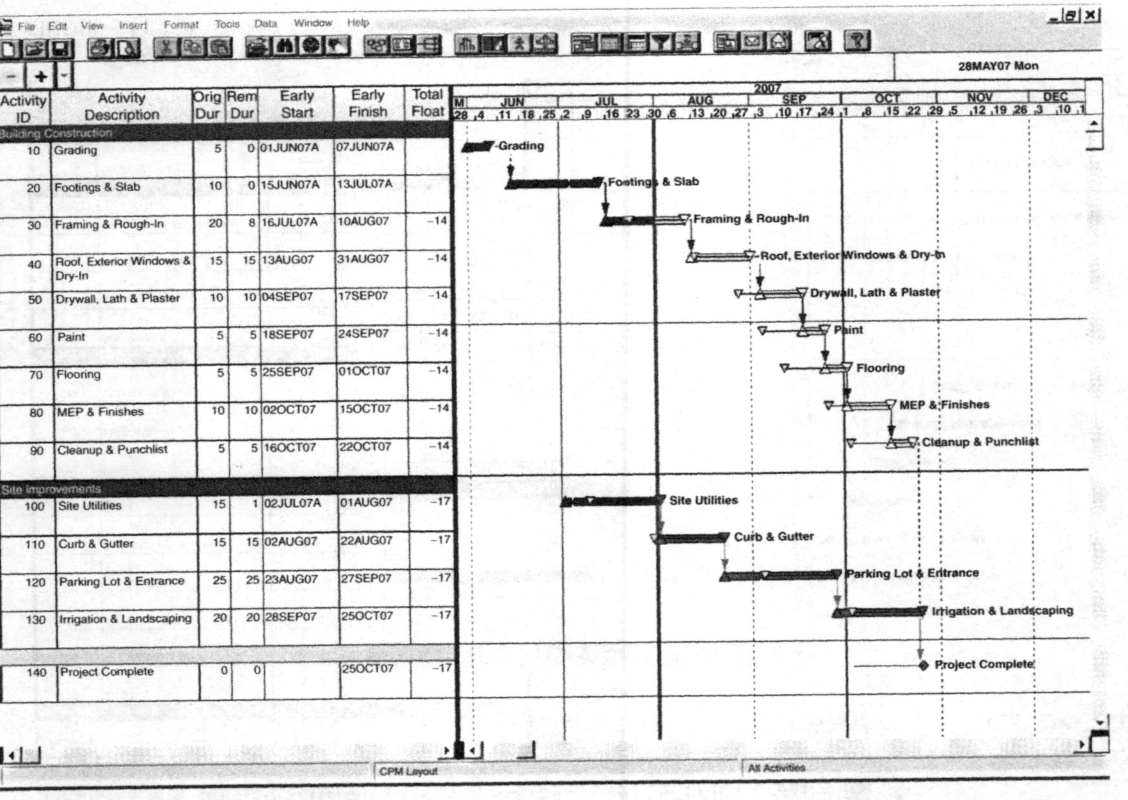

Activity ID	Activity Description	Orig Dur	Rem Dur	Early Start	Early Finish	Total Float
Building Construction						
10	Grading	5	0	01JUN07A	07JUN07A	
20	Footings & Slab	10	0	15JUN07A	13JUL07A	
30	Framing & Rough-In	20	8	16JUL07A	10AUG07	−14
40	Roof, Exterior Windows & Dry-In	15	15	13AUG07	31AUG07	−14
50	Drywall, Lath & Plaster	10	10	04SEP07	17SEP07	−14
60	Paint	5	5	18SEP07	24SEP07	−14
70	Flooring	5	5	25SEP07	01OCT07	−14
80	MEP & Finishes	10	10	02OCT07	15OCT07	−14
90	Cleanup & Punchlist	5	5	16OCT07	22OCT07	−14
Site Improvements						
100	Site Utilities	15	1	02JUL07A	01AUG07	−17
110	Curb & Gutter	15	15	02AUG07	22AUG07	−17
120	Parking Lot & Entrance	25	25	23AUG07	27SEP07	−17
130	Irrigation & Landscaping	20	20	28SEP07	25OCT07	−17
140	Project Complete	0	0		25OCT07	−17

FIGURE 5.12

started as planned on July 16, 2007, and made as-expected progress as of August 1, 2007, reflecting the same August 10, 2007, planned completion as was shown in the July 1, 2007, update. Between July 1, 2007, and August 1, 2007, there was no delay due to the building construction.

The site improvements made slower-than-expected progress between July 1, 2007, and August 1, 2007. The site utilities started as-planned on July 2, 2007. Given the July 2, 2007, start, the utilities should have completed on July 23, 2007. As of August 1, 2007, this work had not completed and had an updated planned finish date of August 1, 2007. The slow progress on the utilities caused this work to become critical and caused the three-day delay to the period.

Summary

Between June 1, 2007, and August 1, 2007, the Project completion date was delayed by 24 days—from October 1, 2007, to October 25, 2007. Of the 24 days, seven were due to the late start of the footings and slab work, 14 were due to the extended duration of the footings and slab work, and three were due to the extended duration of the site utilities.

The Contract completion date was delayed by 23 days, or 17 workdays, as shown by the negative 17 days of total float for the Project completion activity in the August 1, 2007, update. Based on the Contract completion date, the seven-day delay due to the late start of the footings and slab work would have only caused a six-day delay to the Contract completion date.

Example 5-2: Software Features and the Critical Path

With the increased use of CPM schedules in the construction industry, Project schedulers continue to develop an increased understanding of the various features included in the scheduling software in addition to simply adding activities, durations, and logic. These various features can be incorrectly applied or misused and thus affect the critical path of schedules. This, in turn, can result in a schedule reflecting a critical path or delay to a Project other than what is actually occurring on the Project itself. In this example, constraints and multiple calendars will be used to show their effect on the critical path and float reflected in the schedules.

In the previous example, during the construction of a church by XYZ Contracting, Inc., a 24-day delay occurred to the Project between June 1, 2007, and August 1, 2007—21 days of which was due to slow progress on the building and three days of which was due to slow progress on the site utilities. For the purpose of this example, presume that delays to the building construction were the fault of the Contractor, and delays to the site utilities were the responsibility of the Owner.

The Contractor's June 1, 2007, baseline schedule included the same activities as the previous example. As shown in Figure 5.13, however, the critical path ran through the site improvements instead of the building. In the baseline schedule shown in Figure 5.13, the building construction had one day of total float, and the site utilities had zero days of total float. Based on the float alone,

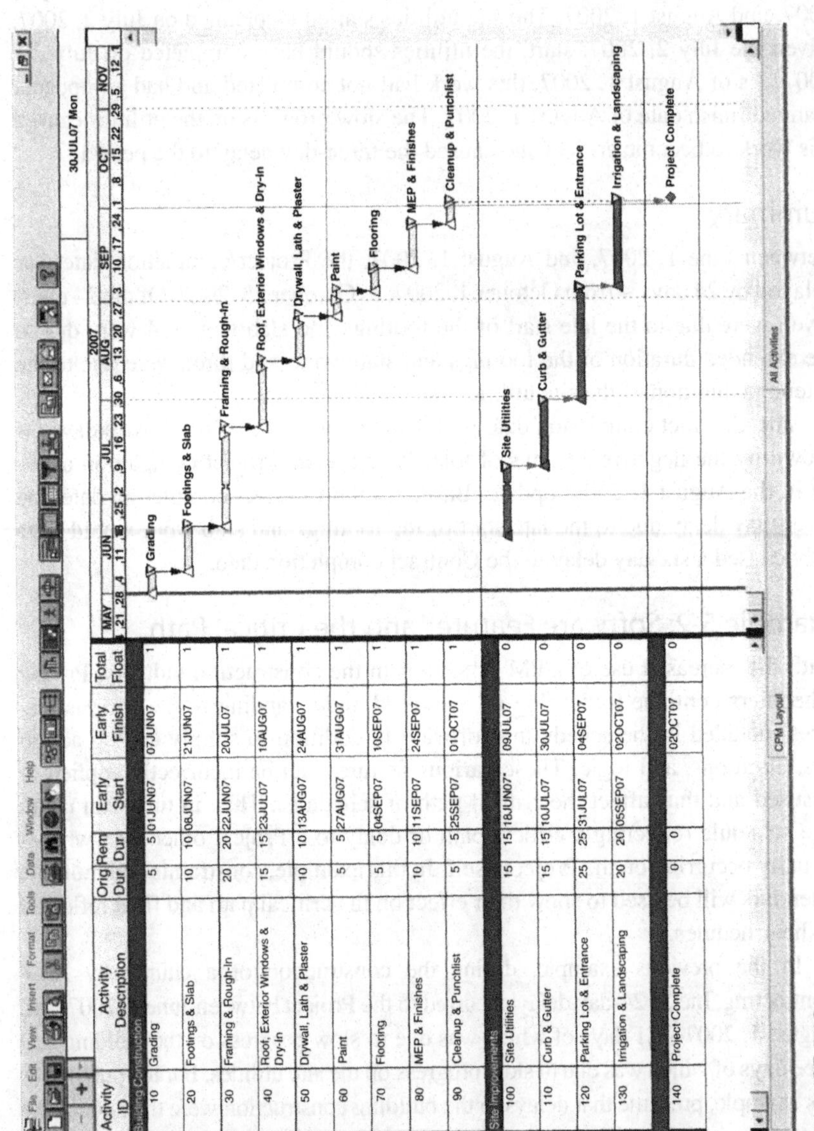
30JUL07 Mon
Activity ID
Activity Description
Orig Dur
Rem Dur
Early Start
Early Finish
Total Float
Building Construction
10 Grading 5 5 01JUN07 07JUN07 1
20 Footings & Slab 10 10 08JUN07 21JUN07 1
30 Framing & Rough-In 20 20 22JUN07 20JUL07 1
40 Roof, Exterior Windows & Dry-In 15 15 23JUL07 10AUG07 1
50 Drywall, Lath & Plaster 10 10 13AUG07 24AUG07 1
60 Paint 5 5 27AUG07 31AUG07 1
70 Flooring 5 5 04SEP07 10SEP07 1
80 MEP & Finishes 10 10 11SEP07 24SEP07 1
90 Cleanup & Punchlist 5 5 25SEP07 01OCT07 1
Site Improvements
100 Site Utilities 15 15 18JUN07 09JUL07 0
110 Curb & Gutter 15 15 10JUL07 30JUL07 0
120 Parking Lot & Entrance 25 25 31JUL07 04SEP07 0
130 Irrigation & Landscaping 20 20 05SEP07 02OCT07 0
140 Project Complete 0 0 02OCT07 0
2007
MAY JUN JUL AUG SEP OCT NOV
Grading
Footings & Slab
Framing & Rough-In
Roof, Exterior Windows & Dry-In
Drywall, Lath & Plaster
Paint
Flooring
MEP & Finishes
Cleanup & Punchlist
Site Utilities
Curb & Gutter
Parking Lot & Entrance
Irrigation & Landscaping
Project Complete
CPM Layout
All Activities

FIGURE 5.13

it would appear that the site utilities were critical. According to the logic in the schedule, however, the site utilities had no predecessors. With no predecessors, the site utilities would normally be expected to start on the first workday of the period: June 1, 2007. In the previous example, the site utilities were planned to start on June 1, 2007 (see Figure 5.8).

A scheduler using constraints may impose on activities—specific planned dates not defined by logic. In this case, the scheduler imposed an early start constraint of June 18, 2007, for the site utilities. This caused the apparent critical path to run through the site improvements instead of the building construction. Note that the asterisk shown next to the June 18, 2007, planned start date identifies that the start date of this work was constrained.

There could be a reasonable explanation for the June 18, 2007, planned start date of the utility work. The Contractor's schedule submittal typically should include a narrative explaining why this constraint was included, especially since it was included on the first activity on the apparent critical path.

Based on the Contractor's baseline schedule shown in Figure 5.13, the critical path ran through the site improvements, starting with the utilities on June 18, 2007. The Project was expected to complete on October 2, 2007, after the completion of the irrigation and landscaping work. The building construction was planned to complete on October 1, 2007, one day earlier than the irrigation and landscaping.

JULY 1, 2007, UPDATE

On July 1, 2007, the Contractor submitted its first monthly update to the Owner, as shown in Figure 5.14. The July 1, 2007, schedule shown in Figure 5.14 reflects a forecast Project completion date on October 16, 2007, 14 days later than the October 2, 2007, forecast Project completion date reflected in the baseline schedule. In Figure 5.14 the site improvements had negative ten days of total float, and the building path had total floats ranging from negative four to negative eight days. Based on the total float, the critical path ran through the site utilities.

According to the baseline schedule (Figure 5.13), the critical path ran through the June 18, 2007, planned start of the site utilities. According to the July 1, 2007, schedule, this work had not started and was still critical. According to the schedules shown in Figures 5.13 and 5.14, a lack of progress on the site utilities caused a 14-day delay to the Project.

In the previous example (Figures 5.8 and 5.9), between June 1, 2007, and July 1, 2007, the footings and slab work started late and made slower-than-expected progress, delaying the planned completion of the building construction and Project completion until October 22, 2007. The July 1, 2007, schedule shown in Figure 5.14 shows the same late start and slow progress on the footings and slab work but reflects a planned completion of the building construction on October 8, 2007.

The planned start and finish dates for the activities following the footings and slab work were earlier in Figure 5.14 than in Figure 5.9. In the previous example,

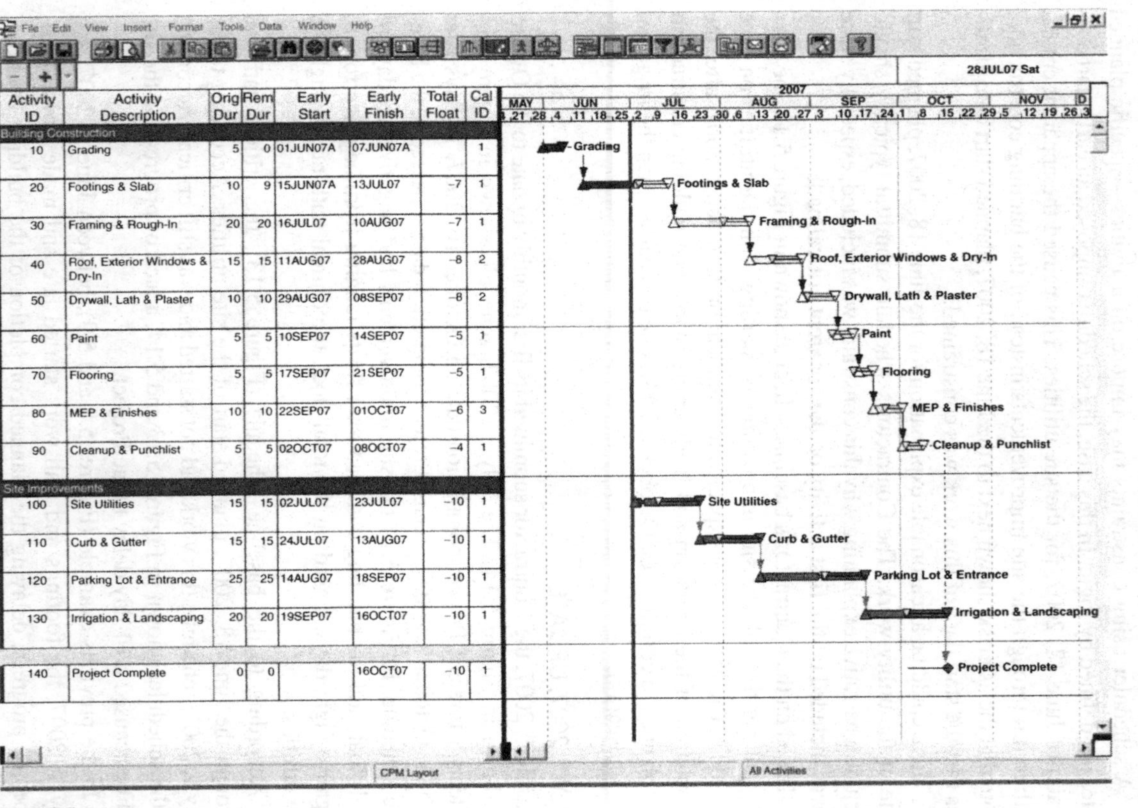

Activity ID	Activity Description	Orig Dur	Rem Dur	Early Start	Early Finish	Total Float	Cal ID
Building Construction							
10	Grading	5	0	01JUN07A	07JUN07A		1
20	Footings & Slab	10	9	15JUN07A	13JUL07	−7	1
30	Framing & Rough-In	20	20	16JUL07	10AUG07	−7	1
40	Roof, Exterior Windows & Dry-In	15	15	11AUG07	28AUG07	−8	2
50	Drywall, Lath & Plaster	10	10	29AUG07	08SEP07	−8	2
60	Paint	5	5	10SEP07	14SEP07	−5	1
70	Flooring	5	5	17SEP07	21SEP07	−5	1
80	MEP & Finishes	10	10	22SEP07	01OCT07	−6	3
90	Cleanup & Punchlist	5	5	02OCT07	08OCT07	−4	1
Site Improvements							
100	Site Utilities	15	15	02JUL07	23JUL07	−10	1
110	Curb & Gutter	15	15	24JUL07	13AUG07	−10	1
120	Parking Lot & Entrance	25	25	14AUG07	18SEP07	−10	1
130	Irrigation & Landscaping	20	20	19SEP07	16OCT07	−10	1
140	Project Complete	0	0		16OCT07	−10	1

FIGURE 5.14

all activities were planned using a standard five-day workweek calendar. In the July 1, 2007, schedule used in this example, however, the scheduler added a six-day workweek calendar and a seven-day workweek calendar. These calendars are identified in Figure 5.14 in the "Cal ID" column, under which the five-, six-, and seven-day workweeks are identified by calendars 1, 2, and 3, respectively.

As of July 1, 2007, both this and the previous examples reflect an October 16, 2007, planned completion date for site improvements due to the site utility delay. This work was not critical in the previous example, however, due to the building construction delays. Due to the calendars added to the July 1, 2007, schedule in this example, the building construction was not critical. As with the use of constraints, had the Contractor planned to work weekends for the selected activities to mitigate the building delays, this information should have been included in the Contractor's schedule narrative.

AUGUST 1, 2007, UPDATE

On August 1, 2007, the Contractor submitted its second monthly update to the Owner, as shown in Figure 5.15. The schedule had a data date of August 1, 2007. The forecast Project completion date as of the August 1, 2007, schedule was October 25, 2007—nine days later than planned in the July 1, 2007, schedule shown in Figure 5.14. As of the Contractor's August 1, 2007, update, the site improvements path had the greatest amount of negative float (negative 17). Based on the total float, the site utilities appeared to be critical.

As of the July 1, 2007, schedule (Figure 5.14), the site utilities were critical and were planned to start on July 2, 2007. This occurred as planned. Given the July 2, 2007, start date, the utilities should have completed on July 23, 2007. As of the August 1, 2007, schedule, however, this work had not completed, and it was expected to complete on August 1, 2007, seven workdays later than July 23, 2007. This delayed the Project completion date by seven workdays, or nine calendar days, from October 16, 2007, to October 25, 2007.

Note that in Figure 5.15 all building construction activities were again planned to occur during a five-day workweek (Cal ID 1), the same calendar used in the baseline schedule. Between July 1, 2007, and August 1, 2007, this delayed the completion of the building construction from October 8, 2007, to October 22, 2007. Due to the site utility delays, however, this did not cause a delay to the Project completion date.

Summary

Between June 1, 2007, and August 1, 2007, a 23-day delay occurred to the Project completion date, from October 2, 2007, to October 25, 2007. According to the Contractor's schedules (Figures 5.13 to 5.15), the 23-day delay was due to the late start and extended duration of the site utilities work. Are the Contractor's schedules reasonable? Maybe. Consider the following questions:

1. In the baseline, why was a June 18, 2007, early start constraint necessary for the site utilities?

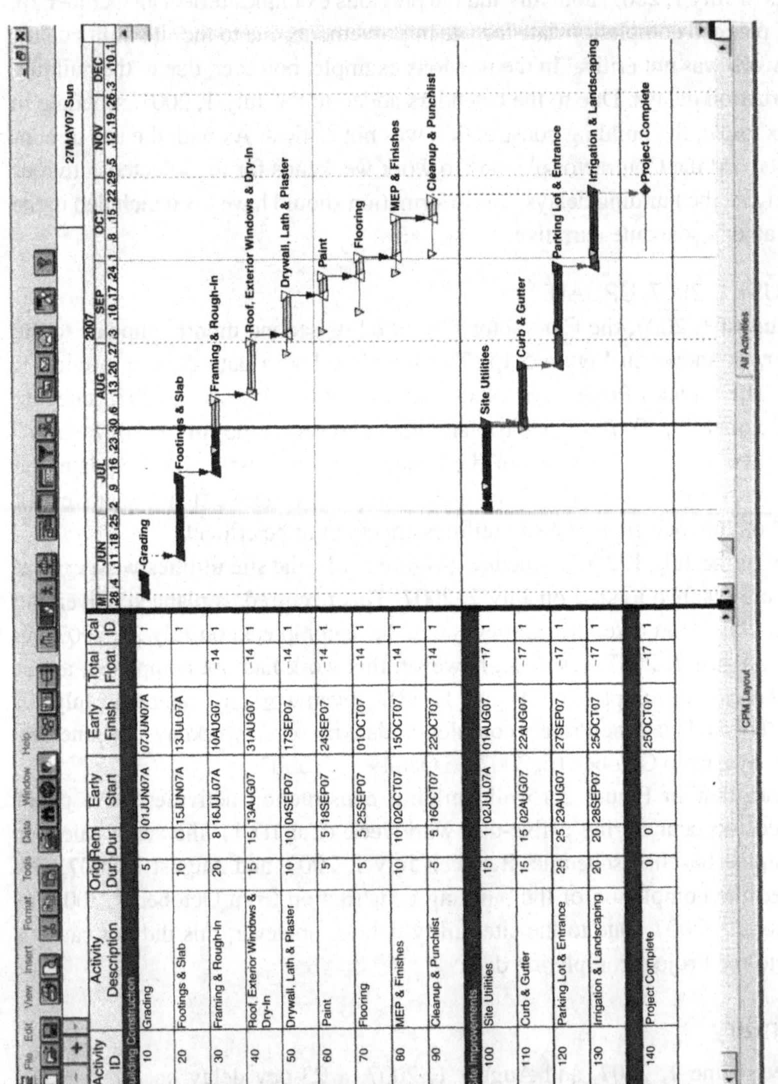

Activity ID	Activity Description	Orig Dur	Rem Dur	Early Start	Early Finish	Total Float	Cal ID
Building Construction							
10	Grading	5	0	01JUN07A	07JUN07A		1
20	Footings & Slab	10	0	15JUN07A	13JUL07A		1
30	Framing & Rough-In	20	8	16JUL07A	10AUG07	−14	1
40	Roof, Exterior Windows & Dry-In	15	15	13AUG07	31AUG07	−14	1
50	Drywall, Lath & Plaster	10	10	04SEP07	17SEP07	−14	1
60	Paint	5	5	18SEP07	24SEP07	−14	1
70	Flooring	5	5	25SEP07	01OCT07	−14	1
80	MEP & Finishes	10	10	02OCT07	15OCT07	−14	1
90	Cleanup & Punchlist	5	5	16OCT07	22OCT07	−14	1
Site Improvements							
100	Site Utilities	15	1	02JUL07A	01AUG07	−17	1
110	Curb & Gutter	15	15	02AUG07	22AUG07	−17	1
120	Parking Lot & Entrance	25	25	23AUG07	27SEP07	−17	1
130	Irrigation & Landscaping	20	20	28SEP07	25OCT07	−17	1
140	Project Complete	0	0		25OCT07	−17	1

FIGURE 5.15

2. Why were the six-day and seven-day workweek calendars added to the July 1, 2007, schedule?
3. Why were the additional calendars then removed in the August 1, 2007, schedule?

The site utilities work did not start until July 2, 2007. It is possible that the early start constraint of June 18, 2007, was based on a reasonable expectation by the Contractor as to when the utilities were expected to start. With the information given in this example, however, we don't know.

In the previous example, the site utilities work was planned to start on June 1, 2007, allowing for a planned completion of the site improvements before the building construction. This caused the building construction to be critical. As of June 18, 2007, the site utilities had not begun. Had the building made as-expected progress through June 18, 2007, this would have caused the critical path to shift to the site utilities. The late start of the footings and slab work, however, would have caused the building construction to remain critical.

Between June 1, 2007, and July 1, 2007, the planned completion of the footings and slab work was delayed by 21 days—from June 21, 2007, to July 13, 2007. In the previous example, this caused a 21-day delay to the Project. Due to the six- and seven-day workweek calendars added to the schedule in this example, however, the overall building construction was delayed by only seven days—from October 1, 2007, to October 8, 2007. This caused the critical path to run through the site utilities.

The removal of the six-day and seven-day calendars in the August 1, 2007, schedule is an indication that these workweeks were never planned to be utilized by the Contractor. Presuming that the utility delays were caused by the Owner, as of August 1, 2007, the site utilities were critical whether or not the additional calendars were used. Had these calendars not been included in the July 1, 2007, schedule (as shown in Figure 5.9), the building construction would have been critical as of July 1, 2007, even with the late start of the utilities. When compared to the previous example, the scheduler's use of constraints and multiple calendars appeared to mask a 21-day delay to the building construction.

Example 5-3: Using Float to Determine Delays and Critical Path

Total float is the number of workdays an activity or path of activities can be delayed without delaying the Project completion date. Total float can be positive, zero, or negative, depending on the planned completion date of the activity or path of activities relative to the required completion date for the Project when the Project end date is constrained:

- *Positive total float*—planned completion before required completion date
- *Zero total float*—planned completion same day as required completion date
- *Negative total float*—planned completion after required completion date

In some cases, the critical path of a schedule can be identified by finding the path of activities with the least positive or greatest negative total float. For example, refer to the June 1, 2007, baseline schedule used in Example 5-1 (Figure 5.8). The building construction path had 1 day of total float, and the site improvements path had 11 days of total float. The building construction path, with the least amount of total float, appears to be critical.

In order to verify whether or not the path with the least amount of total float is the true critical path, or "longest path," the analyst must identify the sequence of activities with the longest duration from the start of the Project through the finish of the Project. As shown in Figure 5.8, both the building construction and site improvements were planned to start on June 1, 2007. The building construction path was planned to complete on October 1, 2007. The site improvements path was planned to complete two weeks earlier—on September 17, 2007. Therefore, the building construction was the longest path and was controlling the completion of the Project.

Although sometimes the critical path can be determined by simply observing the path with the least amount of total float, relying on total float alone can be misleading. A Project scheduler can use various features, such as constraints, that can alter the total float reflected in the schedules. Consider the same sample Project as Figure 5.8, but refer to Figure 5.16 to determine the critical path. In this case, the scheduler included a mandatory finish constraint of September 17, 2007, for Activity 130, *Irrigation and Landscaping*. Because of the mandatory finish constraint on September 17, 2007, the *Irrigation and Landscaping* work could not be delayed, and the total float of this path became zero. Using total float alone to identify the critical path of the Project, the critical path would appear to run through the site improvements. Is this correct?

Again, to verify whether or not the path with the least amount of total float is the critical path, the analyst must identify the sequence of activities with the longest duration from the start of the Project through the finish of the Project. The site improvements were planned to complete on September 17, 2007, but the building construction was not planned to complete until two weeks later—on October 1, 2007. Therefore, even though the site improvements had less total float, the building construction was controlling the Project completion date and was the longest path.

In the July 1, 2007, update, the Contractor removed the mandatory finish constraint from Activity 130, *Irrigation and Landscaping*, but added a mandatory finish constraint of October 22, 2007, for the completion of Activity 90, *Cleanup & Punchlist*. As shown in Figure 5.17, the site improvements continued to have less total float than the building construction path.

In Figure 5.17, Activity 140, *Project Complete*, had the greatest negative total float at 14 workdays. Of the two paths of construction activities required to complete the Project, however, the site improvements path had the greatest negative total float. Between the June 1, 2007, baseline schedule and July 1, 2007, update, Activity 100, *Site Utilities*, made no progress. The forecast

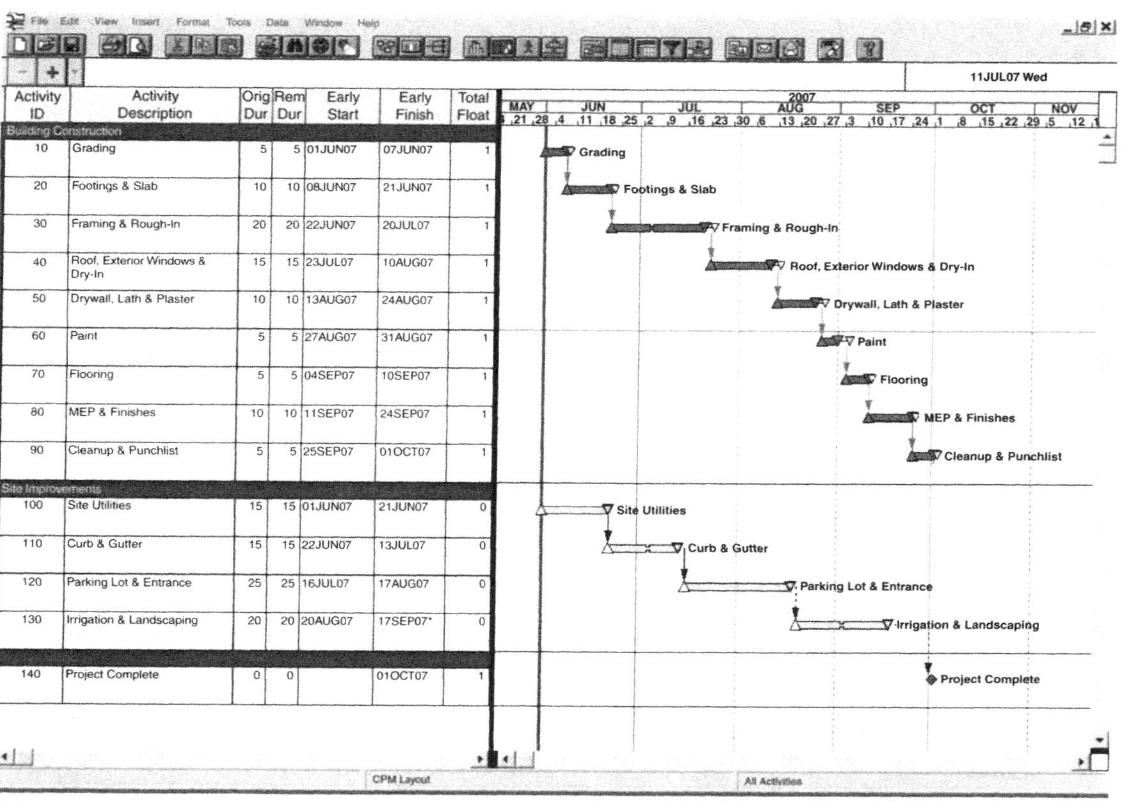

Activity ID	Activity Description	Orig Dur	Rem Dur	Early Start	Early Finish	Total Float
Building Construction						
10	Grading	5	5	01JUN07	07JUN07	1
20	Footings & Slab	10	10	08JUN07	21JUN07	1
30	Framing & Rough-In	20	20	22JUN07	20JUL07	1
40	Roof, Exterior Windows & Dry-In	15	15	23JUL07	10AUG07	1
50	Drywall, Lath & Plaster	10	10	13AUG07	24AUG07	1
60	Paint	5	5	27AUG07	31AUG07	1
70	Flooring	5	5	04SEP07	10SEP07	1
80	MEP & Finishes	10	10	11SEP07	24SEP07	1
90	Cleanup & Punchlist	5	5	25SEP07	01OCT07	1
Site Improvements						
100	Site Utilities	15	15	01JUN07	21JUN07	0
110	Curb & Gutter	15	15	22JUN07	13JUL07	0
120	Parking Lot & Entrance	25	25	16JUL07	17AUG07	0
130	Irrigation & Landscaping	20	20	20AUG07	17SEP07*	0
140	Project Complete	0	0		01OCT07	1

FIGURE 5.16

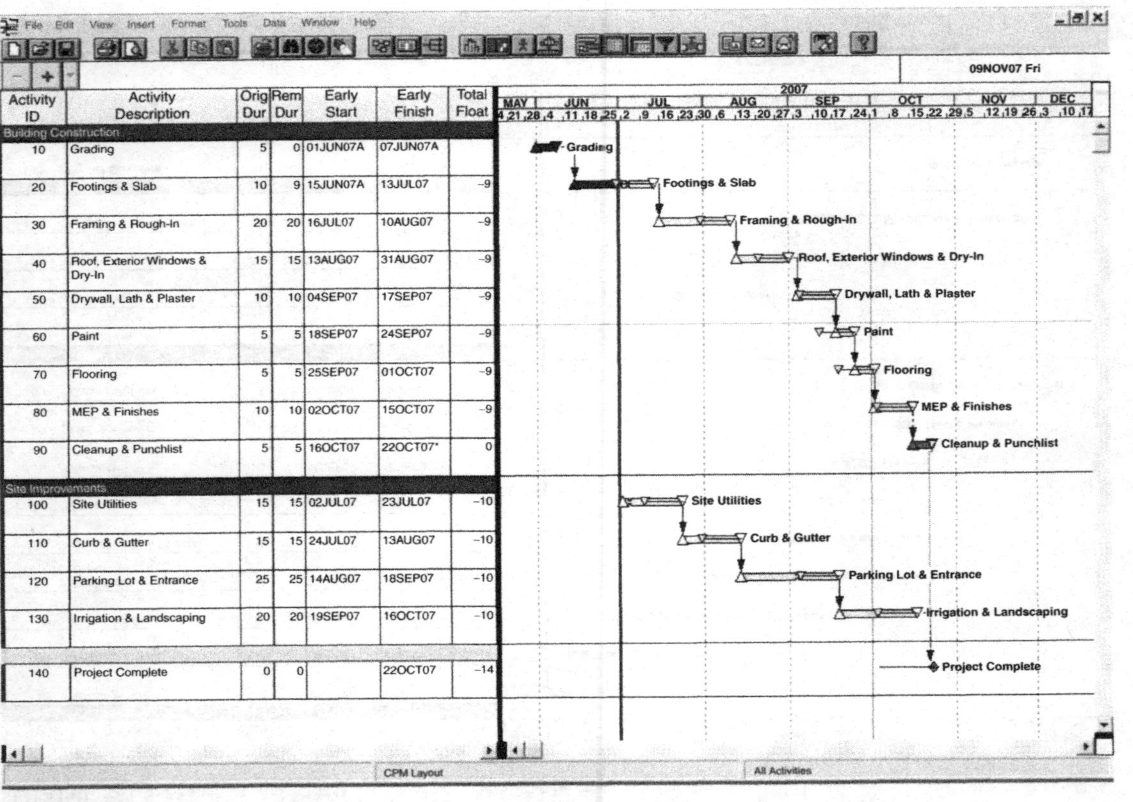

Activity ID	Activity Description	Orig Dur	Rem Dur	Early Start	Early Finish	Total Float
Building Construction						
10	Grading	5	0	01JUN07A	07JUN07A	
20	Footings & Slab	10	9	15JUN07A	13JUL07	–9
30	Framing & Rough-In	20	20	16JUL07	10AUG07	–9
40	Roof, Exterior Windows & Dry-In	15	15	13AUG07	31AUG07	–9
50	Drywall, Lath & Plaster	10	10	04SEP07	17SEP07	–9
60	Paint	5	5	18SEP07	24SEP07	–9
70	Flooring	5	5	25SEP07	01OCT07	–9
80	MEP & Finishes	10	10	02OCT07	15OCT07	–9
90	Cleanup & Punchlist	5	5	16OCT07	22OCT07*	0
Site Improvements						
100	Site Utilities	15	15	02JUL07	23JUL07	–10
110	Curb & Gutter	15	15	24JUL07	13AUG07	–10
120	Parking Lot & Entrance	25	25	14AUG07	18SEP07	–10
130	Irrigation & Landscaping	20	20	19SEP07	16OCT07	–10
140	Project Complete	0	0		22OCT07	–14

FIGURE 5.17

Project completion date was delayed by three weeks—from October 1, 2007, to October 22, 2007. Did the lack of progress on *Site Utilities* cause the delay to the Project?

Based on a review of the total float, the site improvements path had the greatest negative total float in both schedules and would appear to have caused the delay to the Project. As discussed earlier, however, a review of the baseline schedule established that the longest path actually ran through the building construction path. According to the July 1, 2007, schedule, the site improvements path was planned to complete on October 16, 2007. The building construction path was planned to complete about a week later, on October 22, 2007. The building construction path remained critical in the July 1, 2007, schedule, and slow progress on the building construction path between June 1, 2007, and July 1, 2007, caused the three-week delay to the Project completion date. While the site improvements path had the least amount of total float in both schedules, this work never actually controlled the Project completion date.

A review of the total float on the various activities in the July 1, 2007, schedule shown in Figure 5.17 could leave the analyst confused. How could the building construction path start with negative nine days of total float, finish with zero total float, and finish on the same day as the Project completion activity with negative 14 days of total float? Simply put, adding constraints to a schedule can result in the alteration of total float values, causing the total float values to be different from what the planned durations and logic relationships would otherwise dictate. For example, if the mandatory finish constraint on Activity 90, *Cleanup & Punchlist,* were removed from the July 1, 2007, schedule, the resulting schedule reflects a building construction path with a uniform total float of –14, as shown in Figure 5.18.

As shown in Figure 5.18, the building construction path was the longest path when all constraints were removed from the schedule and also had the greatest negative float. While the critical path could be determined in Figure 5.18 by simply observing the path with the least total float, only after determining that the building construction was the longest path of duration between the start and finish of the Project do we know that it was controlling the completion date of the Project. Before removing the constraints in Figure 5.9, total float was meaningless in determining the critical path of the Project.

Summary

Using total float alone as a method of determining the critical path and delays to a Project can be misleading and may result in an incorrect assessment of the activities that are truly controlling the completion of a Project. While they can be used incorrectly, features such as constraints are available to schedulers using CPM software and may be used in the development of the baseline schedule and monthly updates used throughout Projects. To find the correct critical path of the Project, be sure to determine the longest path by running a longest-path filter from the data date of the schedule through the Project completion milestone.

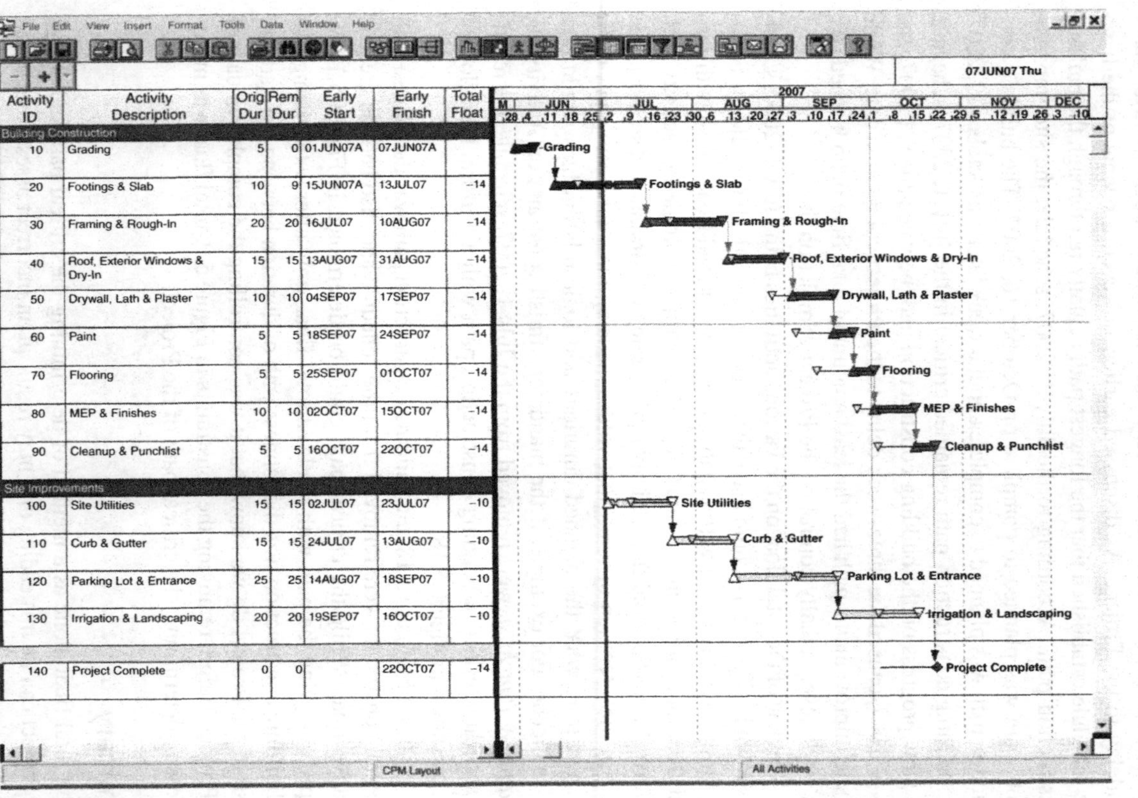

Activity ID	Activity Description	Orig Dur	Rem Dur	Early Start	Early Finish	Total Float
Building Construction						
10	Grading	5	0	01JUN07A	07JUN07A	
20	Footings & Slab	10	9	15JUN07A	13JUL07	−14
30	Framing & Rough-In	20	20	16JUL07	10AUG07	−14
40	Roof, Exterior Windows & Dry-In	15	15	13AUG07	31AUG07	−14
50	Drywall, Lath & Plaster	10	10	04SEP07	17SEP07	−14
60	Paint	5	5	18SEP07	24SEP07	−14
70	Flooring	5	5	25SEP07	01OCT07	−14
80	MEP & Finishes	10	10	02OCT07	15OCT07	−14
90	Cleanup & Punchlist	5	5	16OCT07	22OCT07	−14
Site Improvements						
100	Site Utilities	15	15	02JUL07	23JUL07	−10
110	Curb & Gutter	15	15	24JUL07	13AUG07	−10
120	Parking Lot & Entrance	25	25	14AUG07	18SEP07	−10
130	Irrigation & Landscaping	20	20	19SEP07	16OCT07	−10
140	Project Complete	0	0		22OCT07	−14

FIGURE 5.18

Progress Override Versus Retained Logic

Some Primavera software applications allow the user to customize various options that affect how the software performs its scheduling calculations. The options allow the user to customize how the software addresses the use of leads and lags, the criticality of open ends, and other items that have a significant effect on how the software forecasts the remainder of the Project. Of these options, the most important option, and often the most misunderstood option, is whether to apply Retained Logic or Progress Override when assessing out-of-sequence progress. The scheduler and analyst must have a very clear understanding of these two options and how they affect the Project schedule.

The Primavera Project Planner (P3) Help defines the difference between Retained Logic and Progress Override as follows:

> When you choose Retained Logic, P3 does not schedule the remaining duration of a progressed activity until all of its predecessors are complete. When you choose Progress Override, P3 ignores network logic and allows the activity to progress without delay.

To get a better understanding of how the two options can affect a schedule differently, Figure 5.19 shows a portion of the critical path of the construction of a large ballroom.

The Contractor plans on completing the drywall in seven workdays, then priming for five days, and then painting for five days. All activities are tied using finish-to-start logic with the painting forecasted to finish on Day 17. The drywall actually started on Day 1 and progressed as expected through Day 3. However, on Day 4, the drywall Subcontractor notified the Contractor that it would not be returning to the site until Day 8. To mitigate some of the impending delay from the drywall, the Contractor decided to gain some out-of-sequence progress by priming and painting the portion of the ballroom that had already been drywalled. The painting Subcontractor began priming on Day 5 and began painting the following day on Day 6. Both the prime and paint activities each gained two days of progress before they were stopped because of the lack of progress to the drywall. The as-built diagram of the work accomplished on each activity through the end of Day 7 is shown in Figure 5.20.

Based on the progress accomplished through Day 7, drywall has four workdays remaining, priming has three workdays remaining, and paint has three workdays remaining. The Contractor decided to use the Retained Logic option

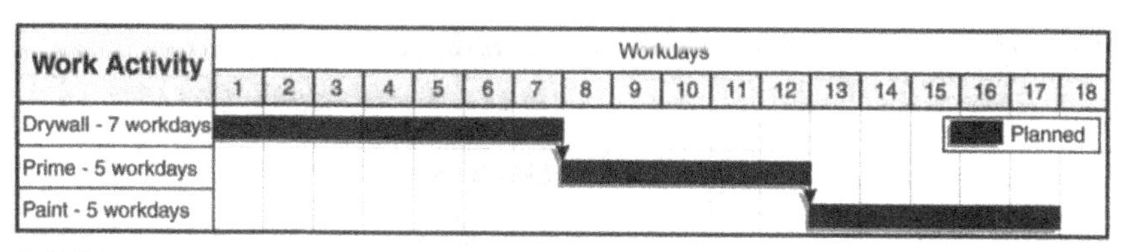

FIGURE 5.19

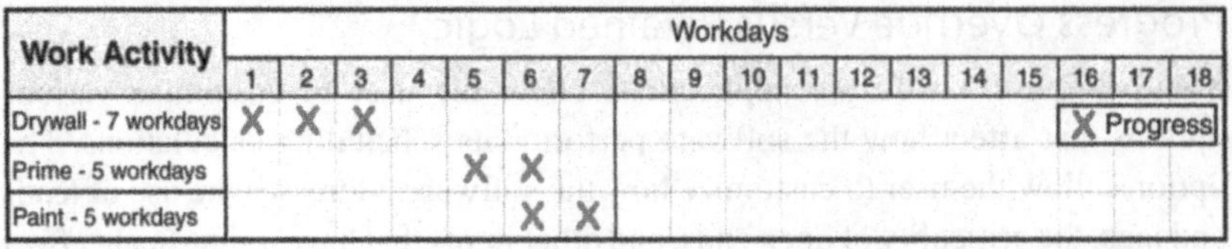

Work Activity	Workdays																	
	1	2	3	4	5	6	7	8	9	10	11	12	13	14	15	16	17	18
Drywall - 7 workdays	X	X	X													X Progress		
Prime - 5 workdays					X	X												
Paint - 5 workdays						X	X											

FIGURE 5.20

Work Activity	Workdays																	
	1	2	3	4	5	6	7	8	9	10	11	12	13	14	15	16	17	18
Drywall - 7 workdays	X	X	X															
Prime - 5 workdays					X	X												
Paint - 5 workdays						X	X											

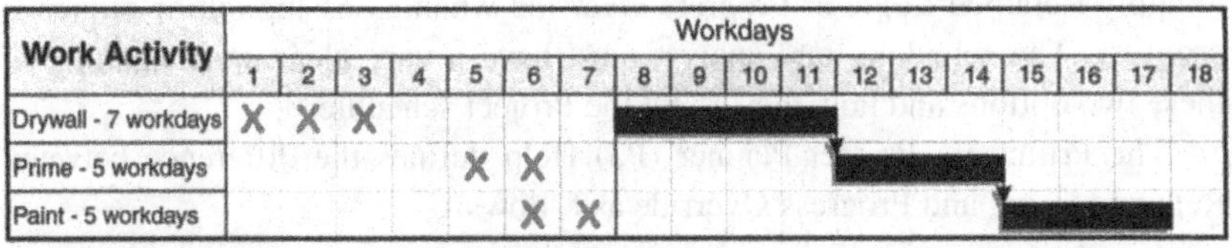

FIGURE 5.21

to forecast the rest of the remaining work because the finish of the drywall was still controlling the remainder of the priming and painting work. Figure 5.21 represents the forecast of the remainder of the work using Retained Logic.

According to Figure 5.21, using the Retained Logic option maintained the finish-to-start relationships between the activities, even though prime and paint made a total of four days of out-of-sequence progress. With Retained Logic, the Contractor still plans on finishing the work on Day 17 and still plans on progressing the remaining work to drywall, prime, and paint using finish-to-start logic relationships.

However, if the Contractor decided to use Progress Override, the software would have ignored the finish-to-start logic tie between drywall and prime, along with the finish-to-start tie between prime and paint. The Progress Override option would forecast the completion of the remainder of the Project assuming that because prime and paint had started without the drywall being finished, then the finish-to-start logic between these activities was not valid. In other words, Progress Override automatically ignores the finish-to-start logic ties because it assumes that the finish of the predecessor activity is no longer controlling the progress on the successor activity.

Figure 5.22 represents the forecast of the remainder of the work using Progress Override. According to Figure 5.22, using the Progress Override option ignored the finish-to-start relationships between the activities and forecasted

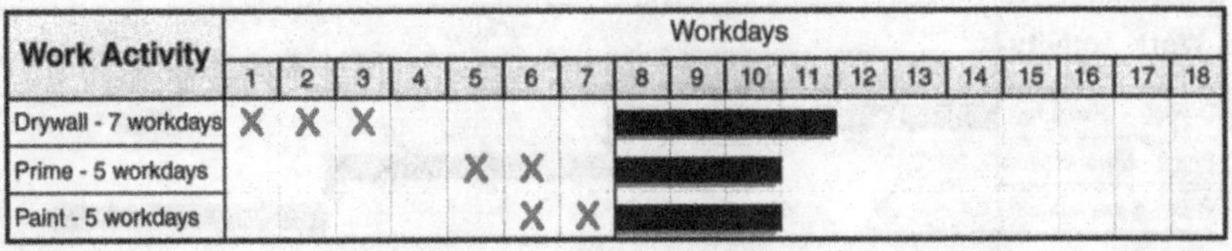

Work Activity	Workdays																	
	1	2	3	4	5	6	7	8	9	10	11	12	13	14	15	16	17	18
Drywall - 7 workdays	X	X	X															
Prime - 5 workdays					X	X												
Paint - 5 workdays						X	X											

FIGURE 5.22

that these activities could progress independently. With Progress Override, the Contractor plans on finishing the work on Day 11, six workdays earlier than when using Retained Logic.

So which option should the scheduler and analyst use? In construction scheduling, Retained Logic is almost always the best option. If a scheduler uses Retained Logic and later decides that previous logic relationships that were input are no longer valid on the Project, then the scheduler can remove or adjust those logic relationships to more accurately represent the current Project conditions. However, logic revisions should be made only to work that has yet to be completed. If a scheduler were to adjust the logic of completed work, his action is pointless because the work has already been completed and will not affect the future of the Project schedule. No matter what option the scheduler uses, he should be consistent.

For the schedule analyst, the decision of which option to use has already been made by the scheduler. Most likely, the analyst is reviewing a schedule that has already been completed, used on the Project, and is now being used to quantify Project delays and improvements. If a scheduler used Retained Logic, and the analyst decided to switch the schedule option to Progress Override, the analyst would essentially be performing an analysis on a different schedule. However, it may be beneficial, for information purposes, for the analyst to view the Project forecast using both options. For example, if Project schedules were consistently being submitted using Retained Logic and midway through the Project the scheduler switched to the Progress Override option, analyzing the schedules in both Progress Override and Retained Logic may be a good source of information. When the switch was made, did the Contractor progress its work sequence following Progress Override or Retained Logic? Such an analysis may provide a more realistic understanding of which schedule option was being used to manage the Project.

Chapter | six

Delay Analysis Using No Schedules

USE OF CONTEMPORANEOUS DOCUMENTS FOR SEQUENCE AND TIMING

The preceding chapters discuss the performance of a delay analysis using a detailed CPM schedule and a bar chart. This chapter addresses the "worst case" situation: the Project with no as-planned schedule. This is the most difficult situation in which to perform a delay analysis. Again, as the available information decreases, the analyst must make more assumptions, and the analysis becomes more subjective. However, while this type of analysis is difficult, it is not impossible.

When there is no as-planned schedule, it is usually because the General Contractor did not prepare a schedule for the Project. In this situation, the analyst should begin by reviewing all available documentation to determine if any information exists that may provide some idea of a proposed sequence of the work or timing for specific activities. The analyst should investigate the following documents:

- *The Contract documents* for any specific sequence, phasing, or staging specified as a requirement on the Project. (Figure 6.1)
- *Correspondence between the General Contractor and the Owner or Owner's representative* for references to sequence or timing, even if only for a portion of the Project. (Figures 6.2 and 6.3)
- *Subcontract agreements* to look for any sequence or timing dictated to the Subcontractors by the General Contractor concerning specific subcontract work.
- *Correspondence with Subcontractors* for any discussion concerning schedules, sequence, or timing. (Figure 6.4)
- *Partial schedules produced during the Project*, which may describe any planned sequence or timing for portions of the Project. (Figure 6.5)

Special Conditions of the Contract-West Street Bridge-Section 1.01, paragraph 5.a:
In staging its work the contractor must complete all abutment construction activities prior to paving of the approaches on both the east and west sides of the bridge.

Special Conditions of the Contract-West Street Bridge-Section 0.01, paragraph 3.b:
TIME FOR COMPLETION: All contract work must be completed by the middle of the 24th week after the Notice to Proceed.

FIGURE 6.1

June 15, 2007

John Lewis
Owner's Representative
Job Trailer
West Street Bridge
Podunk, New York 00001

Dear Mr. Lewis:

As we discussed yesterday during our meeting, we plan to work our construction operations on the West Street Bridge from an east to west direction. In other words, we will construct abutment #1 first and then abutment #2. Our paving of our approaches will follow the same sequence.

Similarly, our operations on the piers will progress from east to west direction, starting with pier #1 and progressing through pier #3.

I trust this answers the question you raised, and you can notify the Township authorities accordingly so they may effect the appropriate detours during the course of the project.

Sincerely,
Joe Super
Ace Construction

FIGURE 6.2

- *Meeting minutes* for any discussions concerning scheduling, particularly the preconstruction meeting minutes. (Figure 6.6)
- *Daily log or diary entries by either the General Contractor's personnel or the Owner's representative.* (Figure 6.7)
- *Purchase orders with suppliers* that show planned dates for delivery of materials or equipment.

The analyst should use all the available information to define some form of an as-planned schedule or at least to establish a general sequence for the work on the Project for different points or portions of the Project.

August 20, 2007

Mr. Joe Super
Ace Construction
Job Trailer/West Street Bridge
Podunk, New York 00001

Dear Mr. Super:

Returned herein are the shop drawings submitted by Ace Construction for the reinforcing steel for the pier caps. As you will note, the drawings are approved as noted. Fabrication of the reinforcing steel can begin in accordance with the corrections noted.

I must note these drawings are being returned to you five weeks later than you requested. There are two reasons why the return of the shop drawings is five weeks beyond your requested date. First, you did not allow the engineer adequate time to review the drawings. As a minimum, the engineer requires three weeks to review, but you allowed only one week if they were to be returned within the time frame requested.

Second, the drawings were not prepared in accordance with normal shop drawing practices and consequently required much more time to review. As submitted, the drawings were extremely difficult to understand. While the engineer could have returned them disapproved and required resubmission in the proper manner, it was decided to take the extra time to correct them on the first submission in order to expedite the process.

If you have any questions, please feel free to call.

Sincerely,
John Lewis
Owner's Representative

FIGURE 6.3

Based on the sample information provided in Figures 6.1 through 6.7, the analyst is unable to determine an as-planned schedule. However, the following information can be determined:

1. The Contract required that all abutment work be completed before any paving was performed on the approaches. The Contract also required that completion of the Project occur by the middle of the 24th week (from Figure 6.1).
2. The Contractor planned to work from east to west, or from abutment #1 to abutment #2, and from pier #1 to pier #3 (from Figure 6.2).
3. The Owner's representative noted that the review and approval of the shop drawings for the reinforcing steel in the pier caps had taken longer than required and were not returned to the Contractor until five weeks later than requested (from Figure 6.3).

June 25, 2007

Mr. Grey Ferrous
Ferrous Steel Erectors
Iron Street
Steeltown, New York 00002

Dear Mr. Ferrous:

In accordance with our previous conversations, Ace Construction will require the delivery and erection of the steel for the West Street Bridge to begin by the beginning of the 13th week of the project. Based on your present fabrication schedule, it appears that this should not be a problem.

Please be advised that time is of the essence on this contract. In the past, we have had problems with steel suppliers promising delivery dates and not adhering to them. Action such as this will not be tolerated on this project. Your delivery and erection of steel is critical to our timely completion of the project.

Sincerely,
John Lewis
Owner's Representative

FIGURE 6.4

4. The Contractor planned on erecting the steel at the beginning of the 13th week of the Project (from Figure 6.4).
5. The Contractor produced at least one ten-week "look-ahead" schedule that addressed the work on the piers and described the specific sequence for this work, including the piles, pile caps, pier columns, and pier caps (from Figure 6.5).
6. The Contractor anticipated a two-week duration for the forming, reinforcing, and placing of each of the concrete deck spans (from Figure 6.6).
7. The Contractor expected four-week durations for each of the abutments and two-week durations for each of the approaches (from Figure 6.7).

Based on this information, the analyst can proceed with the analysis, despite the lack of a complete as-planned schedule.

When no schedule exists for a Project, the analyst may be tempted to create an "after the fact" schedule, reasoning that it will allow the analysis to be more precise. It is best not to create such a schedule. This practice may bias the analysis and not reflect the Contractor's original planned sequence of the work.

USING AN AS-BUILT ANALYSIS TO QUANTIFY DELAYS

Although no as-planned schedule exists for this Project, detailed daily reports do exist. Daily reports allow the analyst to prepare an as-built diagram for the job. The as-built diagram for the example Project based on the daily reports appears in Figure 6.8.

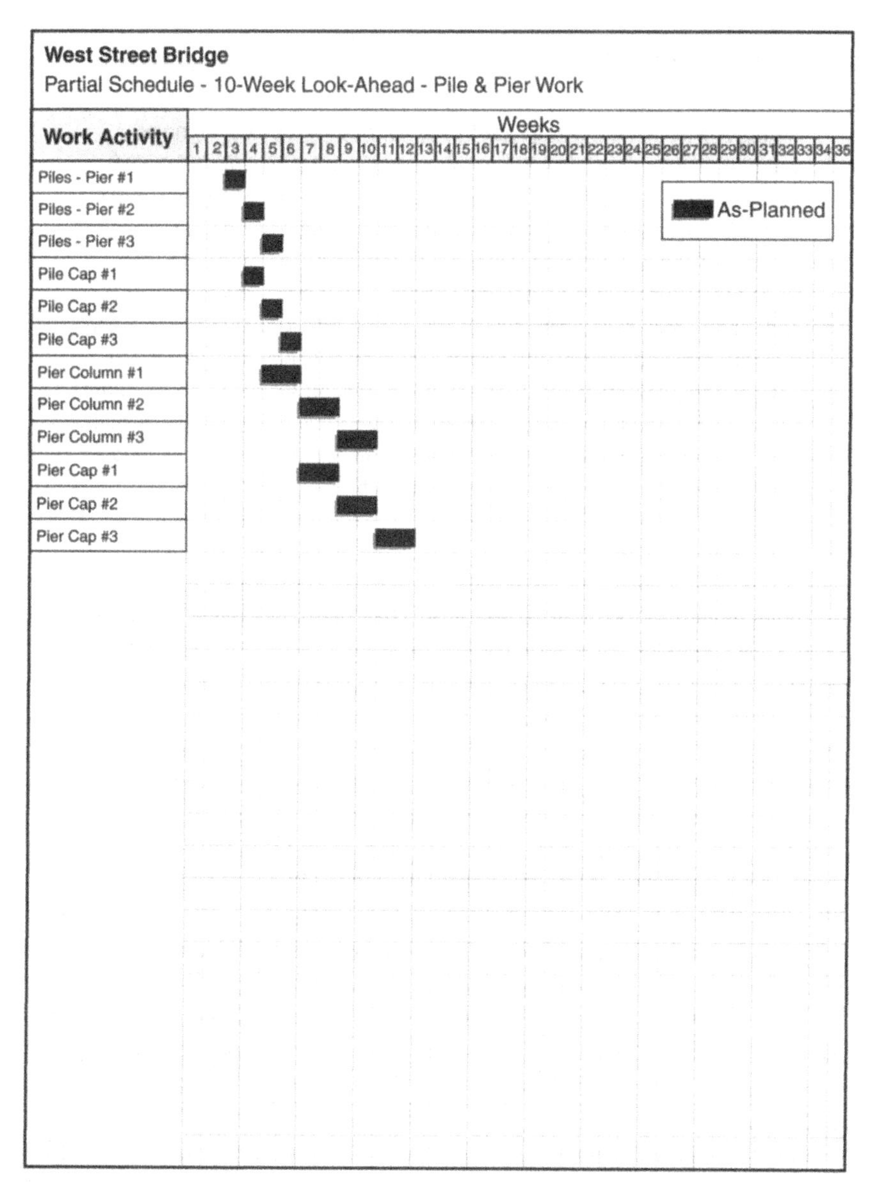

FIGURE 6.5

Preconstruction Meeting Minutes
(Excerpt)

Contractor noted that the forming, reinforcing, and placing of the concrete deck spans will take two weeks per span. Contractor will proceed from east to west constructing span #1 through span #4.

FIGURE 6.6

> Inspector's Daily Diary
> (Excerpt)
>
> During discussion today with Joe Super (Ace Construction), he noted that the abutment work would take about four weeks to perform for each abutment, and the paving of the approaches would take about two weeks for each side of the bridge. Joe noted that the existing abutment sills are acceptable as shown on the contract drawings. Therefore, they can accept the steel as is. The remainder of the work on the abutments can be performed either before the steel is in place or after it is set. He is going to try to have the abutments completed before he sets any steel, since the work will be easier that way. However, he is concerned about the lead time allowed for obtaining the epoxy material called for in the contract. He said that, if necessary, he would set the steel so as not to delay the deck work that he feels controls his completion of the job.

FIGURE 6.7

In reviewing the as-built diagram, the first obvious conclusion is that the Project was delayed seven and one-half weeks. It was to be completed by the middle of the 24th week but was not actually finished until the end of the 31st week. Therefore, the analyst must account for at least seven and one-half weeks of net delay. The delay could, in fact, be greater if the Contractor had planned to finish the Project early. However, no such plans were indicated in the Project correspondence, so this assumption will not be made. The as-built diagram also shows that the Contractor did work from east to west as planned.

Based on the letter to Ferrous Steel Erectors (Figure 6.4), the Contractor planned to start steel erection at the beginning of the 13th week. The as-built diagram (Figure 6.8) shows that steel erection began at the beginning of the 18th week. A five-week delay can be identified for the start of the steel erection.

Following this conclusion, the analyst breaks down the activities that precede the start of steel erection. The analyst uses the ten-week "look-ahead" schedule from Figure 6.5, which the Contractor produced at the end of the second week of the Project. A comparison of this schedule with the as-built diagram for this portion of the Project shows that a delay occurred to the construction of the pier caps. A delay also appears in the construction of pier column #3, which took three weeks instead of two weeks to complete.

The comparison of the partial "look-ahead" schedule with the as-built portion is shown in Figure 6.9. Based on this comparison, it appears that the controlling delay was the delay to the start of the pier caps. This may have been the result of the five-week delay in approval of the shop drawings to which the Owner's representative had referred in correspondence (Figure 6.2). It does not appear that the delay in the completion of pier column #3 affected the completion of the Project, as the flow of activities in the "look-ahead" schedule (and in the logical construction sequence) would have been in the stair-step fashion depicted.

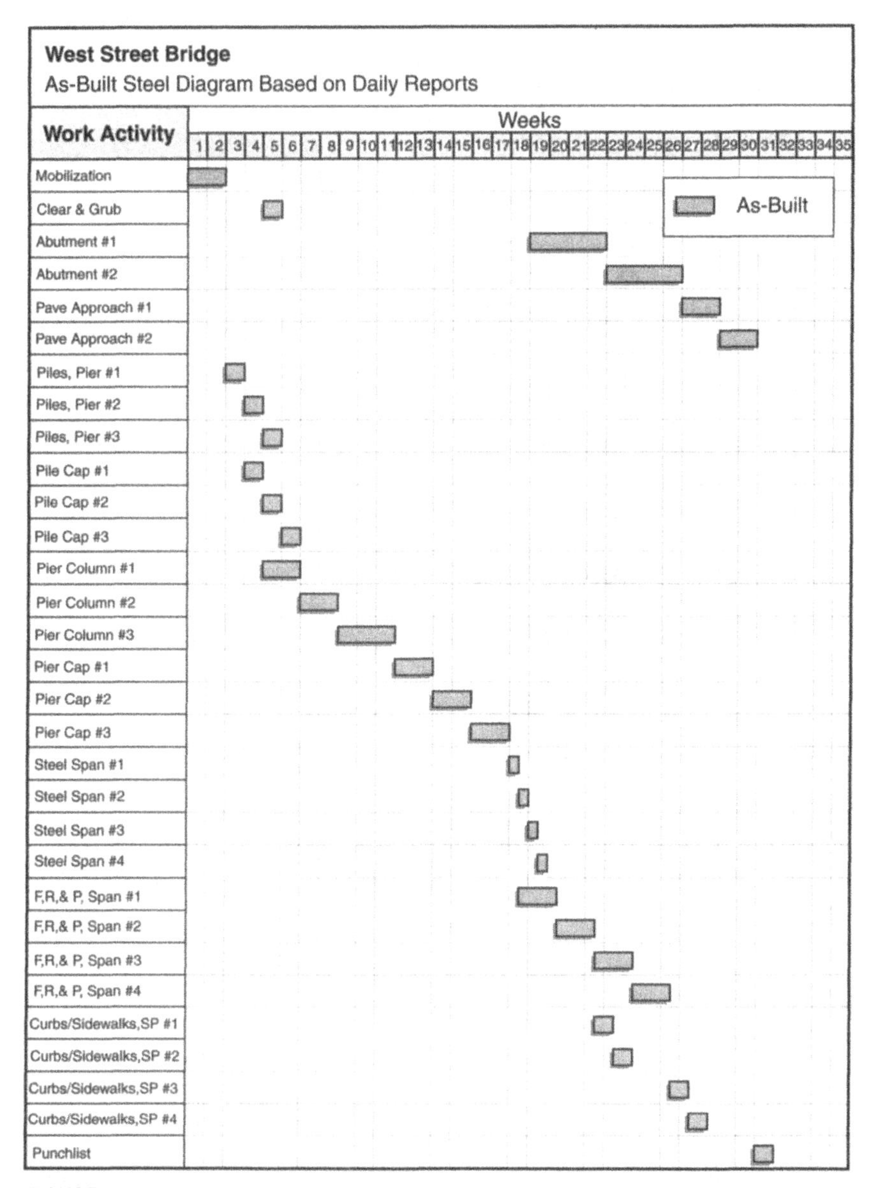

FIGURE 6.8

The remainder of the work along the pier/deck path appears to have been performed in a logical, sequential fashion and does not indicate further delay. There is, however, a gap between the completion of the curbs and sidewalks and the beginning of the punch list work. Based on the as-built diagram, the punch list work did not start until the abutments and approaches were completed. The abutment and approach work occurred in the sequence planned. However, it

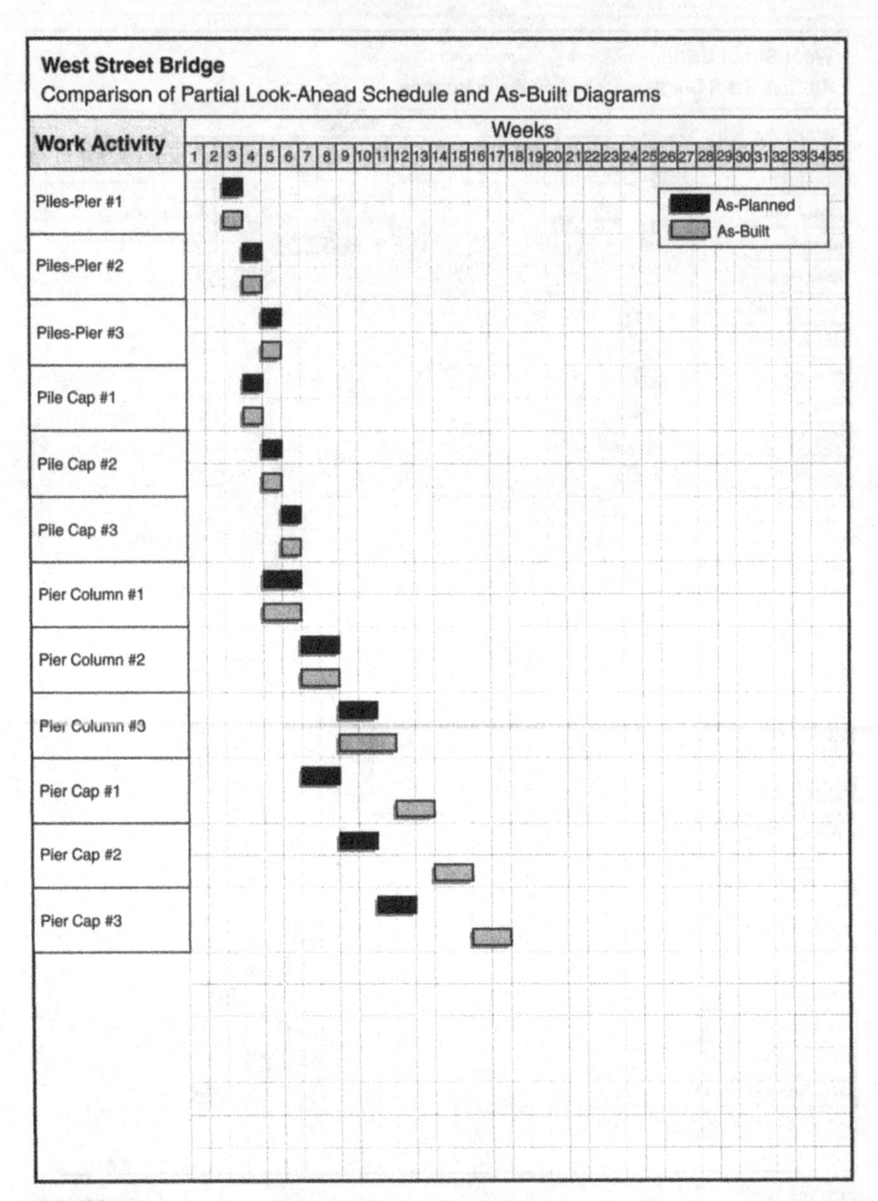

FIGURE 6.9

delayed the Project an additional two and one-half weeks, as this is the gap between the completion of the superstructure work and the start of the punch list work. Based on the documentation available, it can be determined that the abutment/approach work started later than planned, but was performed within the duration planned by the Contractor.

A summary of the delays appears in Figure 6.10. The delay of seven and one-half weeks is from (1) five weeks of delay to the start of the pier cap work

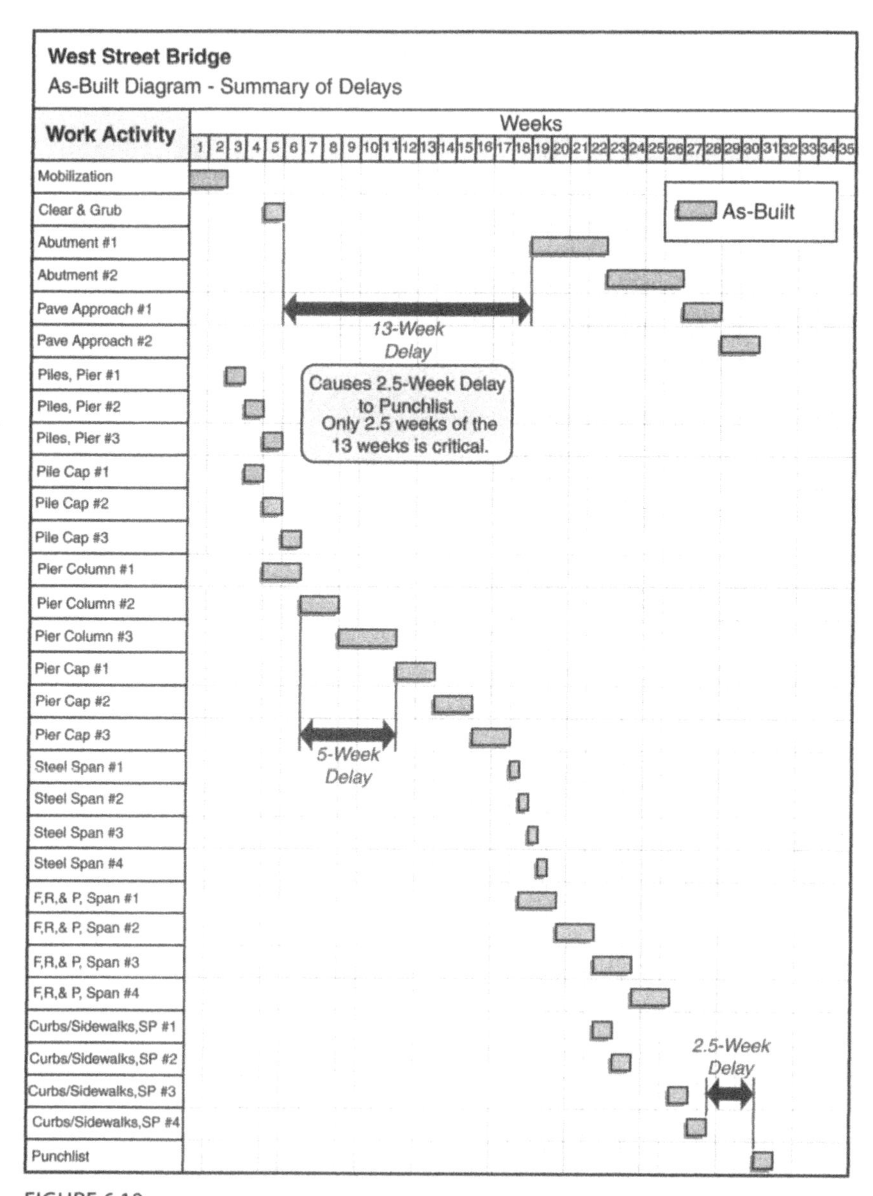

FIGURE 6.10

and (2) two and one-half weeks of delay to the start of the abutment/approach work. At this point in the analysis, the analyst should again carefully review the Project documentation to determine if other supporting information is available to further substantiate the delay analysis. Also at this point, the analyst can begin to assess the reasons or causes for the delay, and thereby establish liability for the delays.

and (2) two and one-half weeks of delay to the start of the abutment approach work. At this point in the analysis, the analyst should again carefully review the project documentation to determine if other supporting information is available to further substantiate the delay analysis. Also at this point the analyst can begin to assess the reasons or causes for the delay, and thereby establish liability for the delays.

Chapter | seven

Other Analysis Techniques—Their Strengths and Weaknesses

There are many types of analysis techniques being used by analysts to measure delays. Many of these fail simply because they express a one-sided view of the Project. Others fail because they are not grounded in the documents used by the participants to manage the Project. A properly performed delay analysis will have certain attributes regardless of the precise methodology used.

To begin with, the analysis must be performed objectively. One way to achieve an objective analysis is for the analyst to focus on determining the source and magnitude of all critical Project delays without regard to the party responsible. For example, an analysis of the schedule may reveal that the late start of foundation excavation caused a critical delay to the Project. This conclusion should be independent of what caused this excavation work to start late. Determining the party responsible for this delay should be a separate task.

This leads to the next attribute. The analysis should account for all Project delays and savings throughout the duration of the Project. The total delay to the Project will be known as the difference between the planned completion date and the actual completion date. The analyst should identify all of the critical delays and savings that total up to the actual total Project delay.

Finally, the delay analysis should rely on the contemporaneous Project schedules as the basis of analysis to the maximum extent possible. Reliance on the contemporaneous Project schedules helps keep the analysis objective and guards against the analyst's drawing erroneous conclusions. As we saw in previous chapters, analyses that stray far from the contemporaneous schedules are usually biased and unpersuasive.

USING FRAGNETS TO QUANTIFY DELAYS

In this section we discuss fragnets, what they are, how they are used to quantify delays, and their advantages and disadvantages. What exactly is a fragnet? In simplest terms, it is a portion of a CPM that represents a specific sequence of work. It is a fragmentary network or a fragment of a network—hence the term *fragnet*. Typically, fragnets are prepared as the result of a change order and consist of the activities and logic necessary to complete the changed work. However, a fragnet can also be a specific portion of the existing CPM—for example, the work associated with the installation of holding tanks at a wastewater treatment facility.

One example of the use of a fragnet is if a Contractor were directed to install an additional wall in an office, the fragnet might include install metal studs, install electric, install drywall, and spackle and paint. This new sequence of activities would be logically tied to one another in a series and then tied logically into the existing CPM (install metal studs might be tied to the existing metal stud installation activity). In addition to identifying added work, fragnets are also prepared to manage distinct features of work within a complex Project, such as construction of a clean room at a new pharmaceutical development and manufacturing installation.

Many private and public Owners require Contractors to use fragnets to express, in a CPM format, the activities associated with change orders and use this to request time extensions. The U.S. Army Corps of Engineers, the Department of Veterans Affairs, and the Florida Department of Transportation are just a few of the public Owners that require Contractors, when appropriate, to use fragnets as part of their requests for time extensions. Typically, the effect of changes on the Project schedule is measured by developing a fragnet for the change and inserting this fragnet into the schedule. The measure of the delay caused by the change is the difference between the scheduled Project completion date before the fragnet is inserted into the schedule and the completion date after the fragnet is inserted.

Returning to the previous example of installing a wall in an office, if the original drywall installation was critical, and the new drywall activity required two workdays and was inserted in series with the existing drywall work, you can measure any additional time required. Prior to inserting the fragnet, the predicted Project completion date was September 19, 2008. After the fragnet is inserted into the CPM, the predicted Project completion date is September 21, 2008. Thus, the added drywall work caused a critical delay to the Project of two calendar days. This delay was quantified by inserting a fragnet and measuring the difference between the predicted completion dates before and after the fragnet was inserted.

Predicted completion date prior to inserting fragnet: September 19, 2008
Predicted completion date after inserting fragnet: September 21, 2008
September 21, 2008 – September 19, 2008 = two calendar days

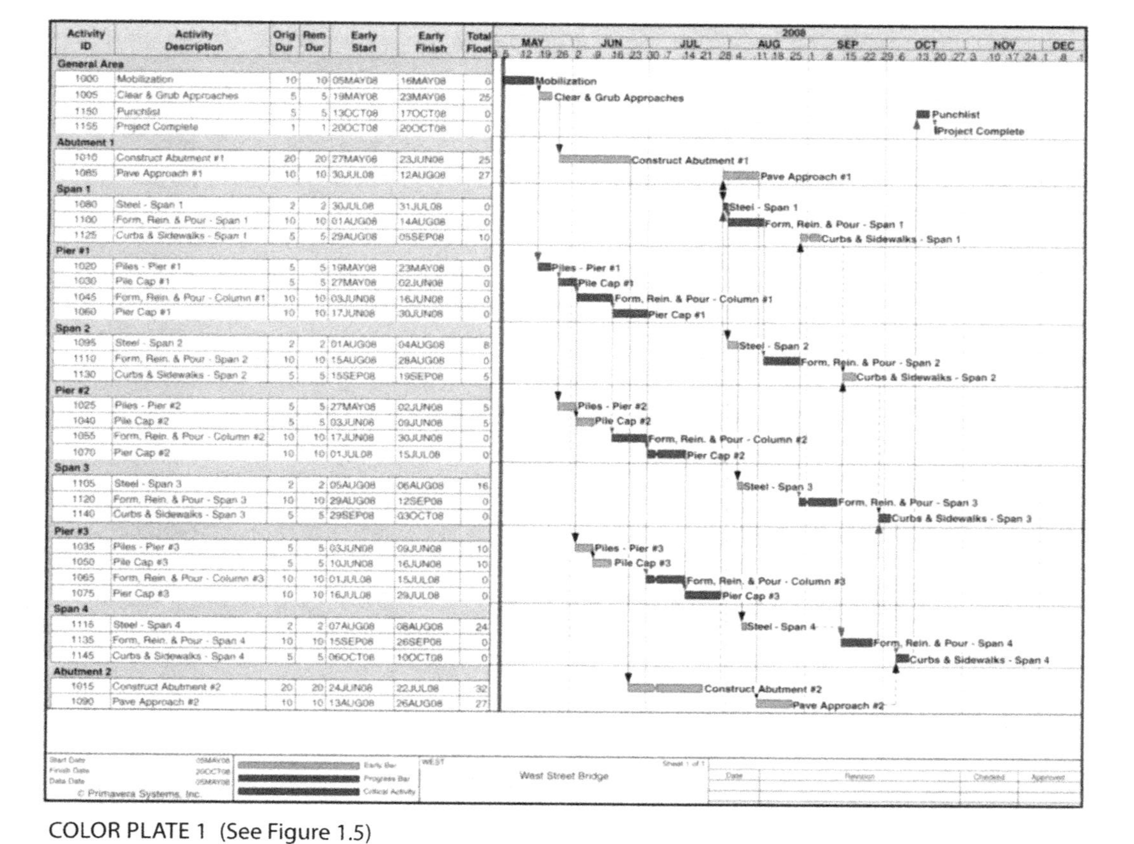

Activity ID	Activity Description	Orig Dur	Rem Dur	Early Start	Early Finish	Total Float
General Area						
1000	Mobilization	10	10	05MAY08	16MAY08	0
1005	Clear & Grub Approaches	5	5	19MAY08	23MAY08	25
1150	Punchlist	5	5	13OCT08	17OCT08	0
1155	Project Complete	1	1	20OCT08	20OCT08	0
Abutment 1						
1010	Construct Abutment #1	20	20	27MAY08	23JUN08	25
1085	Pave Approach #1	10	10	30JUL08	12AUG08	27
Span 1						
1080	Steel - Span 1	2	2	30JUL08	31JUL08	0
1100	Form, Rein. & Pour - Span 1	10	10	01AUG08	14AUG08	0
1125	Curbs & Sidewalks - Span 1	5	5	29AUG08	05SEP08	10
Pier #1						
1020	Piles - Pier #1	5	5	19MAY08	23MAY08	0
1030	Pile Cap #1	5	5	27MAY08	02JUN08	0
1045	Form, Rein. & Pour - Column #1	10	10	03JUN08	16JUN08	0
1060	Pier Cap #1	10	10	17JUN08	30JUN08	0
Span 2						
1095	Steel - Span 2	2	2	01AUG08	04AUG08	8
1110	Form, Rein. & Pour - Span 2	10	10	15AUG08	28AUG08	0
1130	Curbs & Sidewalks - Span 2	5	5	15SEP08	19SEP08	5
Pier #2						
1025	Piles - Pier #2	5	5	27MAY08	02JUN08	5
1040	Pile Cap #2	5	5	03JUN08	09JUN08	5
1055	Form, Rein. & Pour - Column #2	10	10	17JUN08	30JUN08	0
1070	Pier Cap #2	10	10	01JUL08	15JUL08	0
Span 3						
1105	Steel - Span 3	2	2	05AUG08	06AUG08	16
1120	Form, Rein. & Pour - Span 3	10	10	29AUG08	12SEP08	0
1140	Curbs & Sidewalks - Span 3	5	5	29SEP08	03OCT08	0
Pier #3						
1035	Piles - Pier #3	5	5	03JUN08	09JUN08	10
1050	Pile Cap #3	5	5	10JUN08	16JUN08	10
1065	Form, Rein. & Pour - Column #3	10	10	01JUL08	15JUL08	0
1075	Pier Cap #3	10	10	16JUL08	29JUL08	0
Span 4						
1115	Steel - Span 4	2	2	07AUG08	08AUG08	24
1135	Form, Rein. & Pour - Span 4	10	10	15SEP08	26SEP08	0
1145	Curbs & Sidewalks - Span 4	5	5	06OCT08	10OCT08	0
Abutment 2						
1015	Construct Abutment #2	20	20	24JUN08	22JUL08	32
1090	Pave Approach #2	10	10	13AUG08	26AUG08	27

COLOR PLATE 1 (See Figure 1.5)

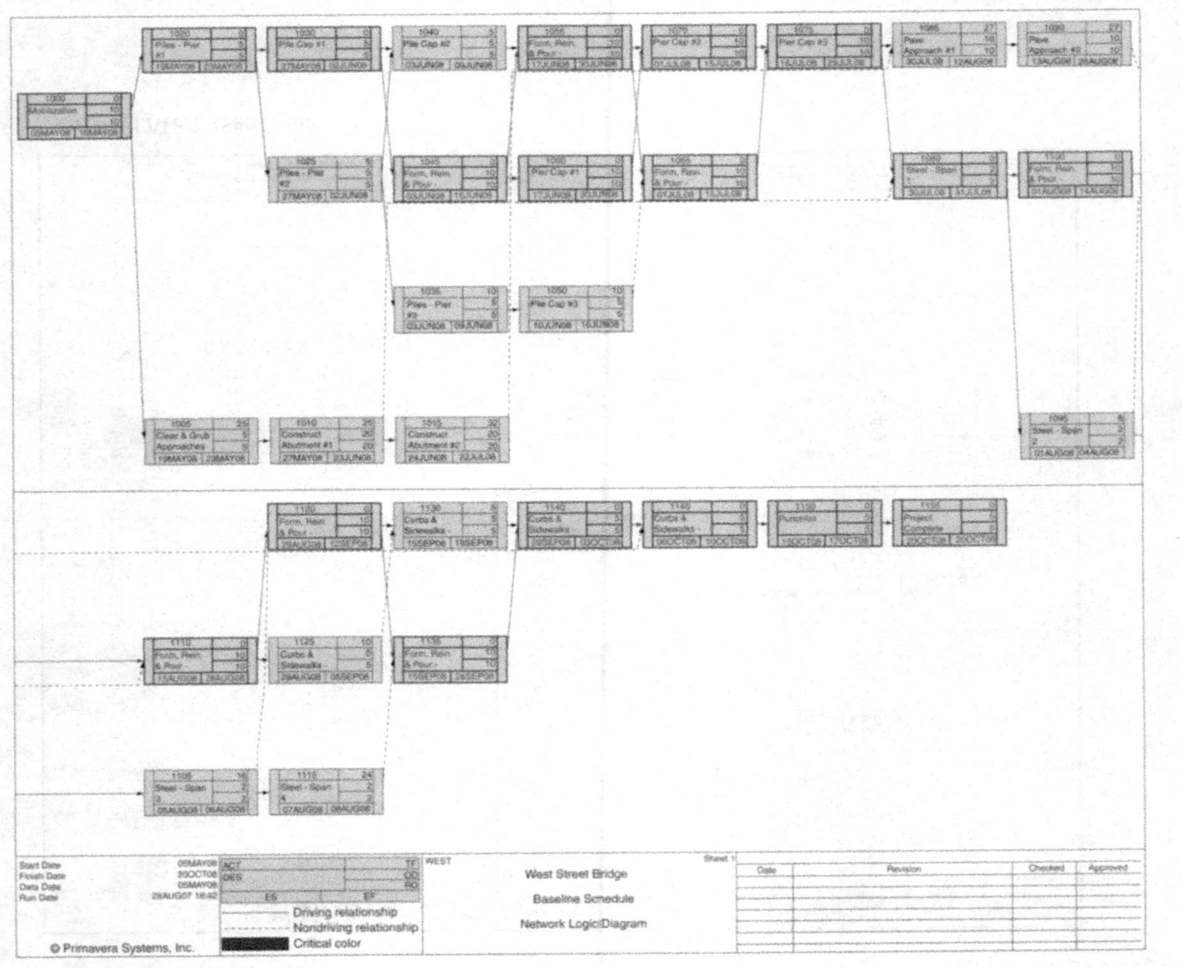

COLOR PLATE 2 (See Figure 5.1)

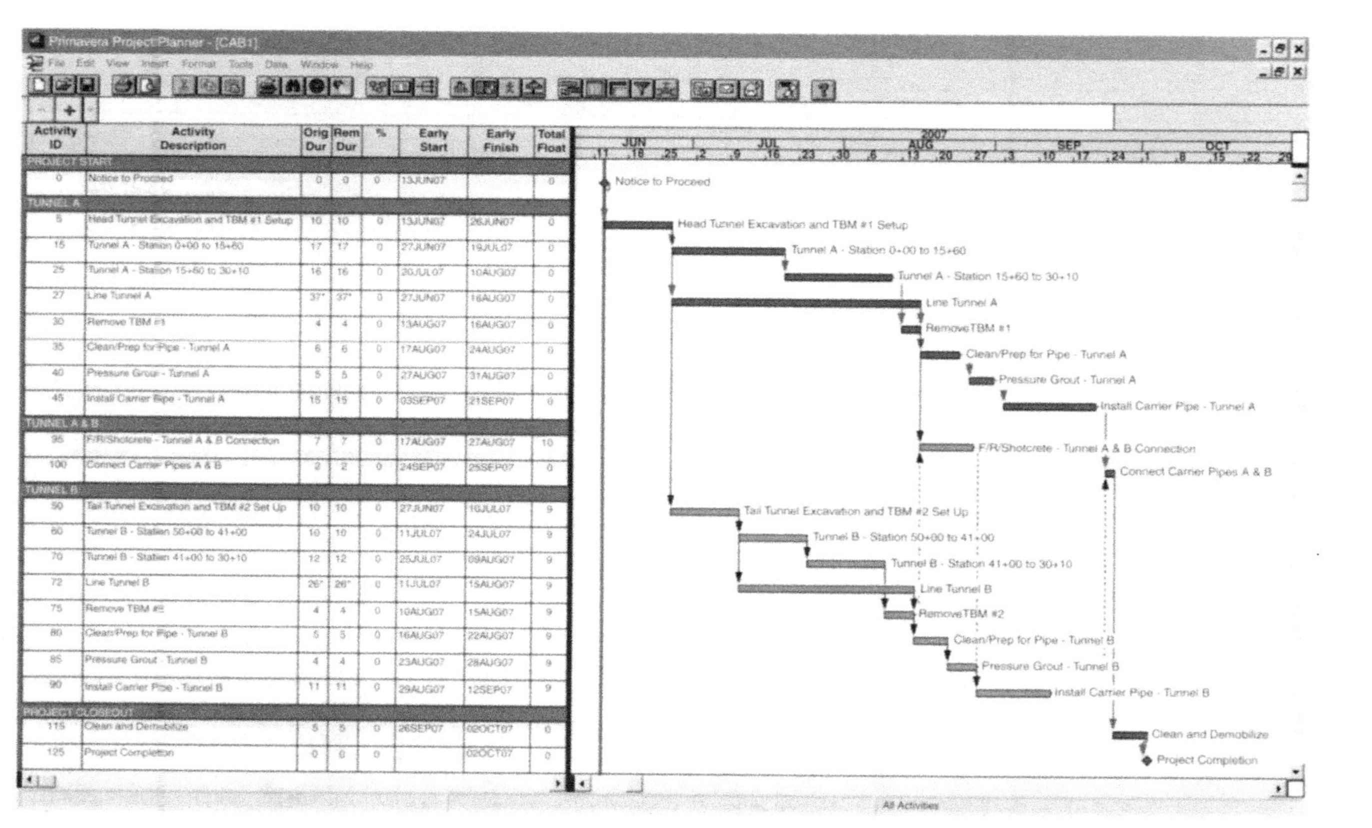

Primavera Project Planner - [CAB1]

Activity ID	Activity Description	Orig Dur	Rem Dur	%	Early Start	Early Finish	Total Float
PROJECT START							
0	Notice to Proceed	0	0	0	13JUN07		0
TUNNEL A							
5	Head Tunnel Excavation and TBM #1 Setup	10	10	0	13JUN07	26JUN07	0
15	Tunnel A - Station 0+00 to 15+60	17	17	0	27JUN07	19JUL07	0
25	Tunnel A - Station 15+60 to 30+10	16	16	0	20JUL07	10AUG07	0
27	Line Tunnel A	37*	37*	0	27JUN07	16AUG07	0
30	Remove TBM #1	4	4	0	13AUG07	16AUG07	0
35	Clean/Prep for Pipe - Tunnel A	6	6	0	17AUG07	24AUG07	0
40	Pressure Grout - Tunnel A	5	5	0	27AUG07	31AUG07	0
45	Install Carrier Pipe - Tunnel A	15	15	0	03SEP07	21SEP07	0
TUNNEL A & B							
95	F/R/Shotcrete - Tunnel A & B Connection	7	7	0	17AUG07	27AUG07	10
100	Connect Carrier Pipes A & B	2	2	0	24SEP07	25SEP07	0
TUNNEL B							
50	Tail Tunnel Excavation and TBM #2 Set Up	10	10	0	27JUN07	10JUL07	9
60	Tunnel B - Station 50+00 to 41+00	10	10	0	11JUL07	24JUL07	9
70	Tunnel B - Station 41+00 to 30+10	12	12	0	25JUL07	09AUG07	9
72	Line Tunnel B	26*	26*	0	11JUL07	15AUG07	9
75	Remove TBM #2	4	4	0	10AUG07	15AUG07	9
80	Clean/Prep for Pipe - Tunnel B	5	5	0	16AUG07	22AUG07	9
85	Pressure Grout - Tunnel B	4	4	0	23AUG07	28AUG07	9
90	Install Carrier Pipe - Tunnel B	11	11	0	29AUG07	12SEP07	9
PROJECT CLOSEOUT							
115	Clean and Demobilize	5	5	0	26SEP07	02OCT07	0
125	Project Completion	0	0	0		02OCT07	0

COLOR PLATE 3 (See Figure 7.1)

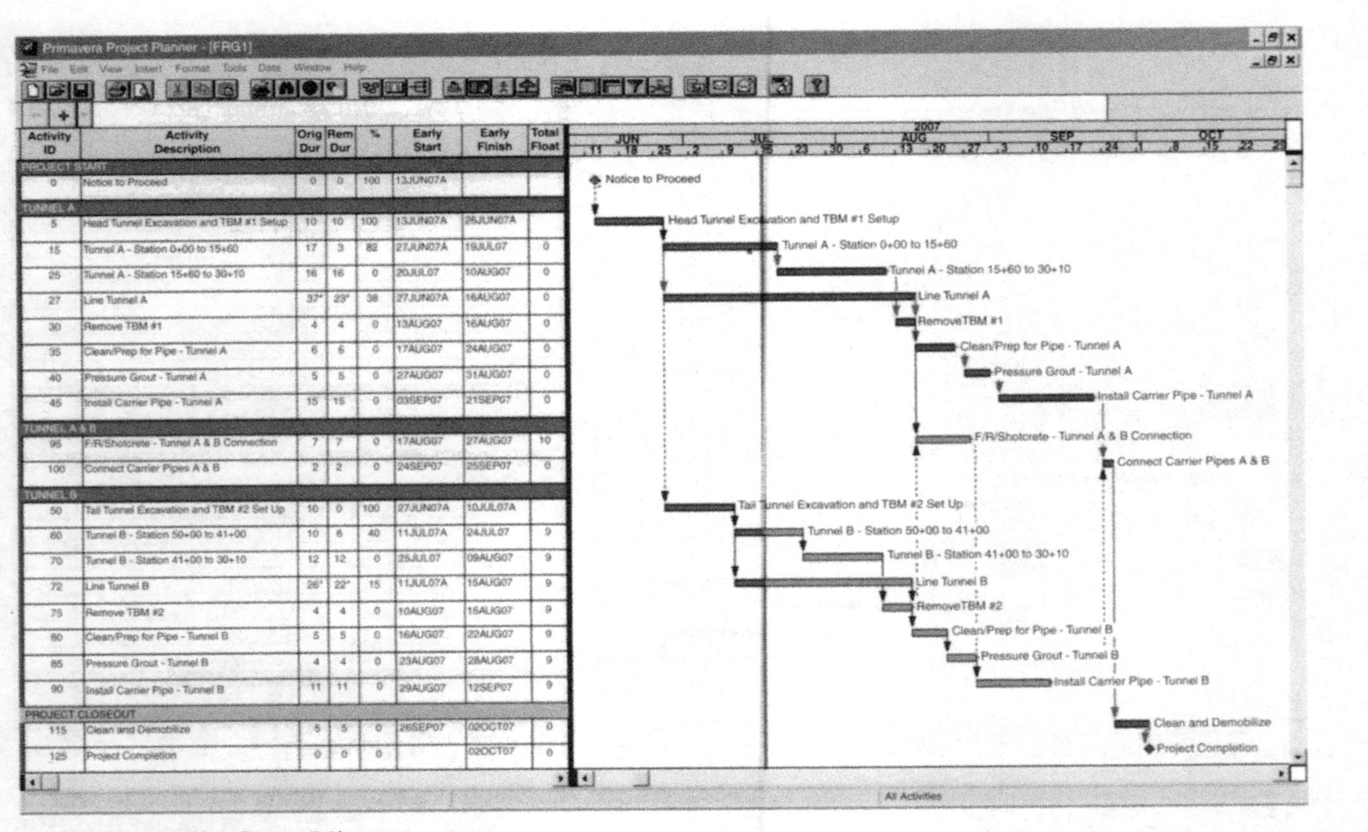

Activity ID	Activity Description	Orig Dur	Rem Dur	%	Early Start	Early Finish	Total Float
PROJECT START							
0	Notice to Proceed	0	0	100	13JUN07A		
TUNNEL A							
5	Head Tunnel Excavation and TBM #1 Setup	10	10	100	13JUN07A	26JUN07A	
15	Tunnel A - Station 0+00 to 15+60	17	3	82	27JUN07A	19JUL07	0
25	Tunnel A - Station 15+60 to 30+10	16	16	0	20JUL07	10AUG07	0
27	Line Tunnel A	37*	23*	38	27JUN07A	16AUG07	0
30	Remove TBM #1	4	4	0	13AUG07	16AUG07	0
35	Clean/Prep for Pipe - Tunnel A	6	6	0	17AUG07	24AUG07	0
40	Pressure Grout - Tunnel A	5	5	0	27AUG07	31AUG07	0
45	Install Carrier Pipe - Tunnel A	15	15	0	03SEP07	21SEP07	0
TUNNEL A & B							
95	F/R/Shotcrete - Tunnel A & B Connection	7	7	0	17AUG07	27AUG07	10
100	Connect Carrier Pipes A & B	2	2	0	24SEP07	25SEP07	0
TUNNEL B							
50	Tail Tunnel Excavation and TBM #2 Set Up	10	0	100	27JUN07A	10JUL07A	
60	Tunnel B - Station 50+00 to 41+00	10	6	40	11JUL07A	24JUL07	9
70	Tunnel B - Station 41+00 to 30+10	12	12	0	25JUL07	09AUG07	9
72	Line Tunnel B	26*	22*	15	11JUL07A	15AUG07	9
75	Remove TBM #2	4	4	0	10AUG07	15AUG07	9
80	Clean/Prep for Pipe - Tunnel B	5	5	0	16AUG07	22AUG07	9
85	Pressure Grout - Tunnel B	4	4	0	23AUG07	28AUG07	9
90	Install Carrier Pipe - Tunnel B	11	11	0	29AUG07	12SEP07	9
PROJECT CLOSEOUT							
115	Clean and Demobilize	5	5	0	26SEP07	02OCT07	0
125	Project Completion	0	0	0		02OCT07	0

COLOR PLATE 4 (See Figure 7.2)

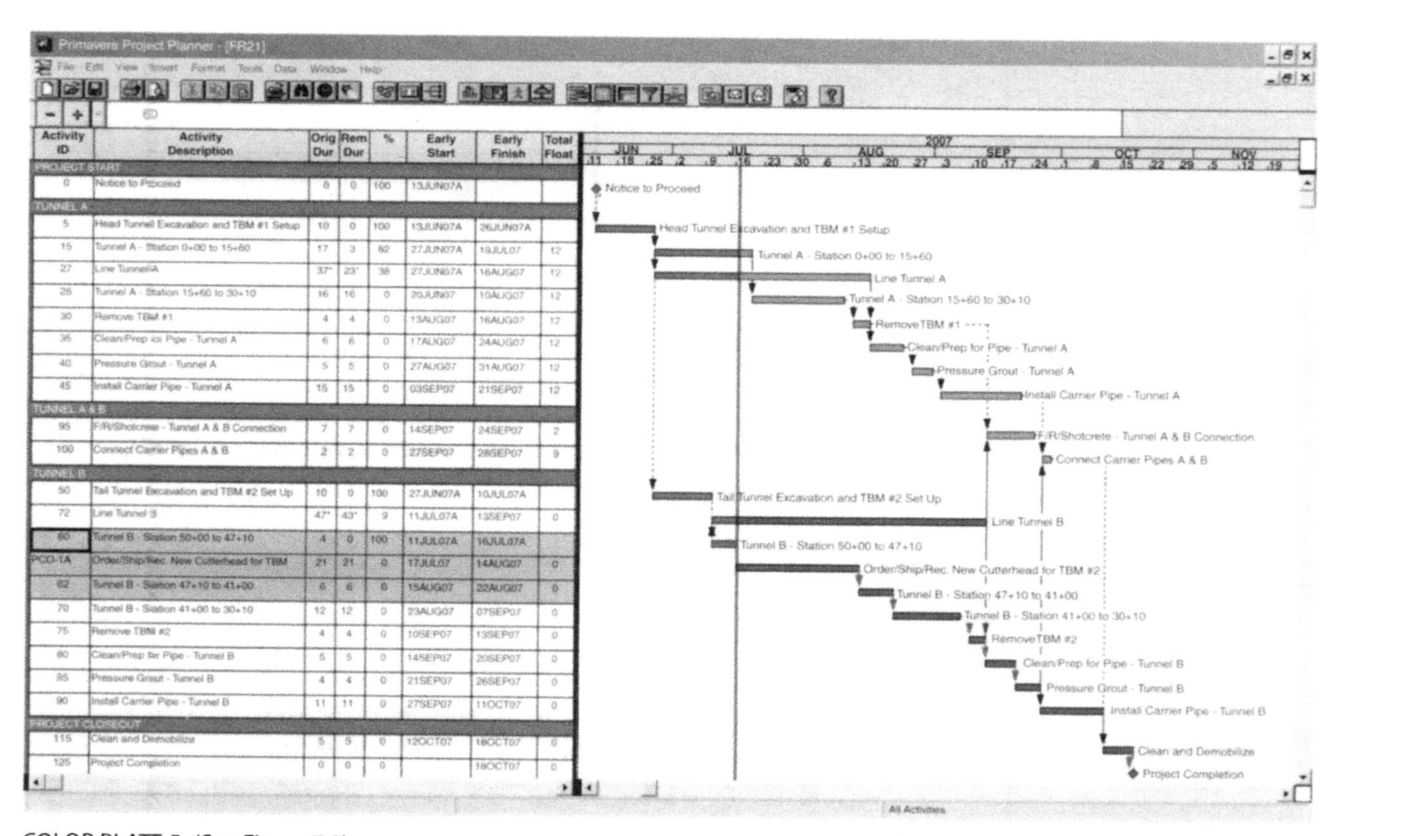

Activity ID	Activity Description	Orig Dur	Rem Dur	%	Early Start	Early Finish	Total Float
PROJECT START							
0	Notice to Proceed	0	0	100	13JUN07A		
TUNNEL A							
5	Head Tunnel Excavation and TBM #1 Setup	10	0	100	13JUN07A	26JUN07A	
15	Tunnel A - Station 0+00 to 15+60	17	3	82	27JUN07A	19JUL07	12
27	Line Tunnel A	37*	23*	38	27JUN07A	16AUG07	12
25	Tunnel A - Station 15+60 to 30+10	16	16	0	20JUL07	10AUG07	12
30	Remove TBM #1	4	4	0	13AUG07	16AUG07	12
35	Clean/Prep for Pipe - Tunnel A	6	6	0	17AUG07	24AUG07	12
40	Pressure Grout - Tunnel A	5	5	0	27AUG07	31AUG07	12
45	Install Carrier Pipe - Tunnel A	15	15	0	03SEP07	21SEP07	12
TUNNEL A & B							
95	F/R/Shotcrete - Tunnel A & B Connection	7	7	0	14SEP07	24SEP07	2
100	Connect Carrier Pipes A & B	2	2	0	27SEP07	28SEP07	9
TUNNEL B							
50	Tail Tunnel Excavation and TBM #2 Set Up	10	0	100	27JUN07A	10JUL07A	
72	Line Tunnel B	47*	43*	9	11JUL07A	13SEP07	0
60	Tunnel B - Station 50+00 to 47+10	4	0	100	11JUL07A	16JUL07A	
PCO-1A	Order/Ship/Rec. New Cutterhead for TBM	21	21	0	17JUL07	14AUG07	0
62	Tunnel B - Station 47+10 to 41+00	6	6	0	15AUG07	22AUG07	0
70	Tunnel B - Station 41+00 to 30+10	12	12	0	23AUG07	07SEP07	0
75	Remove TBM #2	4	4	0	10SEP07	13SEP07	0
80	Clean/Prep for Pipe - Tunnel B	5	5	0	14SEP07	20SEP07	0
85	Pressure Grout - Tunnel B	4	4	0	21SEP07	26SEP07	0
90	Install Carrier Pipe - Tunnel B	11	11	0	27SEP07	11OCT07	0
PROJECT CLOSEOUT							
115	Clean and Demobilize	5	5	0	12OCT07	18OCT07	0
125	Project Completion	0	0	0		18OCT07	0

COLOR PLATE 5 (See Figure 7.3)

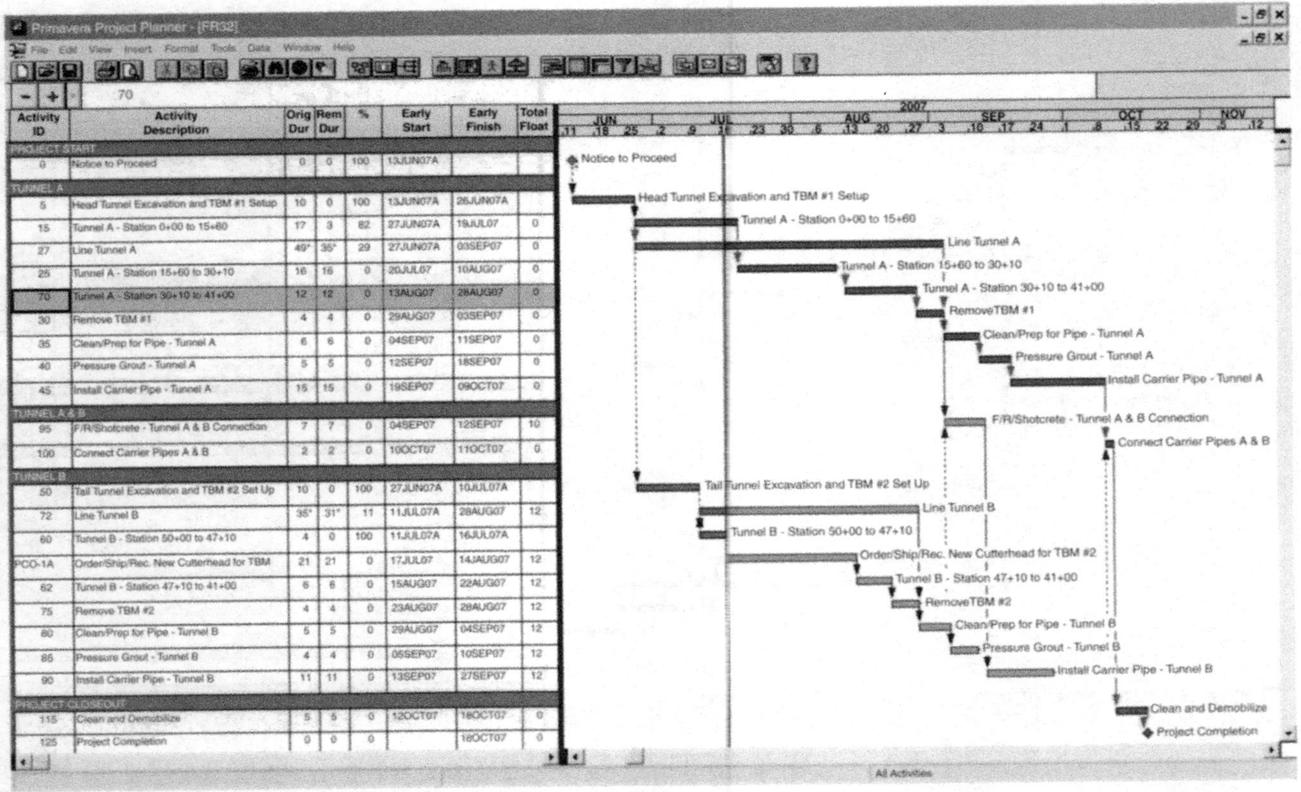
Primavera Project Planner - [FR32]
File Edit View Insert Format Tools Data Window Help
70
Activity ID | Activity Description | Orig Dur | Rem Dur | % | Early Start | Early Finish | Total Float
PROJECT START
0 | Notice to Proceed | 0 | 0 | 100 | 13JUN07A | |
TUNNEL A
5 | Head Tunnel Excavation and TBM #1 Setup | 10 | 0 | 100 | 13JUN07A | 26JUN07A |
15 | Tunnel A - Station 0+00 to 15+60 | 17 | 3 | 82 | 27JUN07A | 19JUL07 | 0
27 | Line Tunnel A | 49* | 35* | 29 | 27JUN07A | 03SEP07 | 0
25 | Tunnel A - Station 15+60 to 30+10 | 16 | 16 | 0 | 20JUL07 | 10AUG07 | 0
70 | Tunnel A - Station 30+10 to 41+00 | 12 | 12 | 0 | 13AUG07 | 28AUG07 | 0
30 | Remove TBM #1 | 4 | 4 | 0 | 29AUG07 | 03SEP07 | 0
35 | Clean/Prep for Pipe - Tunnel A | 6 | 6 | 0 | 04SEP07 | 11SEP07 | 0
40 | Pressure Grout - Tunnel A | 5 | 5 | 0 | 12SEP07 | 18SEP07 | 0
45 | Install Carrier Pipe - Tunnel A | 15 | 15 | 0 | 19SEP07 | 09OCT07 | 0
TUNNEL A & B
95 | F/R/Shotcrete - Tunnel A & B Connection | 7 | 7 | 0 | 04SEP07 | 12SEP07 | 10
100 | Connect Carrier Pipes A & B | 2 | 2 | 0 | 10OCT07 | 11OCT07 | 0
TUNNEL B
50 | Tail Tunnel Excavation and TBM #2 Set Up | 10 | 0 | 100 | 27JUN07A | 10JUL07A |
72 | Line Tunnel B | 35* | 31* | 11 | 11JUL07A | 28AUG07 | 12
60 | Tunnel B - Station 50+00 to 47+10 | 4 | 0 | 100 | 11JUL07A | 16JUL07A |
PCO-1A | Order/Ship/Rec. New Cutterhead for TBM | 21 | 21 | 0 | 17JUL07 | 14AUG07 | 12
62 | Tunnel B - Station 47+10 to 41+00 | 6 | 6 | 0 | 15AUG07 | 22AUG07 | 12
75 | Remove TBM #2 | 4 | 4 | 0 | 23AUG07 | 28AUG07 | 12
80 | Clean/Prep for Pipe - Tunnel B | 5 | 5 | 0 | 29AUG07 | 04SEP07 | 12
85 | Pressure Grout - Tunnel B | 4 | 4 | 0 | 05SEP07 | 10SEP07 | 12
90 | Install Carrier Pipe - Tunnel B | 11 | 11 | 0 | 13SEP07 | 27SEP07 | 12
PROJECT CLOSEOUT
115 | Clean and Demobilize | 5 | 5 | 0 | 12OCT07 | 18OCT07 | 0
125 | Project Completion | 0 | 0 | 0 | | 18OCT07 | 0
2007
JUN JUL AUG SEP OCT NOV
11 18 25 2 9 16 23 30 6 13 20 27 3 10 17 24 1 8 15 22 29 5 12
Notice to Proceed
Head Tunnel Excavation and TBM #1 Setup
Tunnel A - Station 0+00 to 15+60
Line Tunnel A
Tunnel A - Station 15+60 to 30+10
Tunnel A - Station 30+10 to 41+00
RemoveTBM #1
Clean/Prep for Pipe - Tunnel A
Pressure Grout - Tunnel A
Install Carrier Pipe - Tunnel A
F/R/Shotcrete - Tunnel A & B Connection
Connect Carrier Pipes A & B
Tail Tunnel Excavation and TBM #2 Set Up
Line Tunnel B
Tunnel B - Station 50+00 to 47+10
Order/Ship/Rec. New Cutterhead for TBM #2
Tunnel B - Station 47+10 to 41+00
RemoveTBM #2
Clean/Prep for Pipe - Tunnel B
Pressure Grout - Tunnel B
Install Carrier Pipe - Tunnel B
Clean and Demobilize
Project Completion
All Activities

COLOR PLATE 6 (See Figure 7.4)

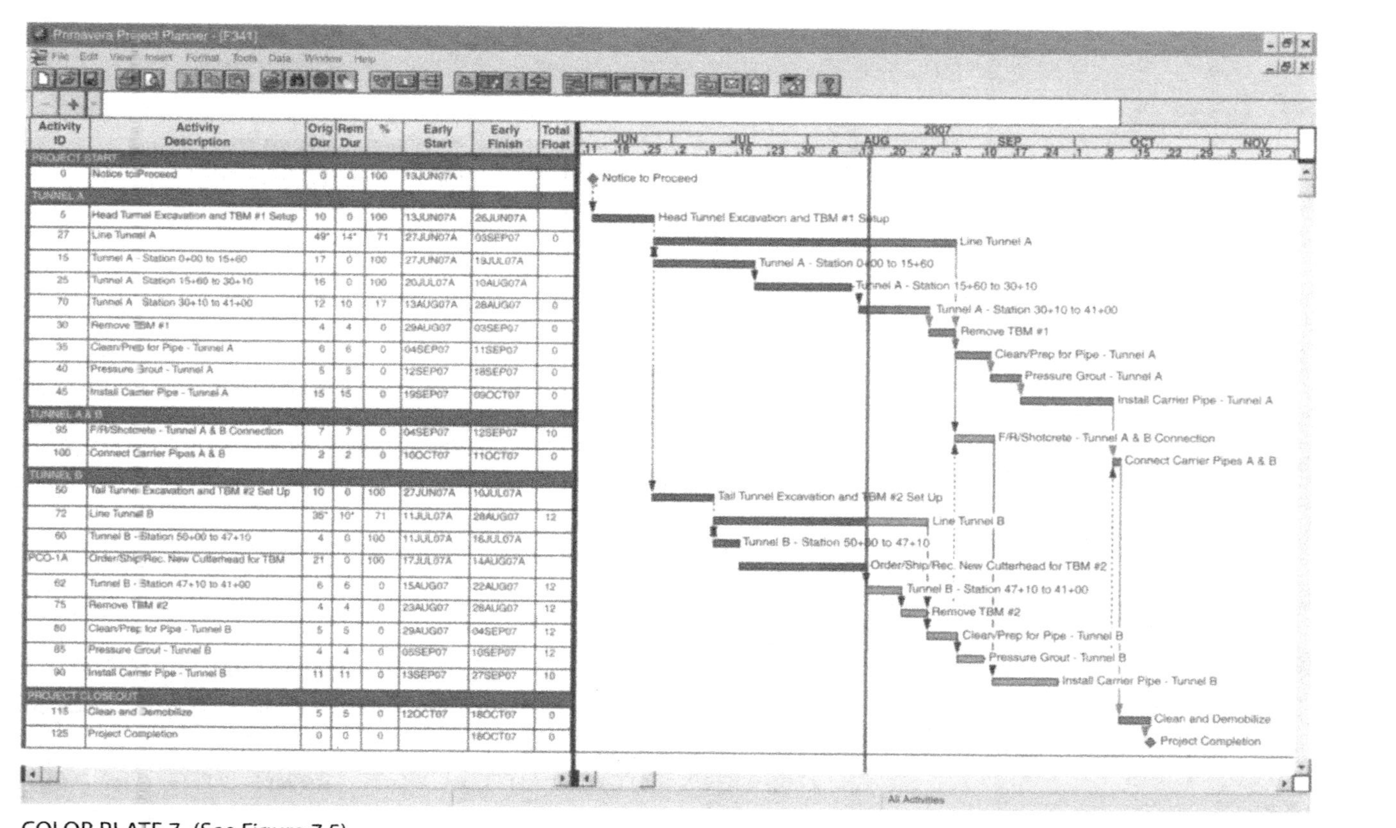

Activity ID	Activity Description	Orig Dur	Rem Dur	%	Early Start	Early Finish	Total Float
PROJECT START							
0	Notice to Proceed	0	0	100	13JUN07A		
TUNNEL A							
5	Head Tunnel Excavation and TBM #1 Setup	10	0	100	13JUN07A	26JUN07A	
27	Line Tunnel A	49*	14*	71	27JUN07A	03SEP07	0
15	Tunnel A - Station 0+00 to 15+60	17	0	100	27JUN07A	19JUL07A	
25	Tunnel A Station 15+60 to 30+10	16	0	100	20JUL07A	10AUG07A	
70	Tunnel A Station 30+10 to 41+00	12	10	17	13AUG07A	28AUG07	0
30	Remove TBM #1	4	4	0	29AUG07	03SEP07	0
35	Clean/Prep for Pipe - Tunnel A	6	6	0	04SEP07	11SEP07	0
40	Pressure Grout - Tunnel A	5	5	0	12SEP07	18SEP07	0
45	Install Carrier Pipe - Tunnel A	15	15	0	19SEP07	09OCT07	0
TUNNEL A & B							
95	F/R/Shotcrete - Tunnel A & B Connection	7	7	0	04SEP07	12SEP07	10
100	Connect Carrier Pipes A & B	2	2	0	10OCT07	11OCT07	0
TUNNEL B							
50	Tail Tunnel Excavation and TBM #2 Set Up	10	0	100	27JUN07A	10JUL07A	
72	Line Tunnel B	35*	10*	71	11JUL07A	28AUG07	12
60	Tunnel B - Station 50+00 to 47+10	4	0	100	11JUL07A	16JUL07A	
PCO-1A	Order/Ship/Rec. New Cutterhead for TBM	21	0	100	17JUL07A	14AUG07A	
62	Tunnel B - Station 47+10 to 41+00	6	6	0	15AUG07	22AUG07	12
75	Remove TBM #2	4	4	0	23AUG07	28AUG07	12
80	Clean/Prep for Pipe - Tunnel B	5	5	0	29AUG07	04SEP07	12
85	Pressure Grout - Tunnel B	4	4	0	05SEP07	10SEP07	12
90	Install Carrier Pipe - Tunnel B	11	11	0	13SEP07	27SEP07	10
PROJECT CLOSEOUT							
115	Clean and Demobilize	5	5	0	12OCT07	18OCT07	0
125	Project Completion	0	0	0		18OCT07	0

COLOR PLATE 7 (See Figure 7.5)

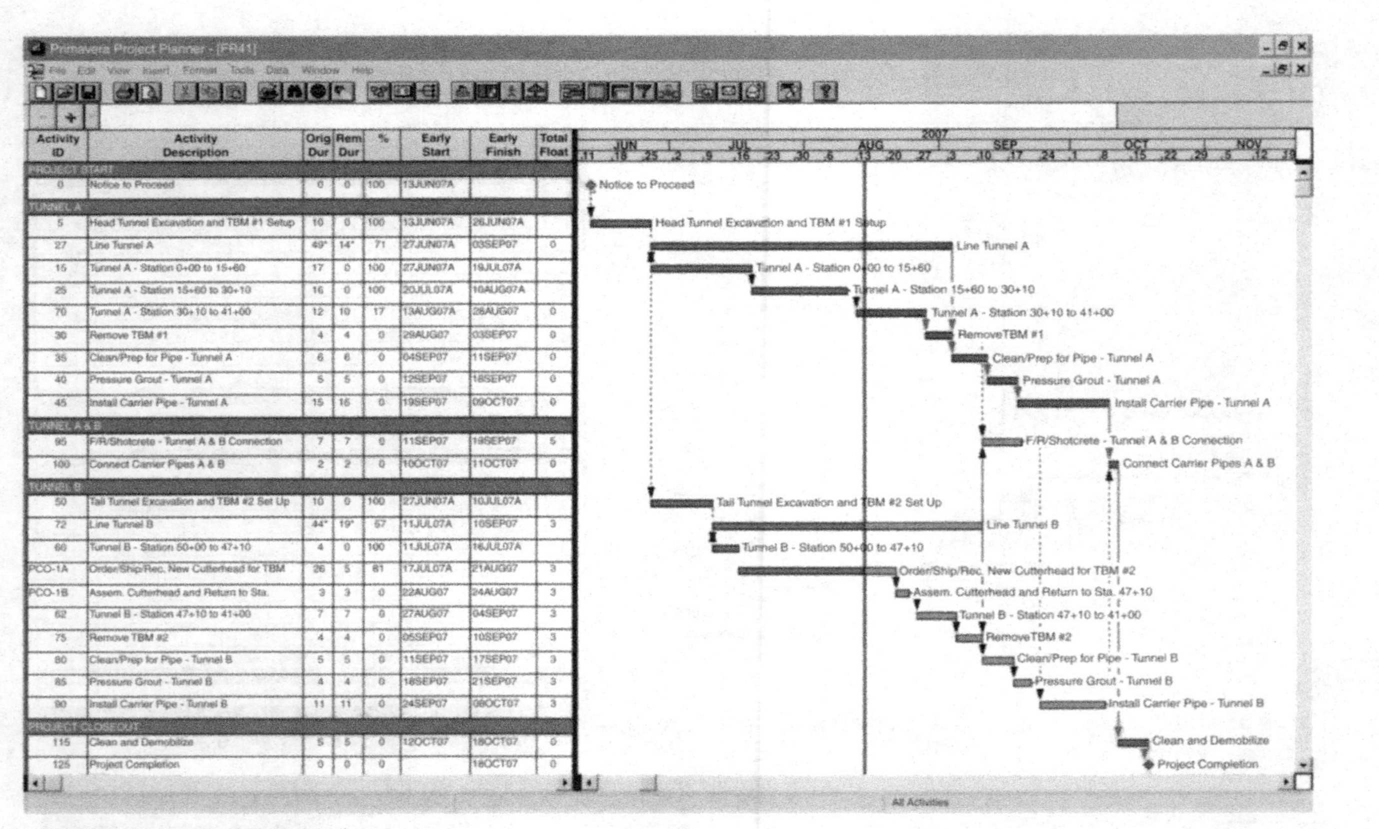

Activity ID	Activity Description	Orig Dur	Rem Dur	%	Early Start	Early Finish	Total Float
PROJECT START							
0	Notice to Proceed	0	0	100	13JUN07A		
TUNNEL A							
5	Head Tunnel Excavation and TBM #1 Setup	10	0	100	13JUN07A	26JUN07A	
27	Line Tunnel A	49*	14*	71	27JUN07A	03SEP07	0
15	Tunnel A - Station 0+00 to 15+60	17	0	100	27JUN07A	19JUL07A	
25	Tunnel A - Station 15+60 to 30+10	16	0	100	20JUL07A	10AUG07A	
70	Tunnel A - Station 30+10 to 41+00	12	10	17	13AUG07A	28AUG07	0
30	Remove TBM #1	4	4	0	29AUG07	03SEP07	0
35	Clean/Prep for Pipe - Tunnel A	6	6	0	04SEP07	11SEP07	0
40	Pressure Grout - Tunnel A	5	5	0	12SEP07	18SEP07	0
45	Install Carrier Pipe - Tunnel A	15	15	0	19SEP07	09OCT07	0
TUNNEL A & B							
95	F/R/Shotcrete - Tunnel A & B Connection	7	7	0	11SEP07	19SEP07	5
100	Connect Carrier Pipes A & B	2	2	0	10OCT07	11OCT07	0
TUNNEL B							
50	Tail Tunnel Excavation and TBM #2 Set Up	10	0	100	27JUN07A	10JUL07A	
72	Line Tunnel B	44*	19*	57	11JUL07A	10SEP07	3
60	Tunnel B - Station 50+00 to 47+10	4	0	100	11JUL07A	16JUL07A	
PCO-1A	Order/Ship/Rec. New Cutterhead for TBM	26	5	81	17JUL07A	21AUG07	3
PCO-1B	Assem. Cutterhead and Return to Sta.	3	3	0	22AUG07	24AUG07	3
62	Tunnel B - Station 47+10 to 41+00	7	7	0	27AUG07	04SEP07	3
75	Remove TBM #2	4	4	0	05SEP07	10SEP07	3
80	Clean/Prep for Pipe - Tunnel B	5	5	0	11SEP07	17SEP07	3
85	Pressure Grout - Tunnel B	4	4	0	18SEP07	21SEP07	3
90	Install Carrier Pipe - Tunnel B	11	11	0	24SEP07	08OCT07	3
PROJECT CLOSEOUT							
115	Clean and Demobilize	5	5	0	12OCT07	18OCT07	0
125	Project Completion	0	0	0		18OCT07	0

COLOR PLATE 8 (See Figure 7.6)

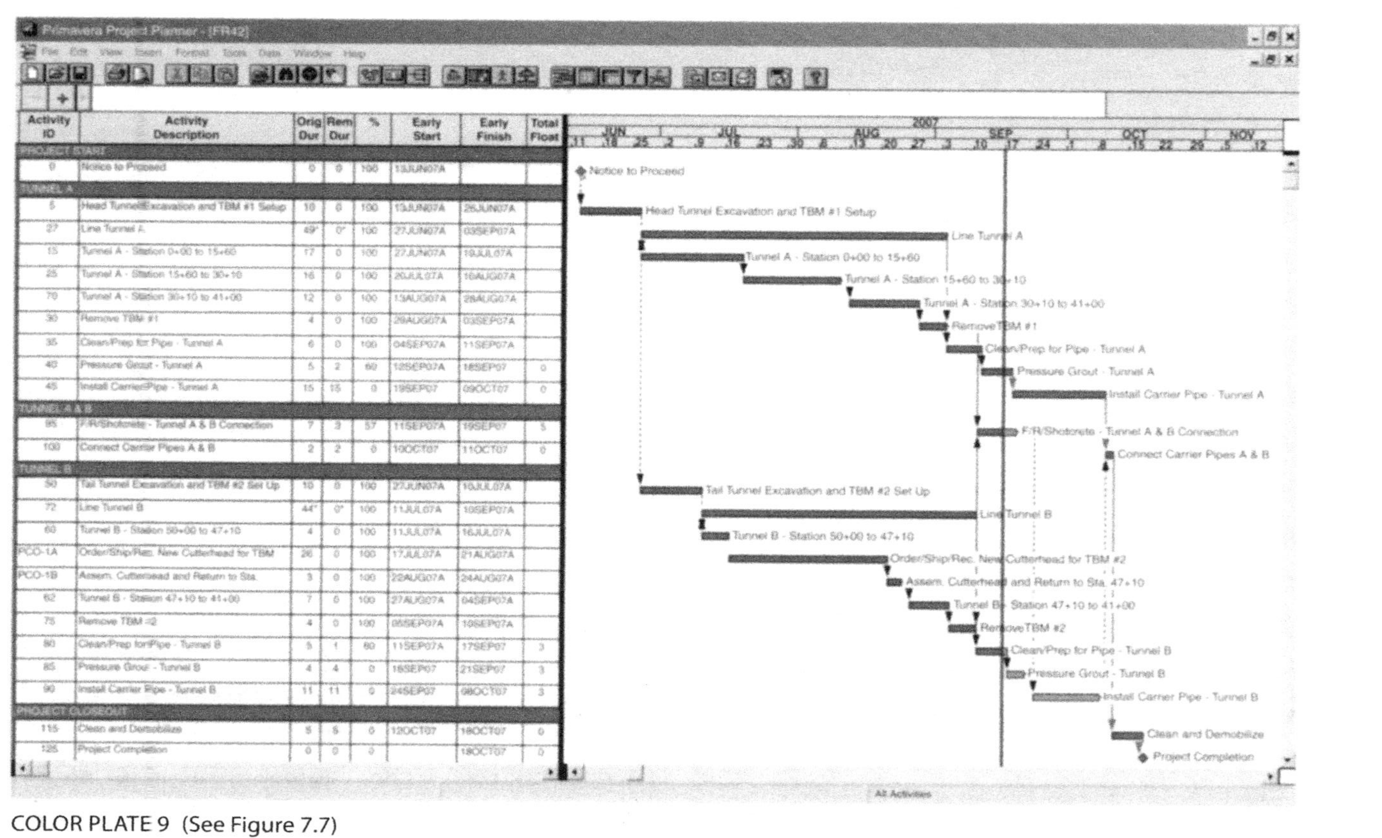

Activity ID	Activity Description	Orig Dur	Rem Dur	%	Early Start	Early Finish	Total Float
PROJECT START							
0	Notice to Proceed	0	0	100	13JUN07A		
TUNNEL A							
5	Head Tunnel Excavation and TBM #1 Setup	10	0	100	13JUN07A	26JUN07A	
27	Line Tunnel A	49*	0*	100	27JUN07A	03SEP07A	
15	Tunnel A - Station 0+00 to 15+60	17	0	100	27JUN07A	19JUL07A	
25	Tunnel A - Station 15+60 to 30+10	16	0	100	20JUL07A	10AUG07A	
70	Tunnel A - Station 30+10 to 41+00	12	0	100	13AUG07A	28AUG07A	
30	Remove TBM #1	4	0	100	29AUG07A	03SEP07A	
35	Clean/Prep for Pipe - Tunnel A	6	0	100	04SEP07A	11SEP07A	
40	Pressure Grout - Tunnel A	5	2	60	12SEP07A	18SEP07	0
45	Install Carrier Pipe - Tunnel A	15	15	0	19SEP07	09OCT07	0
TUNNEL A & B							
95	F/R/Shotcrete - Tunnel A & B Connection	7	3	57	11SEP07A	19SEP07	5
100	Connect Carrier Pipes A & B	2	2	0	10OCT07	11OCT07	0
TUNNEL B							
50	Tail Tunnel Excavation and TBM #2 Set Up	10	0	100	27JUN07A	10JUL07A	
72	Line Tunnel B	44*	0*	100	11JUL07A	10SEP07A	
60	Tunnel B - Station 50+00 to 47+10	4	0	100	11JUL07A	16JUL07A	
PCO-1A	Order/Ship/Rec. New Cutterhead for TBM	26	0	100	17JUL07A	21AUG07A	
PCO-1B	Assem. Cutterhead and Return to Sta.	3	0	100	22AUG07A	24AUG07A	
62	Tunnel B - Station 47+10 to 41+00	7	0	100	27AUG07A	04SEP07A	
75	Remove TBM #2	4	0	100	05SEP07A	10SEP07A	
80	Clean/Prep for Pipe - Tunnel B	5	1	80	11SEP07A	17SEP07	3
85	Pressure Grout - Tunnel B	4	4	0	18SEP07	21SEP07	3
90	Install Carrier Pipe - Tunnel B	11	11	0	24SEP07	08OCT07	3
PROJECT CLOSEOUT							
115	Clean and Demobilize	5	5	0	12OCT07	18OCT07	0
125	Project Completion	0	0	0		18OCT07	0

COLOR PLATE 9 (See Figure 7.7)

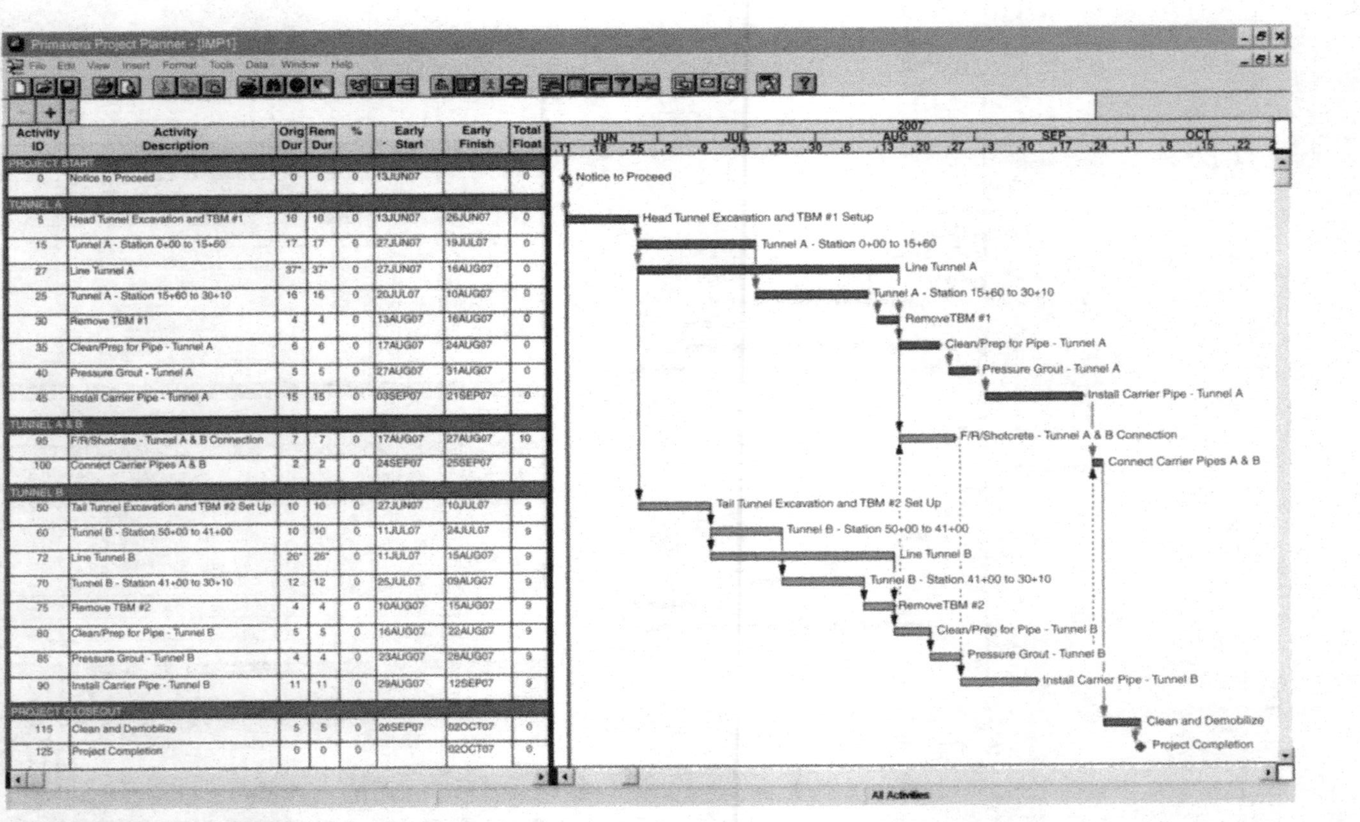

Activity ID	Activity Description	Orig Dur	Rem Dur	%	Early Start	Early Finish	Total Float
PROJECT START							
0	Notice to Proceed	0	0	0	13JUN07		0
TUNNEL A							
5	Head Tunnel Excavation and TBM #1	10	10	0	13JUN07	26JUN07	0
15	Tunnel A - Station 0+00 to 15+60	17	17	0	27JUN07	19JUL07	0
27	Line Tunnel A	37*	37*	0	27JUN07	16AUG07	0
25	Tunnel A - Station 15+60 to 30+10	16	16	0	20JUL07	10AUG07	0
30	Remove TBM #1	4	4	0	13AUG07	16AUG07	0
35	Clean/Prep for Pipe - Tunnel A	6	6	0	17AUG07	24AUG07	0
40	Pressure Grout - Tunnel A	5	5	0	27AUG07	31AUG07	0
45	Install Carrier Pipe - Tunnel A	15	15	0	03SEP07	21SEP07	0
TUNNEL A & B							
95	F/R/Shotcrete - Tunnel A & B Connection	7	7	0	17AUG07	27AUG07	10
100	Connect Carrier Pipes A & B	2	2	0	24SEP07	25SEP07	0
TUNNEL B							
50	Tail Tunnel Excavation and TBM #2 Set Up	10	10	0	27JUN07	10JUL07	9
60	Tunnel B - Station 50+00 to 41+00	10	10	0	11JUL07	24JUL07	9
72	Line Tunnel B	26*	26*	0	11JUL07	15AUG07	9
70	Tunnel B - Station 41+00 to 30+10	12	12	0	25JUL07	09AUG07	9
75	Remove TBM #2	4	4	0	10AUG07	15AUG07	9
80	Clean/Prep for Pipe - Tunnel B	5	5	0	16AUG07	22AUG07	9
85	Pressure Grout - Tunnel B	4	4	0	23AUG07	28AUG07	9
90	Install Carrier Pipe - Tunnel B	11	11	0	29AUG07	12SEP07	9
PROJECT CLOSEOUT							
115	Clean and Demobilize	5	5	0	26SEP07	02OCT07	0
125	Project Completion	0	0	0		02OCT07	0

COLOR PLATE 10 (See Figure 7.15)

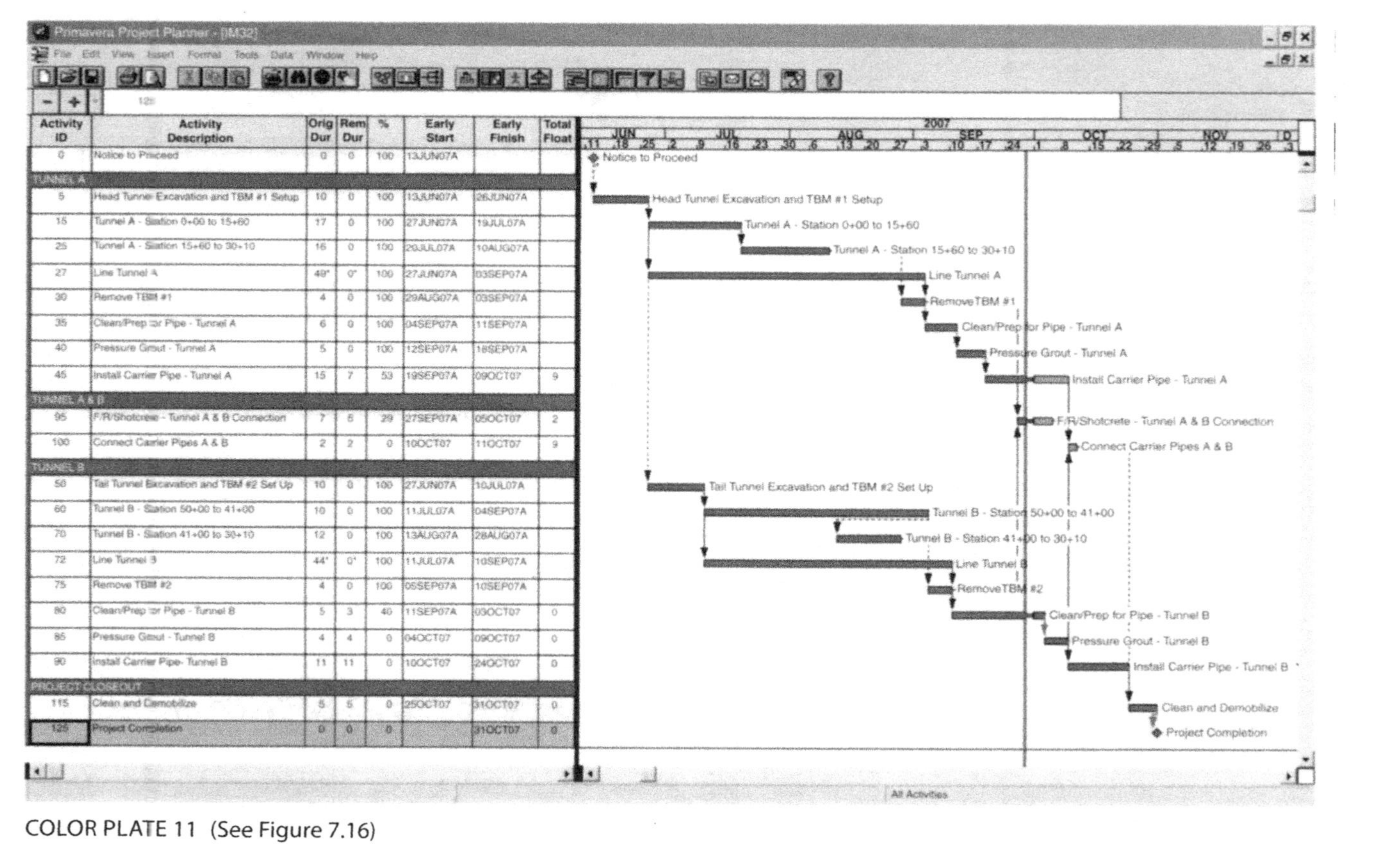

Activity ID	Activity Description	Orig Dur	Rem Dur	%	Early Start	Early Finish	Total Float
0	Notice to Proceed	0	0	100	13JUN07A		
TUNNEL A							
5	Head Tunnel Excavation and TBM #1 Setup	10	0	100	13JUN07A	26JUN07A	
15	Tunnel A - Station 0+00 to 15+60	17	0	100	27JUN07A	19JUL07A	
25	Tunnel A - Station 15+60 to 30+10	16	0	100	20JUL07A	10AUG07A	
27	Line Tunnel A	49*	0*	100	27JUN07A	03SEP07A	
30	Remove TBM #1	4	0	100	29AUG07A	03SEP07A	
35	Clean/Prep for Pipe - Tunnel A	6	0	100	04SEP07A	11SEP07A	
40	Pressure Grout - Tunnel A	5	0	100	12SEP07A	18SEP07A	
45	Install Carrier Pipe - Tunnel A	15	7	53	19SEP07A	09OCT07	9
TUNNEL A & B							
95	F/R/Shotcrete - Tunnel A & B Connection	7	5	29	27SEP07A	05OCT07	2
100	Connect Carrier Pipes A & B	2	2	0	10OCT07	11OCT07	9
TUNNEL B							
50	Tail Tunnel Excavation and TBM #2 Set Up	10	0	100	27JUN07A	10JUL07A	
60	Tunnel B - Station 50+00 to 41+00	10	0	100	11JUL07A	04SEP07A	
70	Tunnel B - Station 41+00 to 30+10	12	0	100	13AUG07A	28AUG07A	
72	Line Tunnel B	44*	0*	100	11JUL07A	10SEP07A	
75	Remove TBM #2	4	0	100	05SEP07A	10SEP07A	
80	Clean/Prep for Pipe - Tunnel B	5	3	40	11SEP07A	03OCT07	0
85	Pressure Grout - Tunnel B	4	4	0	04OCT07	09OCT07	0
90	Install Carrier Pipe- Tunnel B	11	11	0	10OCT07	24OCT07	0
PROJECT CLOSEOUT							
115	Clean and Demobilize	5	5	0	25OCT07	31OCT07	0
125	Project Completion	0	0	0		31OCT07	0

COLOR PLATE 11 (See Figure 7.16)

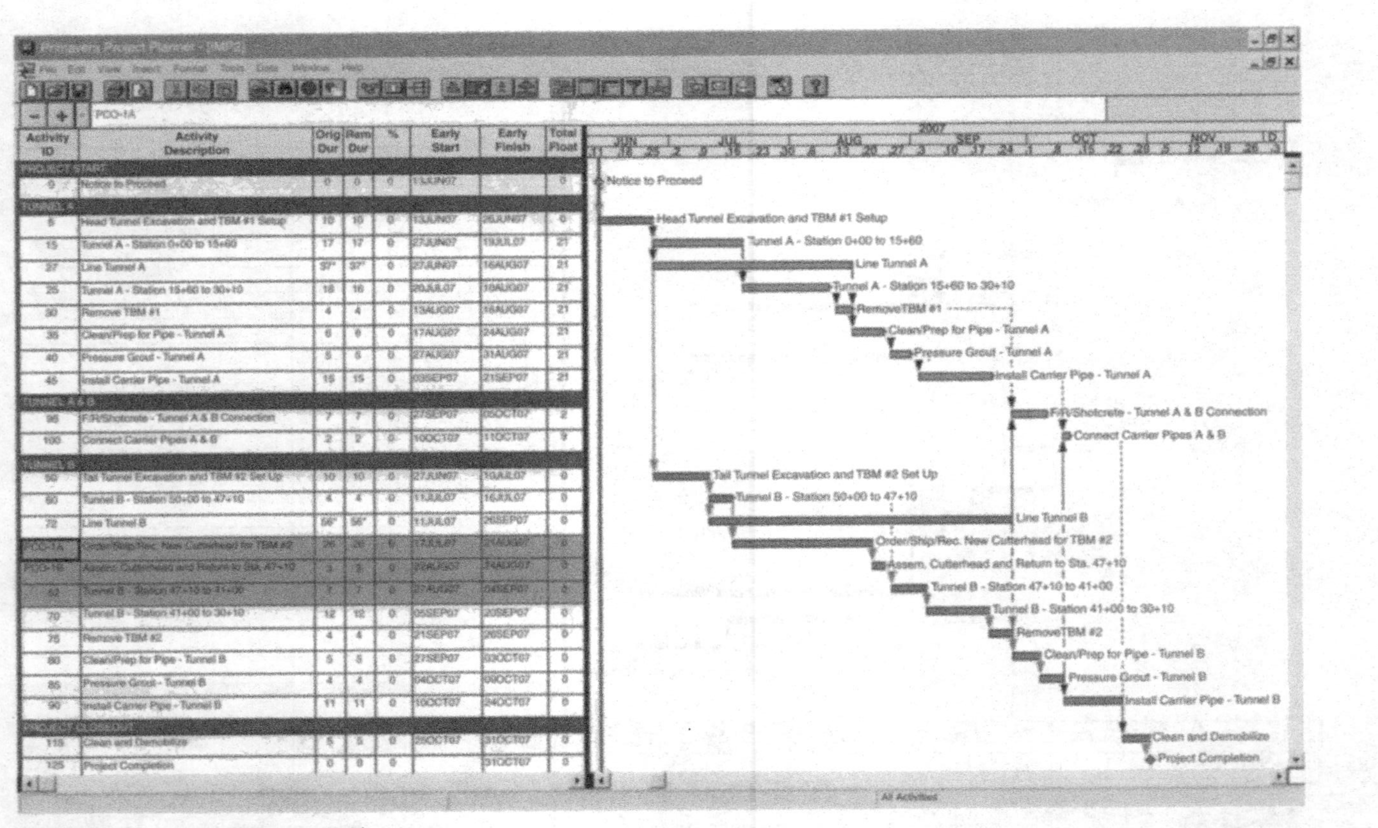

COLOR PLATE 12 (See Figure 7.17)

In the simplest case, this will be the critical Project delay attributable to the added drywall work. In instances where more than one fragnet is inserted into the CPM, each fragnet could contribute to the Project delay. For example, the Owner directed that the new wall be installed in the office, and in a separate change order the Owner directed that an additional toilet be added to each of the bathrooms, so in this case, we have the following:

Predicted completion date prior to inserting fragnets: September 19, 2008
Predicted completion date after inserting fragnets: September 28, 2008
September 28, 2008 – September 19, 2008 = nine calendar days

The additional wall caused two calendar days of delay, and the additional toilets caused seven calendar days of delay.

Using fragnets to measure delays has advantages in that both parties will have agreed to the activities and logic of the fragnet. Typically, the fragnet is required to be submitted as part of a Contractor's change order proposal. The fragnet is negotiated along with the costs for the change. Ideally, the parties will have discussed the crew size and the time required to complete the work and how the fragnet activities are logically tied into the CPM. This negotiation process allows the parties to assure themselves that they fully understand the logic of the fragnet.

While it is advantageous for both parties to understand the fragnet prior to its insertion, there are disadvantages. The two biggest disadvantages are the time it takes to develop and negotiate a fragnet and determining into which CPM update the fragnet should be inserted. Many contracts allow the Contractor 30 calendar days to provide its change order proposal. Once the proposal has been received and reviewed, it may take an Owner several weeks before it is prepared to negotiate the costs and the time. The negotiations themselves may take several weeks depending on the amount in question and the support provided by the Contractor. Therefore, it could be more than two months after the Contractor was directed to perform the work that the parties agree upon the fragnet. In many instances the parties do not agree, and time is not discussed until the end of the Project.

Another issue that arises deals with identifying into which CPM update the fragnet should be inserted. The fragnet should be inserted into the CPM update that was in effect at the time the change was issued, or at the time both parties understood that there was a change, or that the Contractor began its work related to the change, whichever was the earliest. In the case where the Owner issues a directive to change the work, the date of the directive establishes the CPM update that the fragnet is to be inserted. However, lacking a directive, it is not necessarily clear which CPM update should be used. Many times the CPM update to be used must be negotiated along with the fragnet itself. In addition to using a fragnet to analyze change orders, CPM fragnets are also used to evaluate delays to specific features of work or specific Subcontractor's work. For example, a Subcontractor may allege that it was delayed during the installation of a holding tank at a wastewater treatment facility. The CPM may show that the holding tanks are not on the critical path. However, the analyst can extract the holding tank activities as a fragnet and evaluate the alleged time effects.

The fragnet approach is advantageous in that the analysis is focused only on the portion of the work that was changed. One major disadvantage to the fragnet approach is the time it takes to reach agreement on the logic of the fragnet and how it is to be inserted into the overall schedule.

Fragnet Example

D-Tunneling Company (D-Tunneling), a tunneling Subcontractor, was performing the tunneling and drainage piping installation for a large terminal expansion at an airport in New Mexico. In its Contract with the airport authority, D-Tunneling received its notice to proceed on June 13, 2007, and was required to complete the Project by October 2, 2007. Before starting construction, D-Tunneling created a baseline schedule reflecting its original plan for the work. After construction began, D-Tunneling updated its schedule on a monthly basis.

BASELINE SCHEDULE

D-Tunneling's baseline schedule, as shown in Figure 7.1, identifies a notice to proceed on June 13, 2007, and a forecast Project completion of October 2, 2007, the same date as the Contract completion date. D-Tunneling's baseline schedule identifies that it is planning to perform tunneling work on two separate work paths, Tunnels A and B, at the same time. The plans show that Tunnels A and B will combine to form one, continuous, straight drainage tunnel when finished. D-Tunneling has decided to use two tunnel-boring machines (TBMs) to complete its work. The TBMs would be set up at opposite ends of the drainage tunnel and work toward each other, meeting at Station 30+10. TBM #1 will be boring Tunnel A from Station 0+00 to 30+10, and TBM #2 will be boring Tunnel B from Station 50+00 to 30+10.

D-Tunneling's schedule update on the morning of July 17, 2007, identified that between June 13, 2007, and July 16, 2007, the Project had progressed as expected. D-Tunneling's July 17, 2007, Update did not contain any schedule revisions. Figure 7.2 depicts the status of D-Tunneling's work when it submitted the schedule update to the airport authority on the morning of July 17, 2007.

On the afternoon of July 17, TBM #2 encountered rock at Station 47+10 that was harder than the geotechnical report indicated in the Contract documents. In addition, the rock was considered "mixed face" and contained a mixture of rocks with varying compositions and hardnesses. As a result, D-Tunneling stopped work on Activity 60 and was not able to progress work in Tunnel B until it resolved the issue.

D-Tunneling's Contract with the airport authority stated that time extensions must be substantiated by a fragnet analysis, comparing the Project completion date before the fragnet was inserted into the schedule and the Project completion date after the fragnet was inserted into the schedule. On July 25,

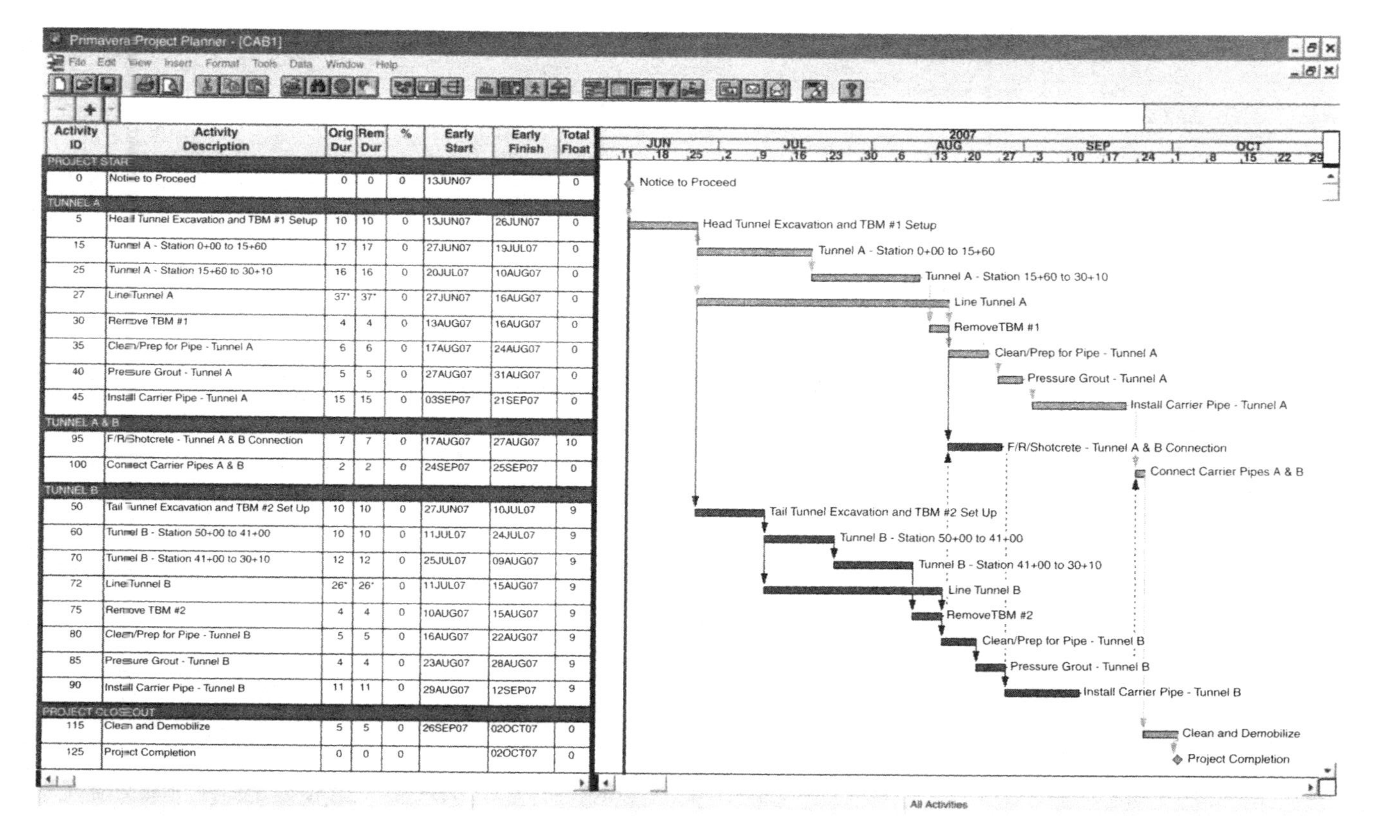

Activity ID	Activity Description	Orig Dur	Rem Dur	%	Early Start	Early Finish	Total Float
PROJECT START							
0	Notice to Proceed	0	0	0	13JUN07		0
TUNNEL A							
5	Head Tunnel Excavation and TBM #1 Setup	10	10	0	13JUN07	26JUN07	0
15	Tunnel A - Station 0+00 to 15+60	17	17	0	27JUN07	19JUL07	0
25	Tunnel A - Station 15+60 to 30+10	16	16	0	20JUL07	10AUG07	0
27	Line Tunnel A	37*	37*	0	27JUN07	16AUG07	0
30	Remove TBM #1	4	4	0	13AUG07	16AUG07	0
35	Clean/Prep for Pipe - Tunnel A	6	6	0	17AUG07	24AUG07	0
40	Pressure Grout - Tunnel A	5	5	0	27AUG07	31AUG07	0
45	Install Carrier Pipe - Tunnel A	15	15	0	03SEP07	21SEP07	0
TUNNEL A & B							
95	F/R/Shotcrete - Tunnel A & B Connection	7	7	0	17AUG07	27AUG07	10
100	Connect Carrier Pipes A & B	2	2	0	24SEP07	25SEP07	0
TUNNEL B							
50	Tail Tunnel Excavation and TBM #2 Set Up	10	10	0	27JUN07	10JUL07	9
60	Tunnel B - Station 50+00 to 41+00	10	10	0	11JUL07	24JUL07	9
70	Tunnel B - Station 41+00 to 30+10	12	12	0	25JUL07	09AUG07	9
72	Line Tunnel B	26*	26*	0	11JUL07	15AUG07	9
75	Remove TBM #2	4	4	0	10AUG07	15AUG07	9
80	Clean/Prep for Pipe - Tunnel B	5	5	0	16AUG07	22AUG07	9
85	Pressure Grout - Tunnel B	4	4	0	23AUG07	28AUG07	9
90	Install Carrier Pipe - Tunnel B	11	11	0	29AUG07	12SEP07	9
PROJECT CLOSEOUT							
115	Clean and Demobilize	5	5	0	26SEP07	02OCT07	0
125	Project Completion	0	0	0		02OCT07	0

FIGURE 7.1 See Color Plate 3.

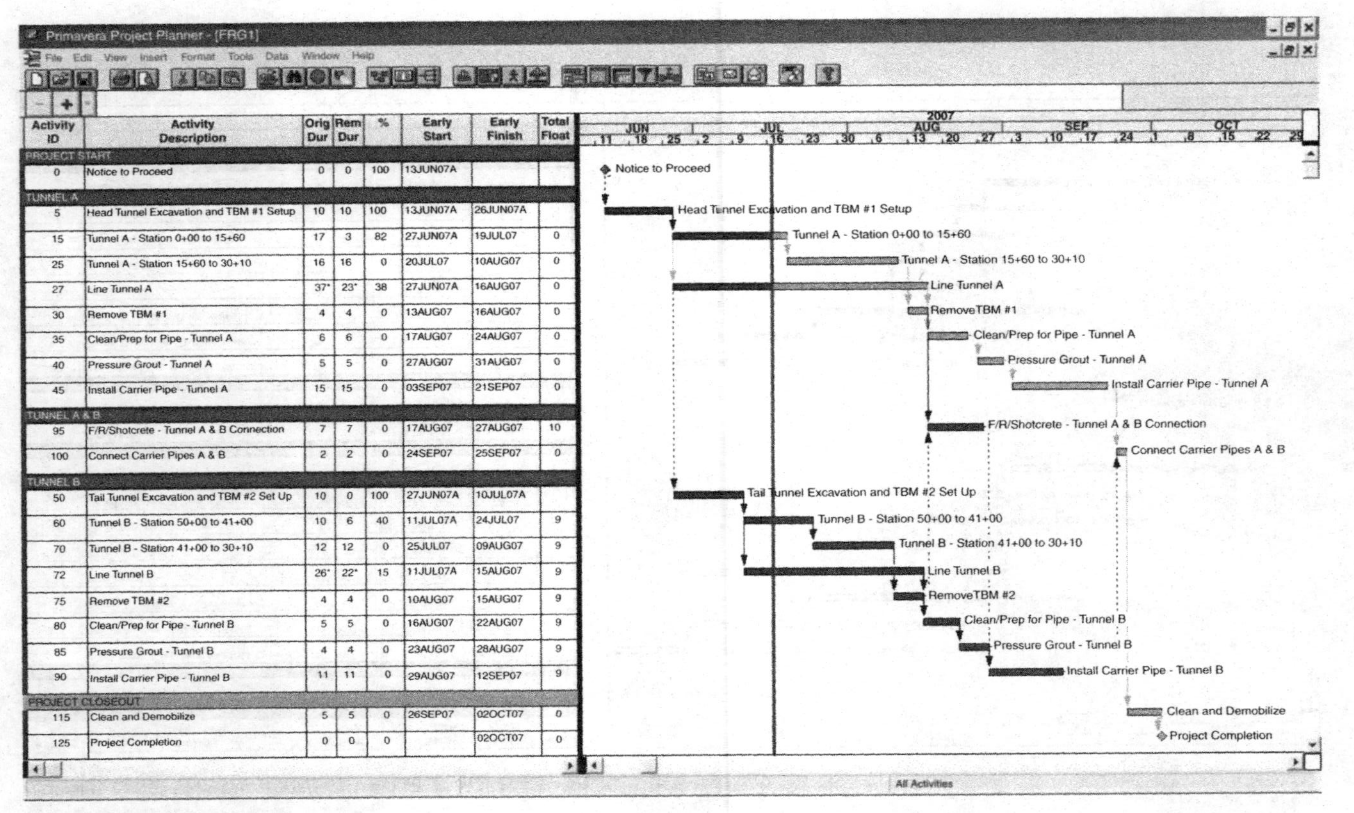

Activity ID	Activity Description	Orig Dur	Rem Dur	%	Early Start	Early Finish	Total Float
PROJECT START							
0	Notice to Proceed	0	0	100	13JUN07A		
TUNNEL A							
5	Head Tunnel Excavation and TBM #1 Setup	10	10	100	13JUN07A	26JUN07A	
15	Tunnel A - Station 0+00 to 15+60	17	3	82	27JUN07A	19JUL07	0
25	Tunnel A - Station 15+60 to 30+10	16	16	0	20JUL07	10AUG07	0
27	Line Tunnel A	37*	23*	38	27JUN07A	16AUG07	0
30	Remove TBM #1	4	4	0	13AUG07	16AUG07	0
35	Clean/Prep for Pipe - Tunnel A	6	6	0	17AUG07	24AUG07	0
40	Pressure Grout - Tunnel A	5	5	0	27AUG07	31AUG07	0
45	Install Carrier Pipe - Tunnel A	15	15	0	03SEP07	21SEP07	0
TUNNEL A & B							
95	F/R/Shotcrete - Tunnel A & B Connection	7	7	0	17AUG07	27AUG07	10
100	Connect Carrier Pipes A & B	2	2	0	24SEP07	25SEP07	0
TUNNEL B							
50	Tail Tunnel Excavation and TBM #2 Set Up	10	0	100	27JUN07A	10JUL07A	
60	Tunnel B - Station 50+00 to 41+00	10	6	40	11JUL07A	24JUL07	9
70	Tunnel B - Station 41+00 to 30+10	12	12	0	25JUL07	09AUG07	9
72	Line Tunnel B	26*	22*	15	11JUL07A	15AUG07	9
75	Remove TBM #2	4	4	0	10AUG07	15AUG07	9
80	Clean/Prep for Pipe - Tunnel B	5	5	0	16AUG07	22AUG07	9
85	Pressure Grout - Tunnel B	4	4	0	23AUG07	28AUG07	9
90	Install Carrier Pipe - Tunnel B	11	11	0	29AUG07	12SEP07	9
PROJECT CLOSEOUT							
115	Clean and Demobilize	5	5	0	26SEP07	02OCT07	0
125	Project Completion	0	0	0		02OCT07	0

FIGURE 7.2 See Color Plate 4.

D-Tunneling sent a letter to the airport authority identifying its new plan to complete the changed work, along with a fragnet schedule to substantiate D-Tunneling's requested time extension for PCO #1, *TBM #2 Shutdown*. D-Tunneling's July 25 letter stated the following:

1. The airport authority/engineer was responsible for all soil borings and site testing in the Contract.
2. D-Tunneling had encountered "mixed face" rock at Station 47+10 that was harder than the tolerance limits identified by the engineers' soil-boring tests in the Contract documents. As a result, D-Tunneling had to shut down its TBM #2 operations in Tunnel B. D-Tunneling has worked and will continue to work in Tunnel A as expected.
3. The engineer has completed additional testing of the area and determined that the out-of-tolerance rock exists between Stations 47+10 and 42+00.
4. D-Tunneling has purchased a new cutterhead that is expected to be delivered on August 21, 2007. The cutterhead will take three workdays to assemble. D-Tunneling will then return to Station 47+10 to resume tunneling. In addition, it will take seven workdays to complete tunneling from Station 47+10 to 41+00, instead of the six remaining workdays shown in the July 15, 2005, Update, due to the hardness of the rock.
5. Using the preceding timeline, D-Tunneling has inserted a fragnet into its July 17, 2007, Update, with the following changes:
   a. Activity 60 received an actual finish of July 16, 2007, and a changed work scope to only cover the tunneling work that had already been completed between Stations 50+00 and 47+10.
   b. Activity 62, Tunnel B—Stations 47+10 to 41+00, was added to the schedule to denote the remaining tunneling work after the new cutterhead is assembled. Also, Activity 62 was given a duration of seven workdays.
   c. Activity PCO-1A, Order/Ship/Rec. New Cutterhead for TBM #2, was added to the schedule to denote the time it will take to receive the new cutterhead. Activity PCO-1A was given a 26-workday duration to take its finish date through August 21, 2007, the date that D-Tunneling expects to receive the new cutterhead.
   d. Activity PCO-1B, Assem. Cutterhead and Return to Sta. 47+10, was added to the schedule to denote the time it will take to assemble the new cutterhead and return it to Station 47+10.
   e. Logic was revised to reflect the changes just identified and the new sequence of work. All logic changes shown are finish-to-start relationships with no leads or lags. No other logic relationships in the schedule were changed, with the exception of the logic relationships between Activity 60, 62, PCO-1A, and PCO-1B.

Figure 7.3 depicts D-Tunneling's new July 17, 2007, Update, including the addition of the new fragnet. D-Tunneling adjusted the schedule to only include

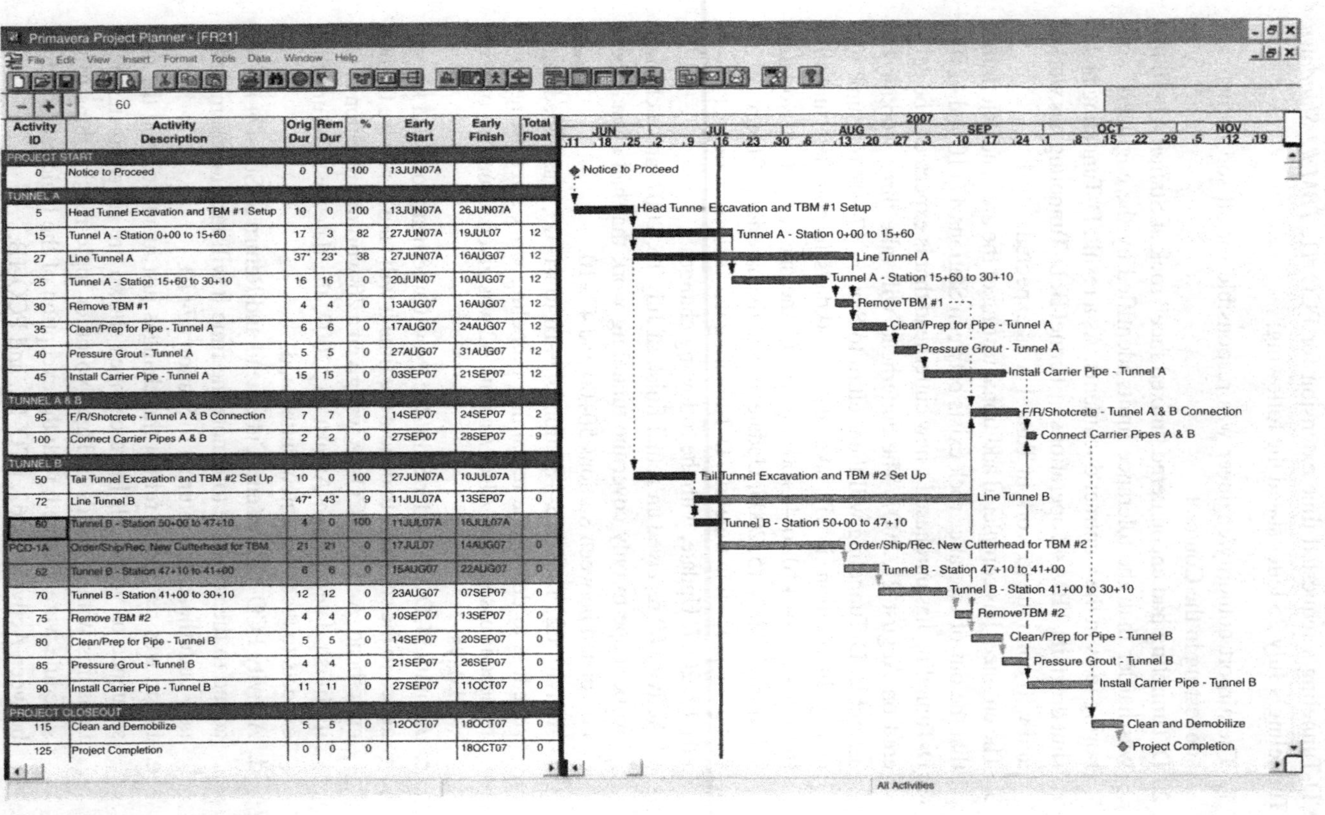

Activity ID	Activity Description	Orig Dur	Rem Dur	%	Early Start	Early Finish	Total Float
PROJECT START							
0	Notice to Proceed	0	0	100	13JUN07A		
TUNNEL A							
5	Head Tunnel Excavation and TBM #1 Setup	10	0	100	13JUN07A	26JUN07A	
15	Tunnel A - Station 0+00 to 15+60	17	3	82	27JUN07A	19JUL07	12
27	Line Tunnel A	37*	23*	38	27JUN07A	16AUG07	12
25	Tunnel A - Station 15+60 to 30+10	16	16	0	20JUN07	10AUG07	12
30	Remove TBM #1	4	4	0	13AUG07	16AUG07	12
35	Clean/Prep for Pipe - Tunnel A	6	6	0	17AUG07	24AUG07	12
40	Pressure Grout - Tunnel A	5	5	0	27AUG07	31AUG07	12
45	Install Carrier Pipe - Tunnel A	15	15	0	03SEP07	21SEP07	12
TUNNEL A & B							
95	F/R/Shotcrete - Tunnel A & B Connection	7	7	0	14SEP07	24SEP07	2
100	Connect Carrier Pipes A & B	2	2	0	27SEP07	28SEP07	9
TUNNEL B							
50	Tail Tunnel Excavation and TBM #2 Set Up	10	0	100	27JUN07A	10JUL07A	
72	Line Tunnel B	47*	43*	9	11JUL07A	13SEP07	0
60	Tunnel B - Station 50+00 to 47+10	4	0	100	11JUL07A	16JUL07A	
PCO-1A	Order/Ship/Rec. New Cutterhead for TBM	21	21	0	17JUL07	14AUG07	0
62	Tunnel B - Station 47+10 to 41+00	6	6	0	15AUG07	22AUG07	0
70	Tunnel B - Station 41+00 to 30+10	12	12	0	23AUG07	07SEP07	0
75	Remove TBM #2	4	4	0	10SEP07	13SEP07	0
80	Clean/Prep for Pipe - Tunnel B	5	5	0	14SEP07	20SEP07	0
85	Pressure Grout - Tunnel B	4	4	0	21SEP07	26SEP07	0
90	Install Carrier Pipe - Tunnel B	11	11	0	27SEP07	11OCT07	0
PROJECT CLOSEOUT							
115	Clean and Demobilize	5	5	0	12OCT07	18OCT07	0
125	Project Completion	0	0	0		18OCT07	0

FIGURE 7.3 See Color Plate 5.

fragnet activities that would affect progress to the Project between July 17, 2007, and August 14, 2007. The remainder of the TBM #2 shutdown delays would be inserted in D-Tunneling's next update on August 15, 2007.

As a result of the fragnet in Figure 7.3, the critical path of the Project shifted to Tunnel B and extended the Project completion date from October 2, 2007, to October 18, 2007. Only fragnet activity PCO-1A was added to the July 17, 2007, Update. The duration for Activity PCO-1A was added to only affect the Project through D-Tunneling's next schedule update that is projected to occur on August 15, 2007. D-Tunneling requested a time extension of 16 days (October 18, 2007 – October 2, 2007 = 16 calendar days) for the differing site condition.

The airport authority requested that D-Tunneling reorganize its work, if possible, to mitigate some of the delays to the new critical path of the Project. D-Tunneling revised its schedule (Figure 7.4) by changing from TBM #2 (Tunnel B) to TBM #1 (Tunnel A) to perform the tunneling for Activity 70, *Tunnel B—Station 41+00 to 30+10.*

Activity 70 has a 12-workday duration, and by resequencing Activity 70 from TBM #2 to TBM #1, Tunnel A became the critical path of the Project. However, Figure 7.4 shows that reorganizing Activity 70 did not result in any improvement to the Project completion date. If this is the case, should D-Tunneling have reassigned the Activity 70 tunneling work to TBM #1? The answer is absolutely, because this more accurately displays D-Tunneling's current plan of the resources (TBM #1) that will be used to complete Activity 70. D-Tunneling has also created 12 days of total float for Tunnel B. By creating 12 days of total float in Tunnel B, D-Tunneling is ensuring that if the delivery of the new cutterhead is late, it will use up some of its total float and not negatively affect the forecast Project completion date as much. In addition, there are further delays to be inserted into D-Tunneling's subsequent August 15, 2007, Schedule that have not been inserted into the July 17, 2007, Schedule.

As a result of the fragnet schedule submitted in the July 17, 2007, Update, D-Tunneling again resubmitted its request for a 16-day time extension due to the differing site condition in Tunnel B. D-Tunneling also reserved its right to request additional days of delay related to the differing site condition in Tunnel B because it had not included all of its fragnet delays in the July 17, 2007, Update. Again, the July 17, 2007, Update only included D-Tunneling's delays between July 17, 2007, and August 14, 2007. Therefore, some of the TBM #2 delays may affect the subsequent schedule update on August 15, 2007, and may require an additional time extension. The airport authority granted D-Tunneling its requested 16-day time extension from the fragnet added to the July 17, 2007, Update.

D-Tunneling submitted its next schedule update on the morning of August 15, 2007. All activities made as-expected progress between July 17, 2007, and August 14, 2007, with no schedule revisions. Figure 7.5 shows D-Tunneling's August 15, 2007, Update, without the additional fragnets

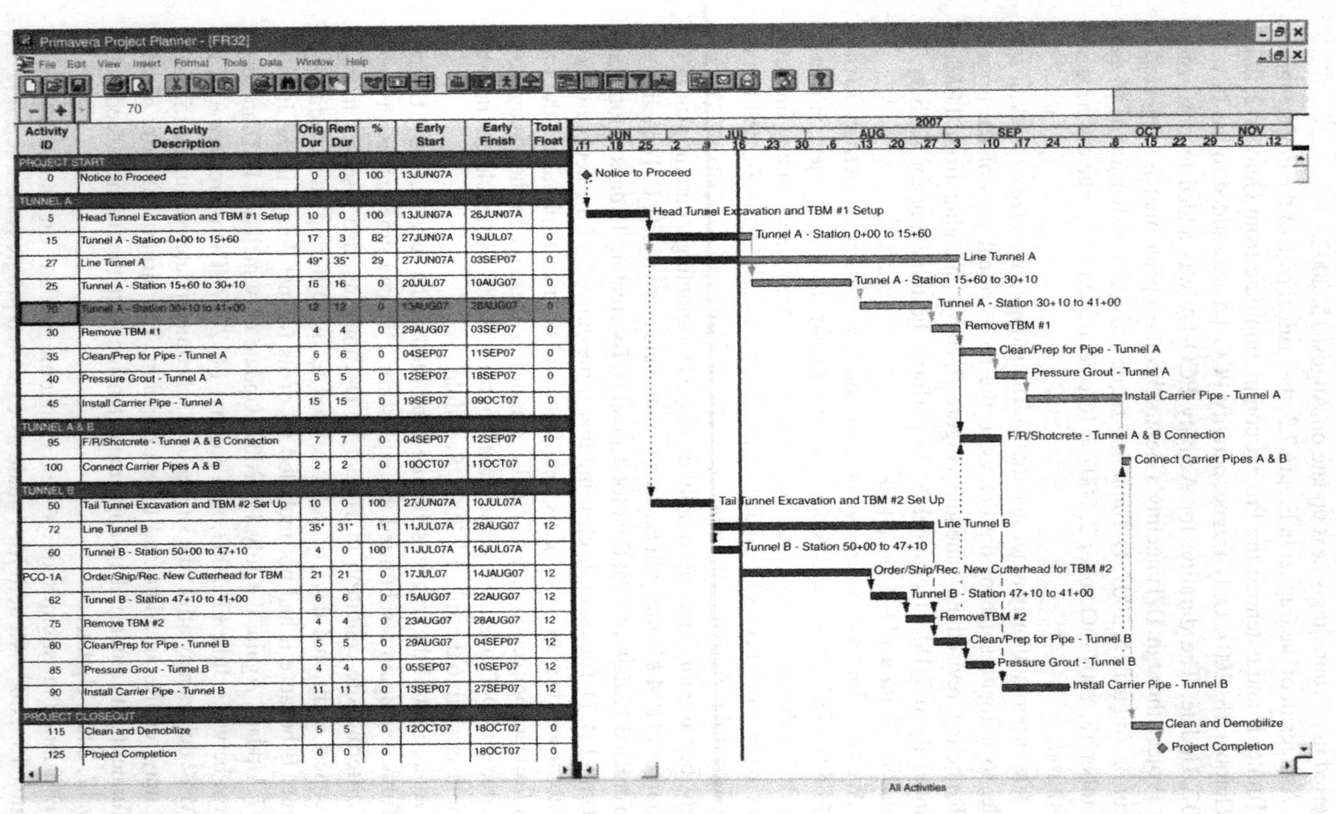

Activity ID	Activity Description	Orig Dur	Rem Dur	%	Early Start	Early Finish	Total Float
PROJECT START							
0	Notice to Proceed	0	0	100	13JUN07A		
TUNNEL A							
5	Head Tunnel Excavation and TBM #1 Setup	10	0	100	13JUN07A	26JUN07A	
15	Tunnel A - Station 0+00 to 15+60	17	3	82	27JUN07A	19JUL07	0
27	Line Tunnel A	49*	35*	29	27JUN07A	03SEP07	0
25	Tunnel A - Station 15+60 to 30+10	16	16	0	20JUL07	10AUG07	0
70	Tunnel A - Station 30+10 to 41+00	12	12	0	13AUG07	28AUG07	0
30	Remove TBM #1	4	4	0	29AUG07	03SEP07	0
35	Clean/Prep for Pipe - Tunnel A	6	6	0	04SEP07	11SEP07	0
40	Pressure Grout - Tunnel A	5	5	0	12SEP07	18SEP07	0
45	Install Carrier Pipe - Tunnel A	15	15	0	19SEP07	09OCT07	0
TUNNEL A & B							
95	F/R/Shotcrete - Tunnel A & B Connection	7	7	0	04SEP07	12SEP07	10
100	Connect Carrier Pipes A & B	2	2	0	10OCT07	11OCT07	0
TUNNEL B							
50	Tail Tunnel Excavation and TBM #2 Set Up	10	0	100	27JUN07A	10JUL07A	
72	Line Tunnel B	35*	31*	11	11JUL07A	28AUG07	12
60	Tunnel B - Station 50+00 to 47+10	4	0	100	11JUL07A	16JUL07A	
PCO-1A	Order/Ship/Rec. New Cutterhead for TBM	21	21	0	17JUL07	14JAUG07	12
62	Tunnel B - Station 47+10 to 41+00	6	6	0	15AUG07	22AUG07	12
75	Remove TBM #2	4	4	0	23AUG07	28AUG07	12
80	Clean/Prep for Pipe - Tunnel B	5	5	0	29AUG07	04SEP07	12
85	Pressure Grout - Tunnel B	4	4	0	05SEP07	10SEP07	12
90	Install Carrier Pipe - Tunnel B	11	11	0	13SEP07	27SEP07	12
PROJECT CLOSEOUT							
115	Clean and Demobilize	5	5	0	12OCT07	18OCT07	0
125	Project Completion	0	0	0		18OCT07	0

FIGURE 7.4 See Color Plate 6.

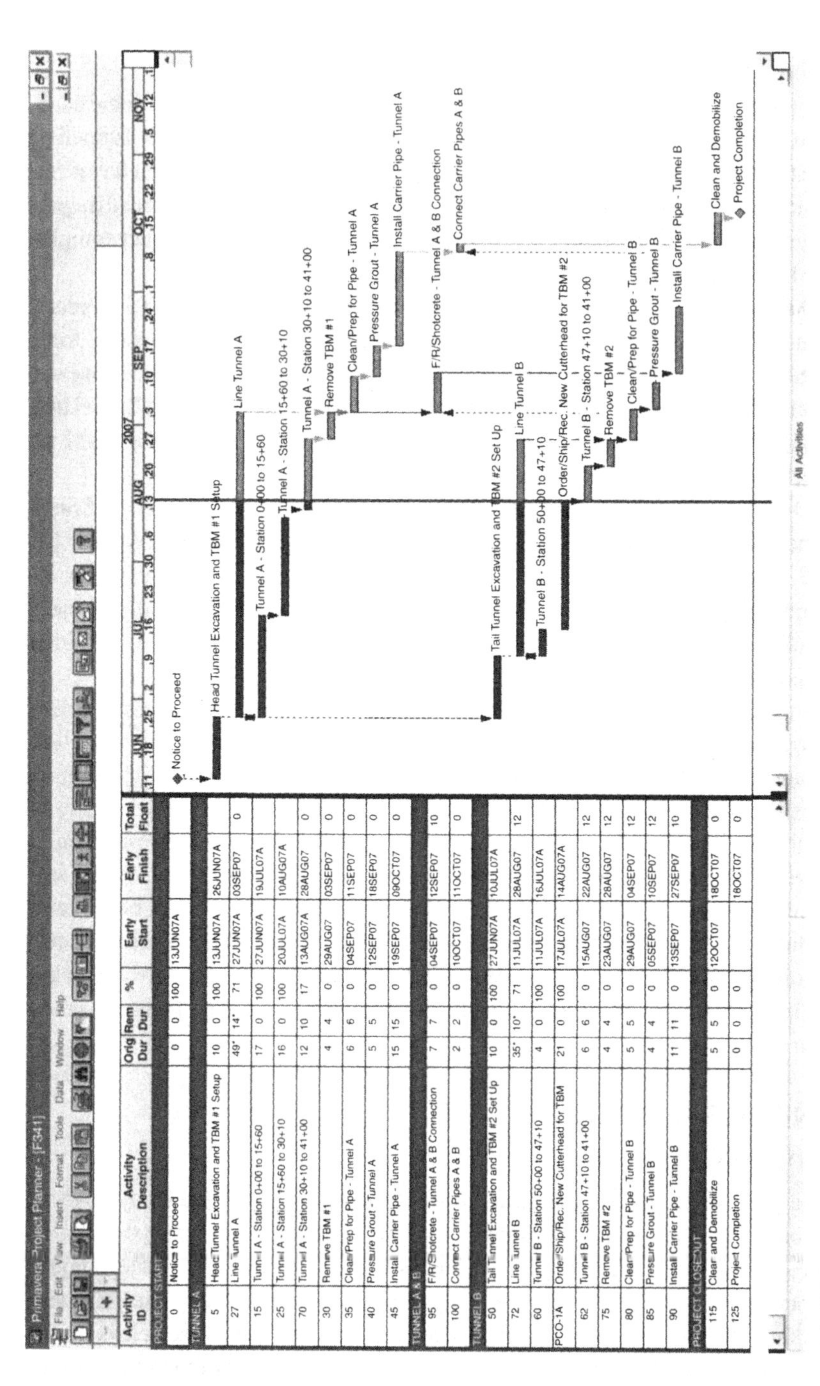

FIGURE 7.5 See Color Plate 7.

for anticipated TBM #2 delays after August 15. As of the August 15, 2007, Update, the Project completion date remained October 18, 2007, without the fragnets for additional delays and work between August 15, 2007, and its subsequent update.

Figure 7.6 represents D-Tunneling's August 15, 2007, Update with the additional fragnet activities and durations beyond August 15, 2007. D-Tunneling adjusted the original August 15, 2007, Update to include the remainder of the fragnet delays that resulted from the TBM #2 shutdown. The additional fragnet delays for TBM #2 in Tunnel B did not result in any delay to the Project completion date of October 18, 2007, and did not cause a critical path shift.

As a result of the fragnet schedule submitted in the August 15, 2007, Update, D-Tunneling did not submit a request for time extension because it forecasted that the fragnets would not affect the Project completion date. However, D-Tunneling also reserved its right to request additional days of delay related to the differing site condition in Tunnel B at a later date, should the TBM #2 cutterhead replacement not progress as expected.

D-Tunneling submitted its next schedule update on the morning of September 17, 2007. All activities made as-expected progress between August 15, 2007, and September 16, 2007, with no schedule revisions. Figure 7.7 shows D-Tunneling's September 17, 2007, Update, which still identified a Project completion date of October 18, 2007, and showed that all fragnet activities related to the TBM #2 shutdown were completed.

In conclusion, this example represents the proper way to address Project changes in a proactive, forward-looking manner using fragnets. D-Tunneling encountered a change in the Project, adjusted its schedule to reflect the Project change, and then filed for a time extension from the airport authority as early as possible. In addition, the airport authority also benefited from swift resolution to Project changes. The airport authority was presented with accurate Project status information and therefore was better prepared to adjust its future budget and planning concerns, alert future tenants of Project changes, and consider acceleration options that reflect the new Project completion date. It is always in the best interests of all parties to resolve Project changes in a timely manner as they occur on the Project.

WINDOWS TECHNIQUES

"Windows" Approach

Some analysts have been measuring critical Project delay with a delay analysis approach that they call a "windows" analysis. Though many analysts classify their analysis technique as a windows analysis, many of these analysis techniques are performed very differently. Typically, the only common feature among these windows analysis techniques is that the analyst chooses to analyze delays in isolated windows, or time periods, during the Project. When comparing many of these windows analyses, both the criteria for selecting the window and the delay

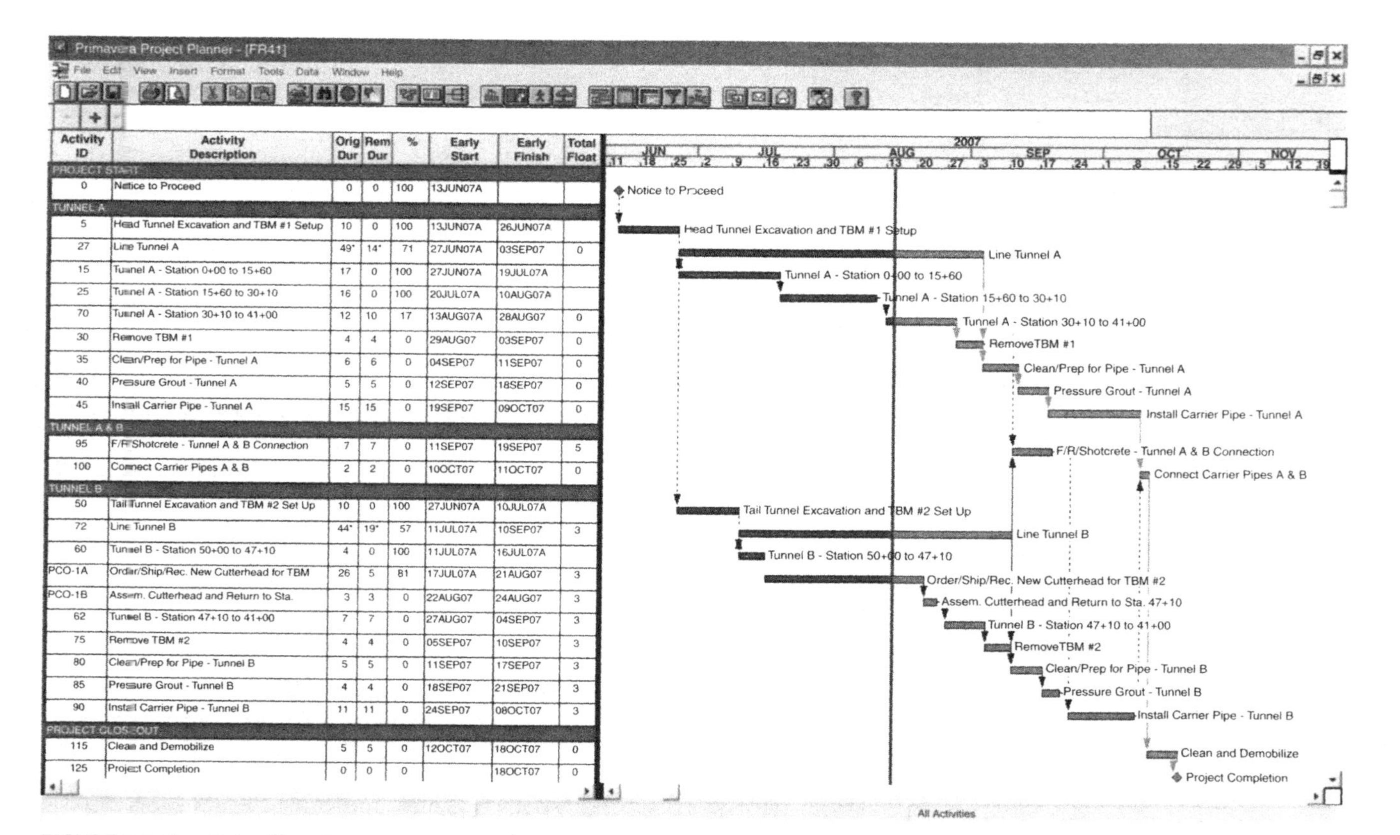

Activity ID	Activity Description	Orig Dur	Rem Dur	%	Early Start	Early Finish	Total Float
PROJECT START							
0	Notice to Proceed	0	0	100	13JUN07A		
TUNNEL A							
5	Head Tunnel Excavation and TBM #1 Setup	10	0	100	13JUN07A	26JUN07A	
27	Line Tunnel A	49*	14*	71	27JUN07A	03SEP07	0
15	Tunnel A - Station 0+00 to 15+60	17	0	100	27JUN07A	19JUL07A	
25	Tunnel A - Station 15+60 to 30+10	16	0	100	20JUL07A	10AUG07A	
70	Tunnel A - Station 30+10 to 41+00	12	10	17	13AUG07A	28AUG07	0
30	Remove TBM #1	4	4	0	29AUG07	03SEP07	0
35	Clean/Prep for Pipe - Tunnel A	6	6	0	04SEP07	11SEP07	0
40	Pressure Grout - Tunnel A	5	5	0	12SEP07	18SEP07	0
45	Install Carrier Pipe - Tunnel A	15	15	0	19SEP07	09OCT07	0
TUNNEL A & B							
95	F/R/Shotcrete - Tunnel A & B Connection	7	7	0	11SEP07	19SEP07	5
100	Connect Carrier Pipes A & B	2	2	0	10OCT07	11OCT07	0
TUNNEL B							
50	Tail Tunnel Excavation and TBM #2 Set Up	10	0	100	27JUN07A	10JUL07A	
72	Line Tunnel B	44*	19*	57	11JUL07A	10SEP07	3
60	Tunnel B - Station 50+00 to 47+10	4	0	100	11JUL07A	16JUL07A	
PCO-1A	Order/Ship/Rec. New Cutterhead for TBM	26	5	81	17JUL07A	21AUG07	3
PCO-1B	Assem. Cutterhead and Return to Sta.	3	3	0	22AUG07	24AUG07	3
62	Tunnel B - Station 47+10 to 41+00	7	7	0	27AUG07	04SEP07	3
75	Remove TBM #2	4	4	0	05SEP07	10SEP07	3
80	Clean/Prep for Pipe - Tunnel B	5	5	0	11SEP07	17SEP07	3
85	Pressure Grout - Tunnel B	4	4	0	18SEP07	21SEP07	3
90	Install Carrier Pipe - Tunnel B	11	11	0	24SEP07	08OCT07	3
PROJECT CLOSE-OUT							
115	Clean and Demobilize	5	5	0	12OCT07	18OCT07	0
125	Project Completion	0	0	0		18OCT07	0

FIGURE 7.6 See Color Plate 8.

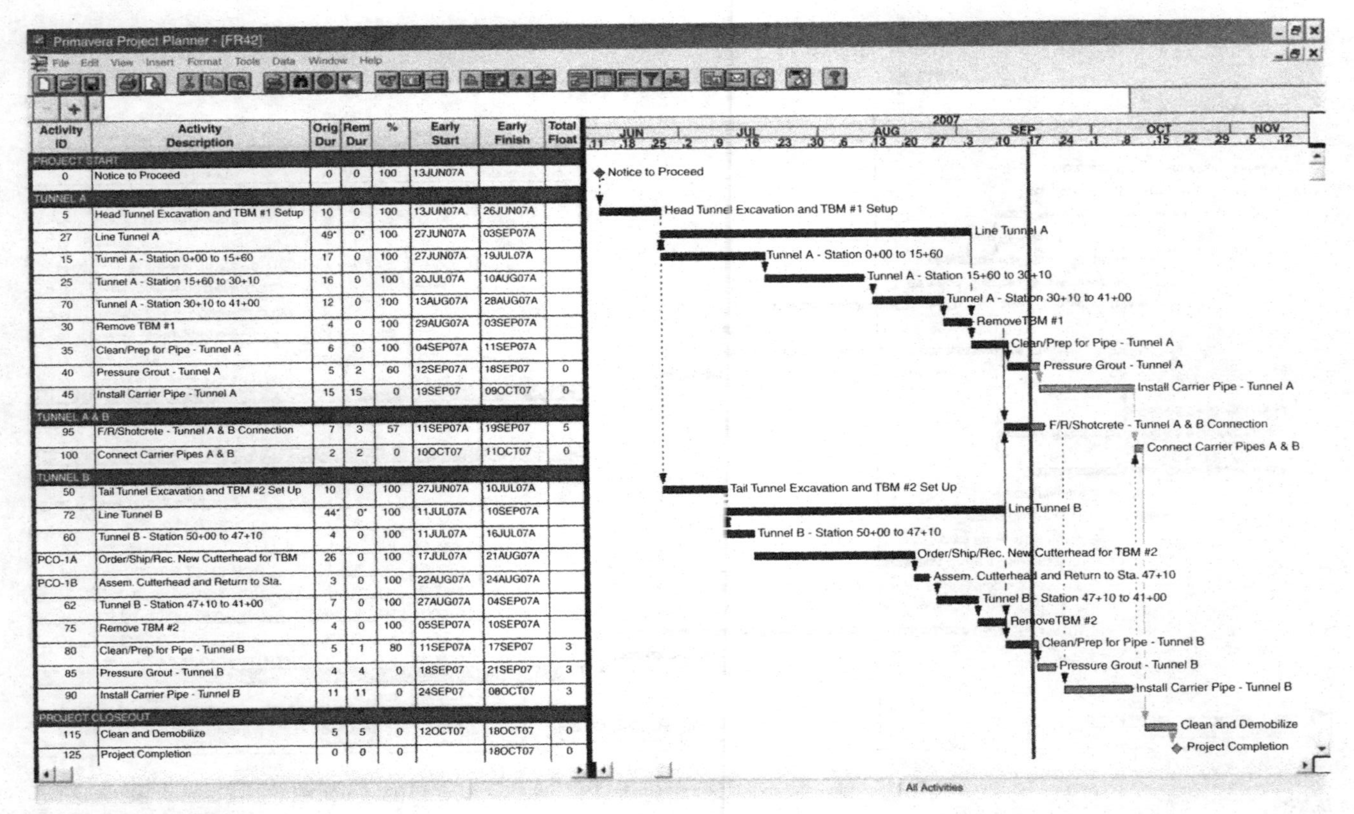

Activity ID	Activity Description	Orig Dur	Rem Dur	%	Early Start	Early Finish	Total Float
PROJECT START							
0	Notice to Proceed	0	0	100	13JUN07A		
TUNNEL A							
5	Head Tunnel Excavation and TBM #1 Setup	10	0	100	13JUN07A	26JUN07A	
27	Line Tunnel A	49*	0*	100	27JUN07A	03SEP07A	
15	Tunnel A - Station 0+00 to 15+60	17	0	100	27JUN07A	19JUL07A	
25	Tunnel A - Station 15+60 to 30+10	16	0	100	20JUL07A	10AUG07A	
70	Tunnel A - Station 30+10 to 41+00	12	0	100	13AUG07A	28AUG07A	
30	Remove TBM #1	4	0	100	29AUG07A	03SEP07A	
35	Clean/Prep for Pipe - Tunnel A	6	0	100	04SEP07A	11SEP07A	
40	Pressure Grout - Tunnel A	5	2	60	12SEP07A	18SEP07	0
45	Install Carrier Pipe - Tunnel A	15	15	0	19SEP07	09OCT07	0
TUNNEL A & B							
95	F/R/Shotcrete - Tunnel A & B Connection	7	3	57	11SEP07A	19SEP07	5
100	Connect Carrier Pipes A & B	2	2	0	10OCT07	11OCT07	0
TUNNEL B							
50	Tail Tunnel Excavation and TBM #2 Set Up	10	0	100	27JUN07A	10JUL07A	
72	Line Tunnel B	44*	0*	100	11JUL07A	10SEP07A	
60	Tunnel B - Station 50+00 to 47+10	4	0	100	11JUL07A	16JUL07A	
PCO-1A	Order/Ship/Rec. New Cutterhead for TBM	26	0	100	17JUL07A	21AUG07A	
PCO-1B	Assem. Cutterhead and Return to Sta.	3	0	100	22AUG07A	24AUG07A	
62	Tunnel B - Station 47+10 to 41+00	7	0	100	27AUG07A	04SEP07A	
75	Remove TBM #2	4	0	100	05SEP07A	10SEP07A	
80	Clean/Prep for Pipe - Tunnel B	5	1	80	11SEP07A	17SEP07	3
85	Pressure Grout - Tunnel B	4	4	0	18SEP07	21SEP07	3
90	Install Carrier Pipe - Tunnel B	11	11	0	24SEP07	08OCT07	3
PROJECT CLOSEOUT							
115	Clean and Demobilize	5	5	0	12OCT07	18OCT07	0
125	Project Completion	0	0	0		18OCT07	0

FIGURE 7.7 See Color Plate 9.

analysis technique used to identify and measure the critical Project delays within windows differ. Perhaps considering that the only common feature among all of these techniques is the selection of time periods within which to analyze delay, it may be more appropriate to more broadly classify these delay analysis techniques as different versions of a windows approach. When encountering a delay analysis that uses the windows approach, there are two questions that an analyst must answer when assessing the strengths and weaknesses of the delay analysis technique: "How were the time periods, or windows, established?" and "How were the delays determined within the window?"

The windows can be established in many ways. Selection of the time period can range from using the actual submission of the Project schedule updates to identify the windows to instances when the analyst chooses to arbitrarily identify the time periods. When using the Project schedules to identify the time periods, the analyst should rely on the frequency of the schedule updates to determine the intervals. Some analysts believe it is acceptable to base the window on the time period during which a particular controlling work activity is critical to determine the duration of the time period.

Allowing the controlling critical activities to determine the time periods means that each window should analyze a time period during which a specific work activity is the initial activity on the critical path. A window should begin when a specific work activity first became critical, either by its predecessor finishing or the critical path shifts making it critical. Similarly, the window should finish when the specific work activity is no longer critical, either when it finishes or the critical path shifts from it to another work activity. Realistically, selecting the window based on an activity or activities being critical still requires the analyst to determine throughout the specific time period if that activity or those activities remained critical throughout every day of the period. Consequently, the analyst still must go through a day-by-day analysis to determine this. Therefore, the optimum selection of time frames for a window would be the contemporaneous schedule updates that existed on the Project.

If you encounter a windows technique where the analyst selects the windows, you should ask the analyst to clearly explain the reasoning for selecting those specific windows of time. Any time an analyst makes a decision that is not directly supported by the contemporaneous Project documents, such as the Project schedules, or grounded by the facts, then the analyst's opinion is inserted and the analysis becomes more subjective.

The other question that must be answered is "How were the delays determined within the window?" The analyst should also clearly explain how the critical Project delays were identified and measured during the window. First and foremost, the analyst should focus his or her attention on the critical path because only delays to the critical path can delay the Project. Therefore, the analyst must demonstrate that the change identified actually delayed the contemporaneous critical path. If the analyst cannot credibly show that the change it identified did delay the critical path during the window, then its findings are questionable.

It is interesting to note that if an analysis that the analyst chooses to call a windows analysis is performed correctly, then it is no different from the contemporaneous delay analysis explained in other sections of this book. To put it plainly, a "windows analysis" is just window dressing. It has no analytical meaning and deserves no special significance or recognition.

IMPACTED AS-PLANNED ANALYSES

Some analysts prefer to present a delay analysis in the form of an impacted as-planned analysis. In this approach, the analyst identifies the as-planned schedule and then inserts into this schedule the changes that it believes may have caused Project delays. The source for these changes is only the perceived delays noted during construction that may have affected Project time. The impacted as-planned approach was discussed in Chapter 3 when we addressed perspectives.

The major weaknesses of this approach are that the "impacted" schedule does not depict actual Project events as they occurred, the decision as to what changes or impacts to place in the schedule is highly subjective, and, most significantly, it does not consider the dynamic nature of a construction Project and the critical path.

The first major weakness is that this method does not update the schedule using actual as-built information or rely on the contemporaneous schedule updates. Therefore, there is no comparison between events depicted in the impacted schedule and the actual Project events. This technique assumes that the Project would have progressed exactly as it was planned except for the inserted changes. In most instances, this assumption ignores what really occurred on the Project and does not consider changes in sequence or changes in the critical path that may have occurred.

Additionally, the analyst cannot always identify from the Project documentation all circumstances that affected the schedule. Events that are never mentioned in any Project documentation can and often do affect the schedule. For example, a specific activity might have a longer-than-planned duration. If the extended duration of this activity is not noted in the Project documentation as an "impact" or a change, then its effects will not be measured. In the methodology described in the preceding chapters, an extended duration would be accounted for by updating the schedule with the actual as-built durations from contemporaneous documentation, such as the daily reports.

Another major weakness of this analysis technique is that it is often a subjective or one-sided representation of the Project delays. In most cases, when this technique is used, the analyst only inserts the delays caused by the other party. By only inserting the delay of one party, the analyst attempts to identify the schedule impacts resulting from the opposing party's delays. Unfortunately, if all the changes that affected the schedule are not inserted into the impacted schedule analysis, then the results will at best be incomplete.

Last but not least, some analysts like the impacted as-planned approach because it is simple and "clean," but it does not consider the dynamic nature of construction projects or the critical path. It allows the analyst to present an analysis showing one impact at a time so he can "demonstrate" how the Project completion date is being delayed. He then proceeds to each succeeding delay, inserting a new impact each time to show how the Project end date is extended by each impact.

While this method is indeed "clean," it is also highly inaccurate. By using the first schedule, this method "freezes" the critical path at the beginning of the Project, and, thus, the actual shifts in the critical path will not be recognized.

A simple Project with a series of consecutive activities appears in Figure 7.8 to demonstrate this method. Activities A through F must be performed in sequence and are scheduled for the durations shown. At the same time, but not until after the completion of Activity B, there is another sequence of consecutive activities consisting of G through I with the respective durations shown. The total Project duration is 35 days. On day 15, the Contractor modifies its work and adds another activity, J, immediately after Activity I. Activity J has a duration of three days, shown in Figure 7.9.

From the initial schedule perspective, the critical path has not changed; it still depends on activities A through F. The Project is actually completed in 45 days—five days late. The as-built information for the Project appears in Figure 7.10, plotted against the original as-planned schedule. A review of the as-built data shows that Activity A started three days late. Activity C started six days early and out of sequence. Activity F took eight days longer to complete than planned. Activity J took four days longer than planned. All other activities started in sequence and were performed in the time frames originally scheduled.

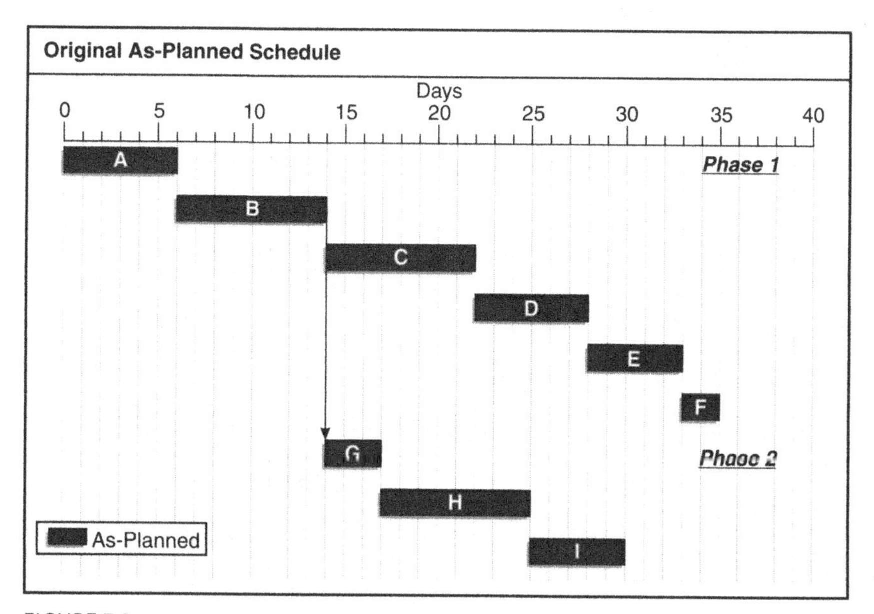

FIGURE 7.8

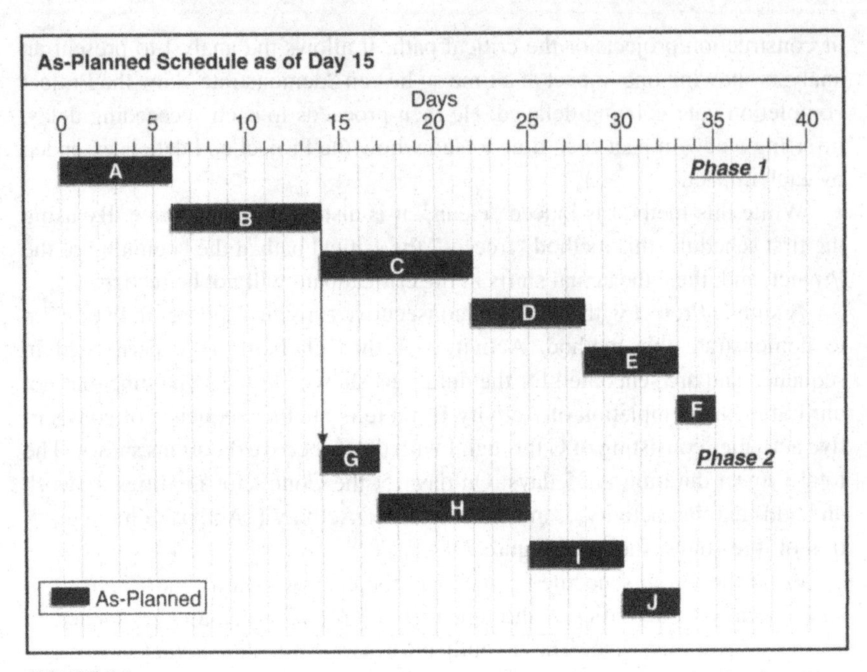

FIGURE 7.9

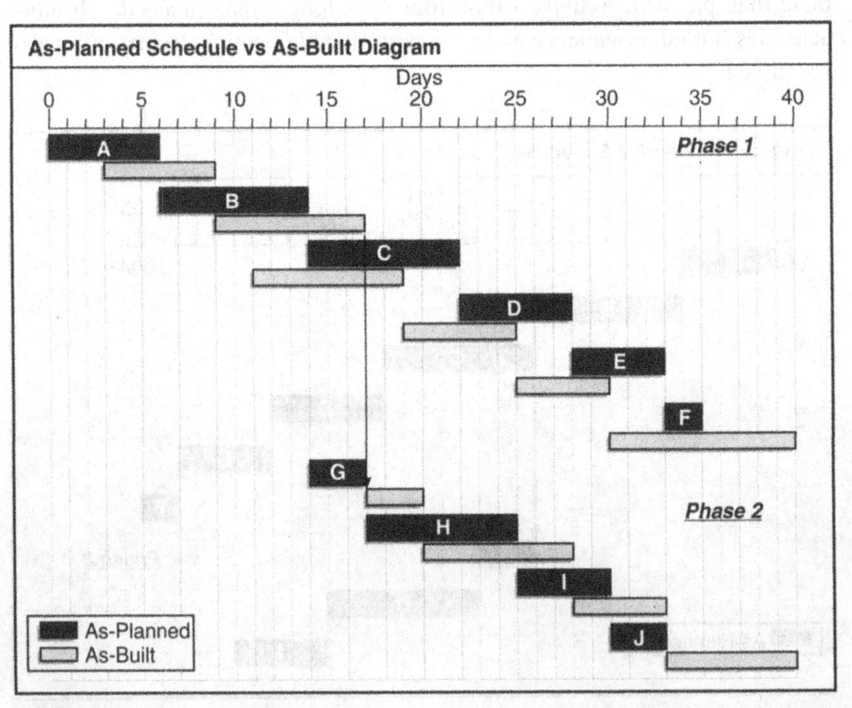

FIGURE 7.10

The impacted as-planned approach would show the following results:

1. Figure 7.11: Project delayed by three days because of the late start of Activity A.
2. Figure 7.12: Project ahead of schedule by three days because of the early start of Activity C.
3. Figure 7.13: Project delayed a total of five days due to the extended duration of Activity F.

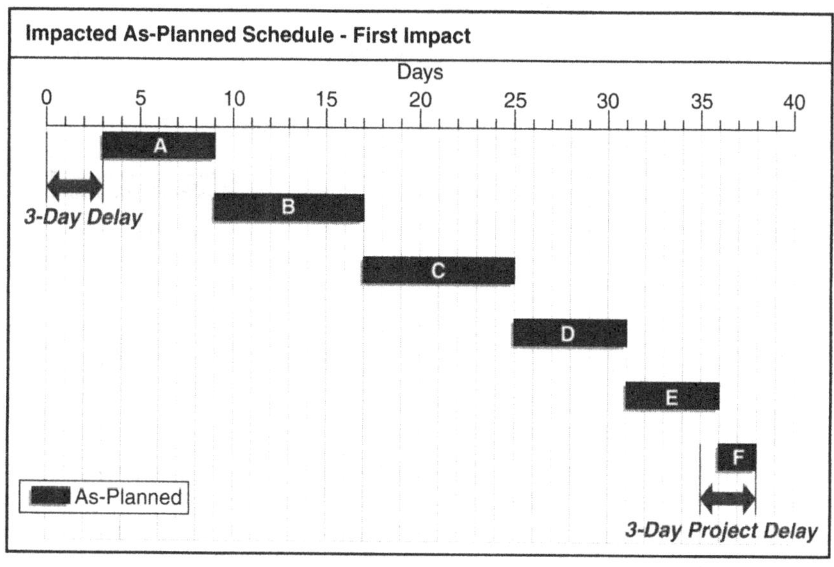

FIGURE 7.11

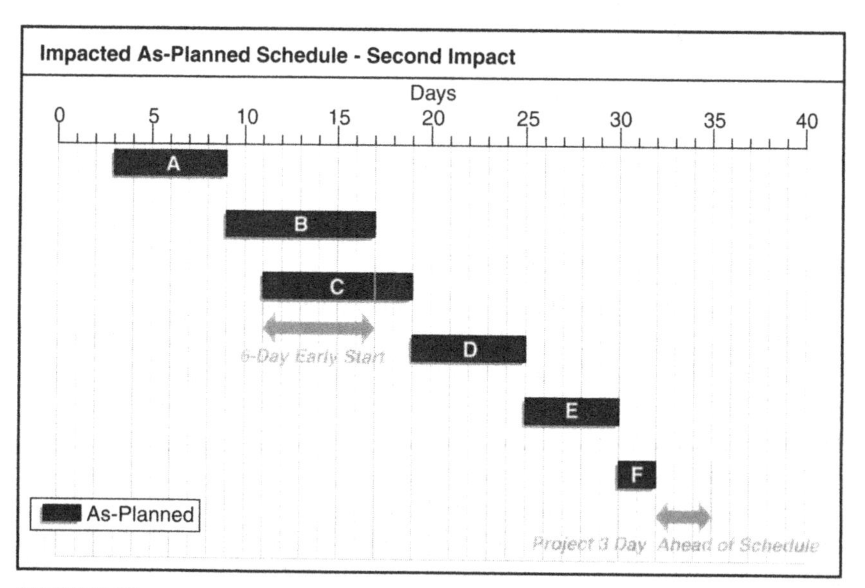

FIGURE 7.12

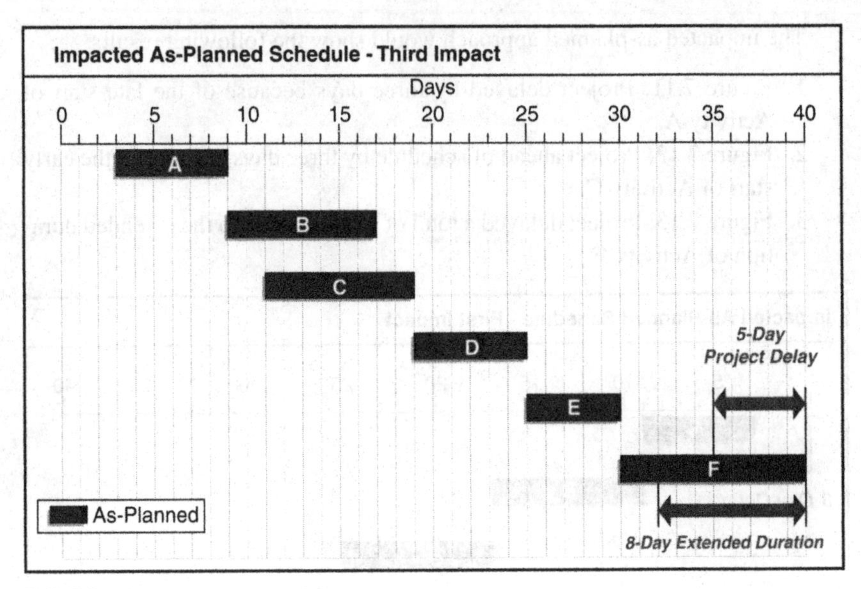

FIGURE 7.13

Therefore, the overall conclusion is that the Project was delayed by five days, primarily because of the late finish of activity F. This conclusion is erroneous. Using the analysis method presented in Chapters 4 and 5, the schedule is instead updated as of day 15, the day the Contractor added activity J. The results appear in Figure 7.14. This update shows that the Project is behind schedule by one day because of the delay to activity A and that the critical path has shifted to the Phase 2 path of activities, G through J. This path then controls the remainder of the Project and all five days of delay. The original critical path never becomes critical again. In this case, the impacted as-planned approach for the entire Project clearly leads to the wrong conclusion. The following additional example is presented to show how grossly inaccurate the impacted as-planned approach can be.

Impacted As-Planned Example

In a previous example of a Fragnet Analysis (Forward-Looking), D-Tunneling properly submitted its request for a 16-day time extension for the shutdown of TBM #2 in Tunnel B. The airport authority approved D-Tunneling's request for a 16-day time extension based on the clarity of PCO #1 and because of D-Tunneling's subsequent fragnet schedule submissions. D-Tunneling chose to be proactive in assessing the projected delays caused by the shutdown on TBM #2. However, D-Tunneling could have chosen a different methodology to assess the time impacts caused by this shutdown. For the next example, let's assume that D-Tunneling never submitted a request for a time extension with PCO #1 but instead stated the following:

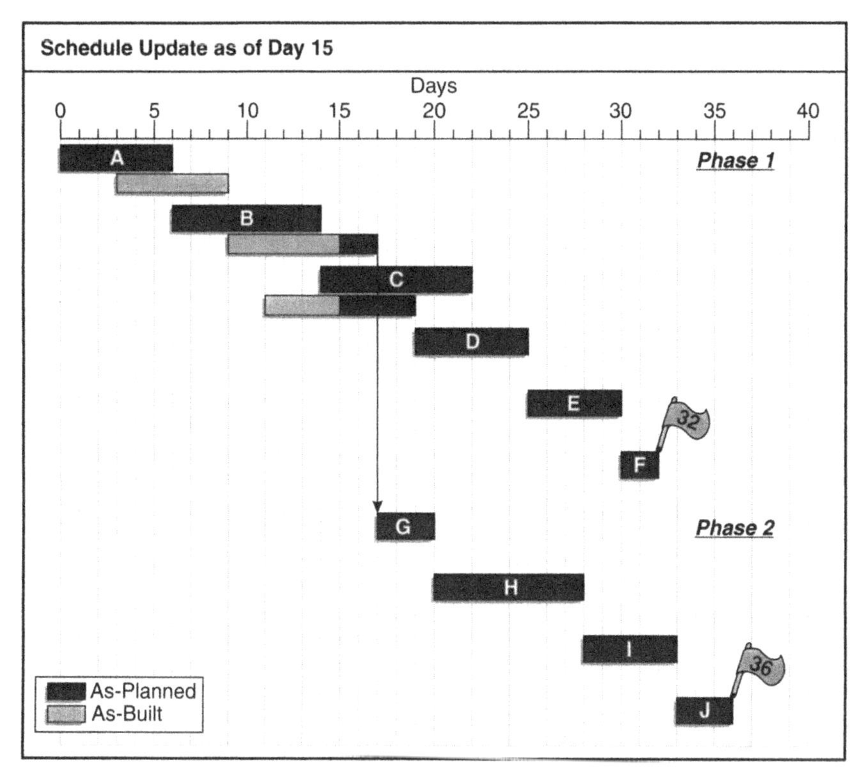

FIGURE 7.14

> On July 17, 2007, D-Tunneling encountered a rock that was harder than the rock specified in the geotechnical report provided by the airport authority in the Contract documents. As a result, all tunneling operations in Tunnel B have stopped due to the shutdown of TBM #2.
>
> D-Tunneling has ordered a new cutterhead for TBM #2 that will allow TBM #2 to complete the work in Tunnel B. D-Tunneling believes this shutdown will cause a significant delay to the Project completion date, but it is not possible at this time for D-Tunneling Company to quantify the magnitude of delay that will result from PCO #1. However, D-Tunneling Company specifically reserves the right to submit a claim for any cost incurred as a result of construction schedule impacts due to the changes in PCO #1.

The airport authority directed D-Tunneling to progress the remaining work and agreed to address the delays related to PCO #1 at a later date.

On September 30, 2007, D-Tunneling submitted a letter to the airport authority along with a 29-day time extension request for delays related to PCO #1. In its time extension request, D-Tunneling supplied a copy of its baseline schedule, or as-planned schedule, from June 13, 2007. D-Tunneling's as-planned schedule is shown in Figure 7.15. According to its as-planned schedule, on June 13,

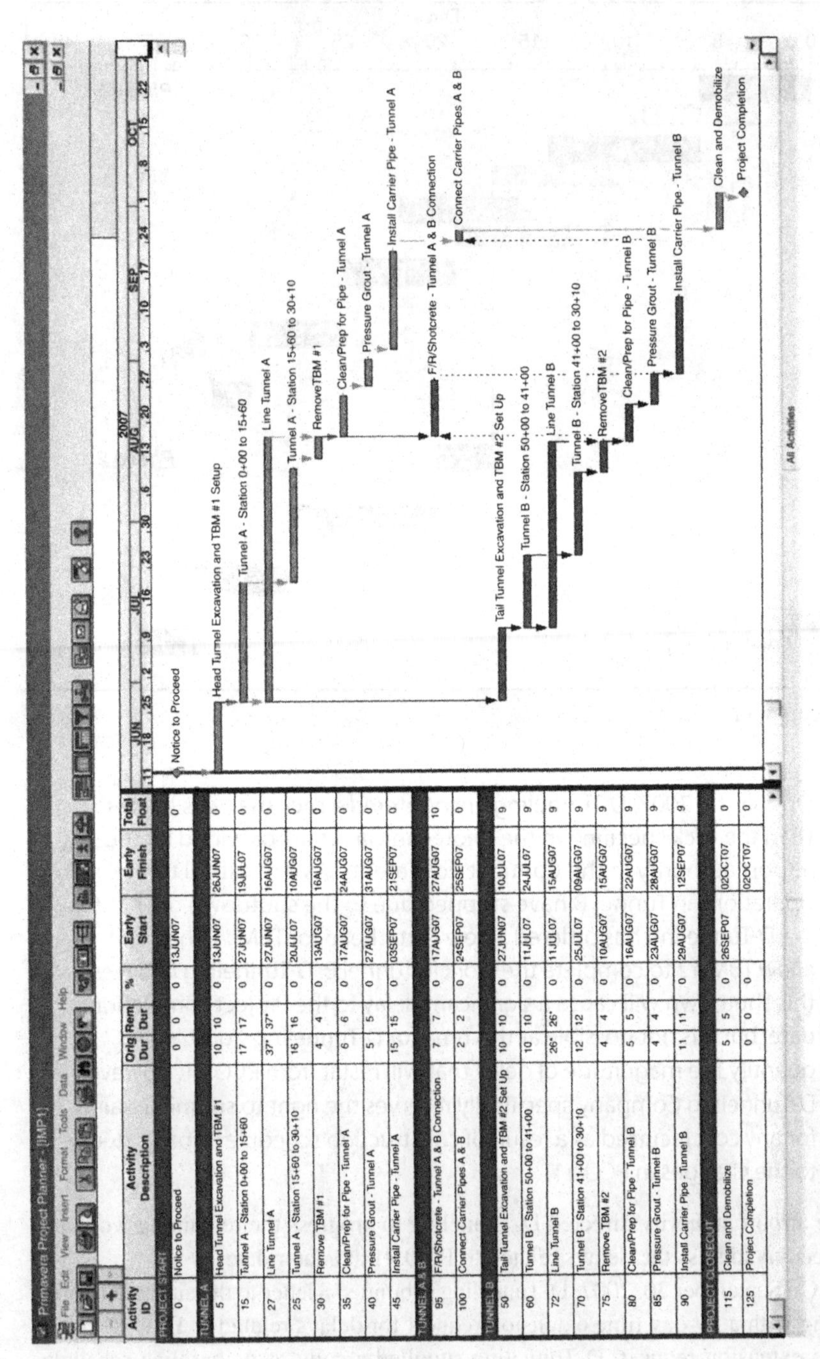

FIGURE 7.15 See Color Plate 10.

2007, D-Tunneling had planned on completing the Project on October 2, 2007. D-Tunneling's request for a 29-day time extension stated the following:

> 1. The airport authority/engineer was responsible for all soil borings and testing in the Contract and provided D-Tunneling with a geotechnical report of the expected soil conditions on the Project.
> 2. D-Tunneling encountered "mixed face" rock at Station 47+10 that was harder than the tolerance limits identified in the geotechnical report. As a result, D-Tunneling had to shut down its TBM #2 operations in Tunnel B. D-Tunneling continued work on Tunnel A while Tunnel B was shut down.
> 3. The engineer completed additional testing of the area and determined that the out-of-tolerance rock existed between Station 47+10 and Station 42+00.
> 4. D-Tunneling purchased a new cutterhead that was delivered on August 21, 2007. It took D-Tunneling three workdays to complete the assembly of the new cutterhead. D-Tunneling returned to Station 47+10 and resumed tunneling operations following the assembly of the cutterhead.
> 5. As a result of the differing site condition, it also took D-Tunneling seven workdays to complete tunneling from Station 47+10 to 41+00.
>
> D-Tunneling completed the added work that was associated with the differing site condition and has provided its September 29, 2007, schedule that includes all of the actual start and finish dates on the Project, thus far.

D-Tunneling's September 29, 2007, schedule is shown in Figure 7.16. D-Tunneling went on to state the following:

> As it can be seen, D-Tunneling incurred substantial delays to Tunnel B and the critical path of the Project. In addition, D-Tunneling performed its work with significant added cost as a result of the differing site condition not identified by the airport authority in the Contract documents.
>
> D-Tunneling inserted the delays incurred as a result of the differing site condition into its June 13, 2007, schedule, which was the baseline schedule submitted to the airport authority.

D-Tunneling's impacted as-planned schedule is shown in Figure 7.17. D-Tunneling's September 30, 2007, schedule also stated the following:

> As shown in the impacted as-planned, D-Tunneling's forecast Project completion date was delayed 29 calendar days as a result of the differing site condition. Had the airport authority's geotechnical report identified the differing site condition at Station 47+10, D-Tunneling could have possibly avoided this delay. However, the airport authority did not identify the differing site condition, and, therefore, D-Tunneling's Project schedule was delayed 29 days.

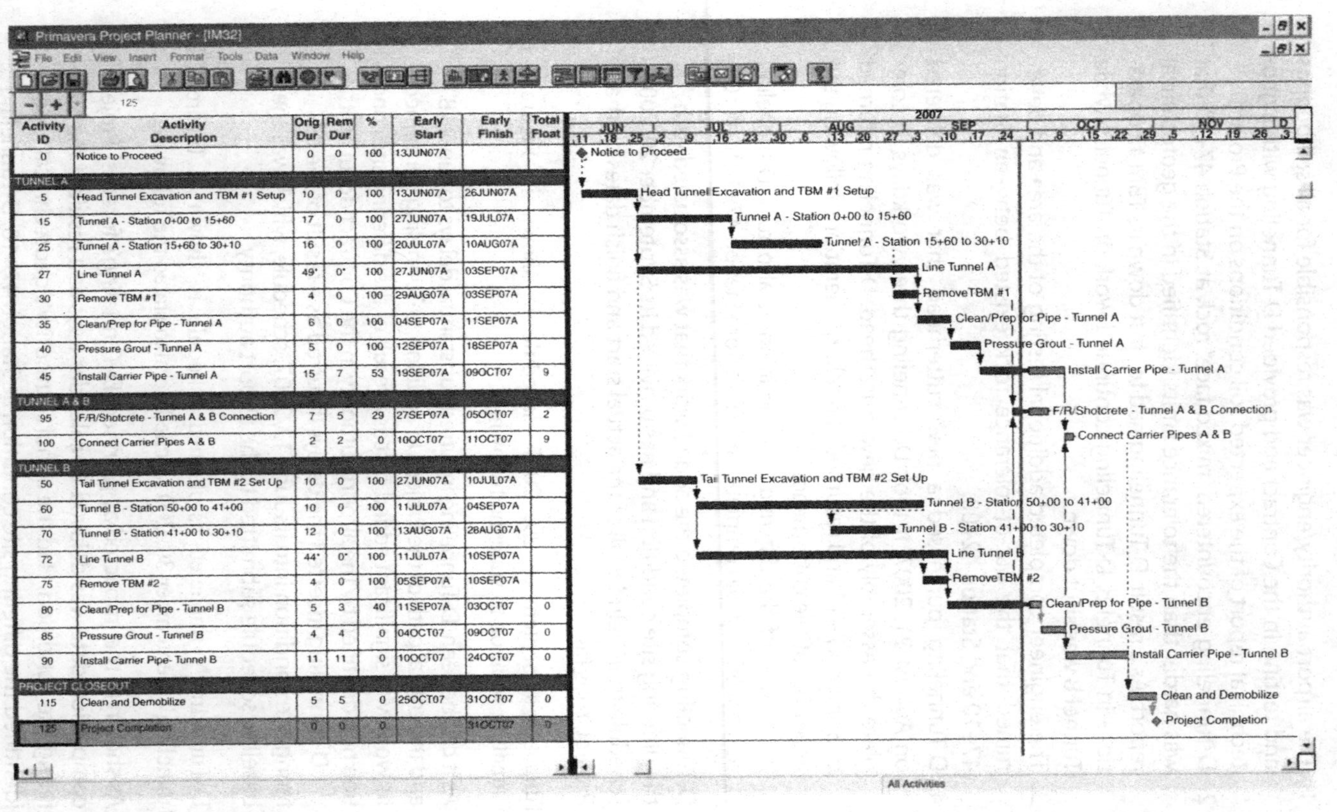

Activity ID	Activity Description	Orig Dur	Rem Dur	%	Early Start	Early Finish	Total Float
0	Notice to Proceed	0	0	100	13JUN07A		
TUNNEL A							
5	Head Tunnel Excavation and TBM #1 Setup	10	0	100	13JUN07A	26JUN07A	
15	Tunnel A - Station 0+00 to 15+60	17	0	100	27JUN07A	19JUL07A	
25	Tunnel A - Station 15+60 to 30+10	16	0	100	20JUL07A	10AUG07A	
27	Line Tunnel A	49*	0*	100	27JUN07A	03SEP07A	
30	Remove TBM #1	4	0	100	29AUG07A	03SEP07A	
35	Clean/Prep for Pipe - Tunnel A	6	0	100	04SEP07A	11SEP07A	
40	Pressure Grout - Tunnel A	5	0	100	12SEP07A	18SEP07A	
45	Install Carrier Pipe - Tunnel A	15	7	53	19SEP07A	09OCT07	9
TUNNEL A & B							
95	F/R/Shotcrete - Tunnel A & B Connection	7	5	29	27SEP07A	05OCT07	2
100	Connect Carrier Pipes A & B	2	2	0	10OCT07	11OCT07	9
TUNNEL B							
50	Tail Tunnel Excavation and TBM #2 Set Up	10	0	100	27JUN07A	10JUL07A	
60	Tunnel B - Station 50+00 to 41+00	10	0	100	11JUL07A	04SEP07A	
70	Tunnel B - Station 41+00 to 30+10	12	0	100	13AUG07A	28AUG07A	
72	Line Tunnel B	44*	0*	100	11JUL07A	10SEP07A	
75	Remove TBM #2	4	0	100	05SEP07A	10SEP07A	
80	Clean/Prep for Pipe - Tunnel B	5	3	40	11SEP07A	03OCT07	0
85	Pressure Grout - Tunnel B	4	4	0	04OCT07	09OCT07	0
90	Install Carrier Pipe- Tunnel B	11	11	0	10OCT07	24OCT07	0
PROJECT CLOSEOUT							
115	Clean and Demobilize	5	5	0	25OCT07	31OCT07	0
125	Project Completion	0	0	0		31OCT07	0

FIGURE 7.16 See Color Plate 11.

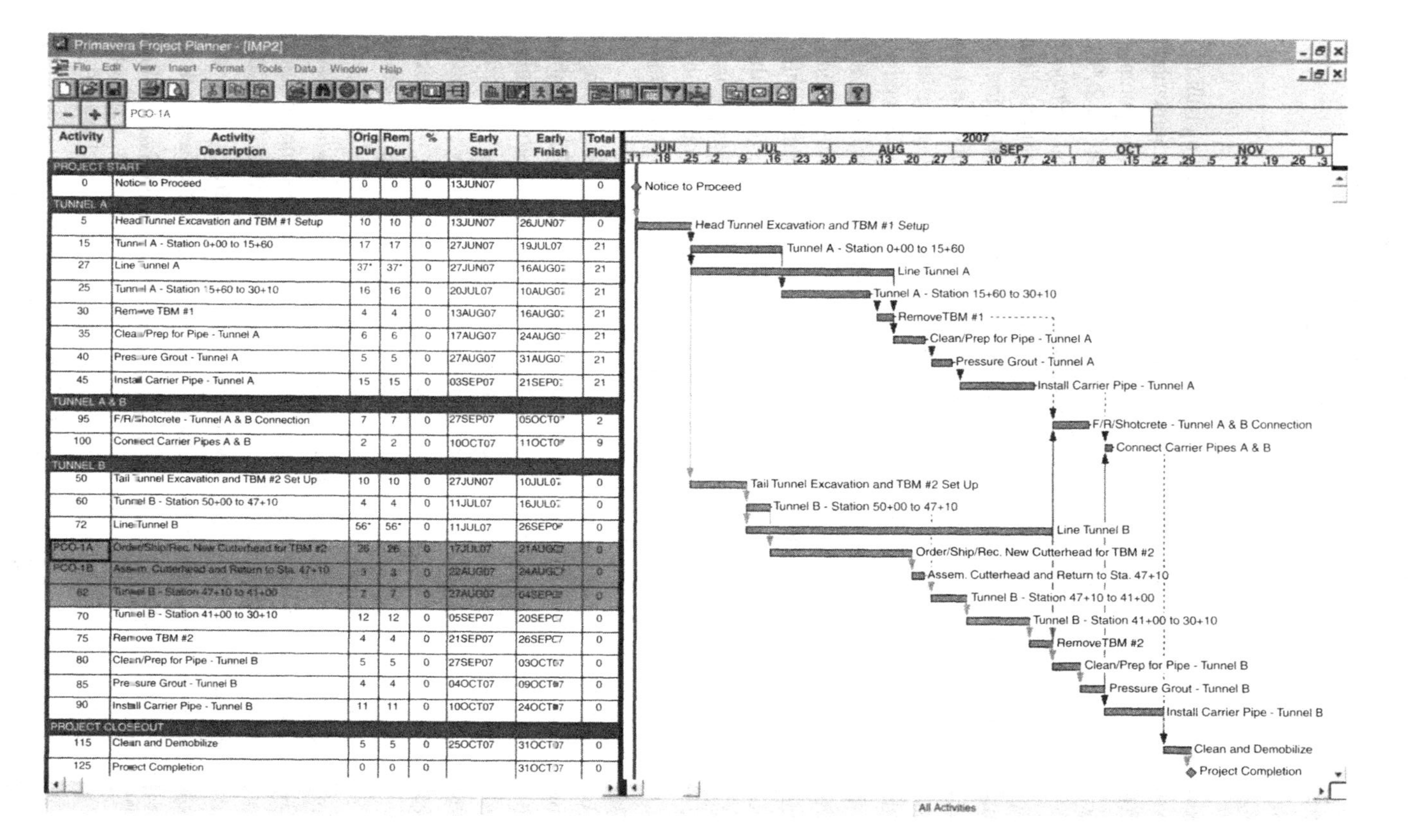

Activity ID	Activity Description	Orig Dur	Rem Dur	%	Early Start	Early Finish	Total Float
PROJECT START							
0	Notice to Proceed	0	0	0	13JUN07		0
TUNNEL A							
5	Head Tunnel Excavation and TBM #1 Setup	10	10	0	13JUN07	26JUN07	0
15	Tunnel A - Station 0+00 to 15+60	17	17	0	27JUN07	19JUL07	21
27	Line Tunnel A	37*	37*	0	27JUN07	16AUG07	21
25	Tunnel A - Station 15+60 to 30+10	16	16	0	20JUL07	10AUG07	21
30	Remove TBM #1	4	4	0	13AUG07	16AUG07	21
35	Clean/Prep for Pipe - Tunnel A	6	6	0	17AUG07	24AUG07	21
40	Pressure Grout - Tunnel A	5	5	0	27AUG07	31AUG07	21
45	Install Carrier Pipe - Tunnel A	15	15	0	03SEP07	21SEP07	21
TUNNEL A & B							
95	F/R/Shotcrete - Tunnel A & B Connection	7	7	0	27SEP07	05OCT07	2
100	Connect Carrier Pipes A & B	2	2	0	10OCT07	11OCT07	9
TUNNEL B							
50	Tail Tunnel Excavation and TBM #2 Set Up	10	10	0	27JUN07	10JUL07	0
60	Tunnel B - Station 50+00 to 47+10	4	4	0	11JUL07	16JUL07	0
72	Line Tunnel B	56*	56*	0	11JUL07	26SEP07	0
PCO-1A	Order/Ship/Rec. New Cutterhead for TBM #2	26	26	0	17JUL07	21AUG07	0
PCO-1B	Assem. Cutterhead and Return to Sta. 47+10	3	3	0	22AUG07	24AUG07	0
62	Tunnel B - Station 47+10 to 41+00	7	7	0	27AUG07	04SEP07	0
70	Tunnel B - Station 41+00 to 30+10	12	12	0	05SEP07	20SEP07	0
75	Remove TBM #2	4	4	0	21SEP07	26SEP07	0
80	Clean/Prep for Pipe - Tunnel B	5	5	0	27SEP07	03OCT07	0
85	Pressure Grout - Tunnel B	4	4	0	04OCT07	09OCT07	0
90	Install Carrier Pipe - Tunnel B	11	11	0	10OCT07	24OCT07	0
PROJECT CLOSEOUT							
115	Clean and Demobilize	5	5	0	25OCT07	31OCT07	0
125	Project Completion	0	0	0		31OCT07	0

FIGURE 7.17 See Color Plate 12.

So is D-Tunneling correct in stating that it is due a 29-day time extension from PCO #1? Possibly, but let's look at the details of D-Tunneling's schedule submissions. D-Tunneling submitted three of its schedules with time extension requests:

June 13, 2007, schedule—the as-planned schedule
September 29, 2007, schedule—the as-built schedule
June 13, 2007, impacted schedule—the impacted as-planned

Looking at the data dates of those schedules, it appears that D-Tunneling submitted two Project schedules: the June 13, 2007, Project schedule and the September 29, 2007, Project schedule. These are schedules that D-Tunneling contemporaneously submitted to the airport authority as Project updates. The June 13, 2007, impacted as-planned schedule was not submitted by D-Tunneling as a Project update to the airport authority, and, thus, it is a schedule that was submitted by D-Tunneling in an effort to substantiate its request for a 29-day time extension.

If D-Tunneling's first schedule submission was on June 13, 2007, and its last schedule submission was on September 29, 2007, did D-Tunneling perform monthly updates between these two schedules? In other words, did D-Tunneling base its analysis on all of the contemporaneous Project schedules, or did it selectively choose which schedules to use for its analysis in order to show the largest impact to the schedule? Since there are three and one-half months between June 13, 2007, and September 29, 2007, and D-Tunneling stated that it would perform monthly updates, it appears that D-Tunneling has not used all of the contemporaneous schedules to perform its analysis of the delays that occurred as a result of PCO #1. Therefore, D-Tunneling's method for substantiating its time extension may be flawed.

According to the two contemporaneous schedule submissions, D-Tunneling's planned work sequence did not change between June 13, 2007, and September 29, 2007. Is this true? The analyst would need to verify that D-Tunneling maintained the same work sequence between the schedule updates by reviewing the Project documents. Did D-Tunneling perform out-of-sequence work on the critical path or possibly reorganize its resources to mitigate some of the critical path delay? Remember, in the previous fragnet example, at the airport authority's request, D-Tunneling reorganized its work sequence and used TBM #1 to perform the required tunneling work on Activity 70, *Tunnel B—Station 41+00 to 30+10*. Did D-Tunneling reorganize its work sequence in the same manner? According to the work sequence shown in D-Tunneling's September 29, 2007, schedule (Figure 7.16), the answer would appear to be no. Activity 70 is still classified as work associated with Tunnel B. However, take a closer look at Activity 70 and its actual start and finish dates. Activity 70 received an actual start date of August 13, 2007, and an actual finish date of August 28, 2007, and thus Activity 70 was completed during the shutdown of TBM #2 and Tunnel B. The actual start of Activity 70 on August 13, 2007, also follows the completion of Activity 25, *Tunnel A—Station 15+60 to 30+10*. Also, it appears that the actual finish date of Activity 70 on August 28, 2007, is the reason that Activity 30,

Remove TBM #1, is not started until August 29, 2007. As a result, it appears that D-Tunneling performed out-of-sequence work by using TBM #1 to complete Activity 70, which would have mitigated some of the delay resulting from the shutdown on TBM #2.

In examining the critical path for delays, the analyst must also identify if other activities that are unrelated to the shutdown of TBM #2 were delaying the critical path. Are there other critical path delays besides the differing site condition that are delaying the Project completion date? All activities progressed as expected, besides the resequencing of Activity 70, with the exception of Activity 80, *Clean/Prep for Pipe—Tunnel B*. Activity 80 received an actual start date of September 11, 2007, but only gained two days of progress between its start on September 11, 2007, and the schedule update on September 29, 2007. D-Tunneling had 14 available workdays between September 11, 2007, and September 28, 2007—the day before the September 29, 2007, schedule update. D-Tunneling received only two days of progress between September 11, 2007, and September 28, 2007, resulting in 12 workdays of delay to the Project completion date. Since Activity 80 was delayed 12 workdays, and it was controlling the critical path of the Project, it pushed the Project completion date out an additional 12 workdays from October 15, 2007, to October 31, 2007, a delay of 16 calendar days.

So is D-Tunneling entitled to its requested 29-day time extension? A critique of D-Tunneling's impacted as-planned analysis of the Project delays indicates that D-Tunneling overstated the critical path delays that resulted from the shutdown of TBM #2. It appears that D-Tunneling was delayed as a result of the shutdown to TBM #2, but the shutdown would have extended the Project completion date from October 2, 2007, to October 15, 2007. The reason the Project completion date was delayed from October 15, 2007, to October 31, 2007, was a result of the delays to Activity 80. The delays that resulted from the shutdown of TBM #2 would have been greater if D-Tunneling did not resequence Activity 70 to be completed by TBM #1. However, D-Tunneling, like most Contractors, has an obligation to mitigate Project delays by resequencing work or by adjusting its resources. D-Tunneling may have incurred some added cost while performing out-of-sequence work on Activity 70, but it would not be due time-related damages associated with Activity 70, as shown in its impacted as-planned schedule (Figure 7.17). Although it looks logical, the impacted as-planned analysis is flawed in the details. It often neglects Project changes, such as reorganizing Activity 70 to be performed by TBM #1, and hides other critical path delays that are unrelated to the activities for which the Contractor is seeking time extension/delay costs, such as the delays to Activity 80. It is too easy to manipulate the schedule and manufacture outcomes that are beneficial to the party making the claim. An impacted as-planned analysis does not use all of the Project schedules to assess the critical path delays and, therefore, ignores the changing nature of construction projects and Project schedule changes related to the critical path.

COLLAPSED AS-BUILT ANALYSES

Another method of analysis is the collapsed as-built analysis, also called the subtractive as-built. This approach is perhaps best described as the reverse of the impacted as-planned analysis described previously. In the collapsed as-built, the analyst studies all of the contemporaneous Project documentation and constructs as detailed an as-built schedule as possible. In this case it truly is referred to as a schedule and not an as-built diagram.

Normally, this as-built schedule is a CPM network drawn on a time scale. Next, the analyst "subtracts" or removes activities or shortens actual durations of delayed work items, which he has determined have impacted the Project. If the removal of these activities or the shortening of these durations affects the new schedule's end date, then the difference in days between the as-built and the collapsed as-built end dates is considered to be the delay caused by the specific activities that were removed. There are a few different variations of this approach. It is worthwhile to mention two of them and then note the weaknesses of this approach.

Unit Subtractive As-Built

For simplicity, the first variation will be called a unit subtractive as-built approach. This method starts with the preparation of the overall as-built schedule and subtracts one "impact" at a time in an attempt to correlate a measurable number of days attributable to each item removed or each delay. Once the alleged "impacts" are removed, the argument is made that the Project would have finished earlier were it not for the impacts noted.

Gross Subtractive As-Built

The second variation is the gross subtractive as-built approach. This method starts with the preparation of the overall as-built schedule and removes all possible impacts that may have caused a delay to the Project. These potential impacts are both Owner and Contractor caused. The resulting schedule and overall Project duration supposedly represent the time the Project would have taken if no problems had occurred.

During the next step in the process, the analyst adds specific problems back into the now "collapsed" as-built, one at a time. As the analyst reinserts each impact, he or she attributes the corresponding increase in the Project duration or delay to the respective impact items.

Basic Flaws of the Collapsed or Subtractive As-Built Method

This overall method, regardless of the variation used, has several major problems. These are its three primary flaws:

1. It requires the analyst to construct a CPM network diagram based on as-built information.
2. It is extremely subjective and highly amenable to manipulation.
3. With very little effort, the analyst can create an as-built schedule that supports a predisposed conclusion.

The method also requires the analyst to determine specific "impacts" before performing any analysis. Not only is this subjective, but it is impossible. The method forces the analyst to first reach a conclusion about what caused a delay and then uses the analysis to prove its conclusion. This is the proverbial "tail wagging the dog."

Finally, the method "freezes" the critical path based on the as-built critical path constructed by the analyst. The critical path then changes depending on the "subtraction" made; these changes normally do not agree with what was actually taking place on the Project. This method ignores the fact that the CPM schedule is dynamic.

While there are numerous approaches used to analyze delays, care must be exercised throughout the process. The method used must incorporate the available contemporaneous information, recognize the dynamic nature of a construction schedule and the critical path, and avoid after-the-fact hypotheses that do not reflect all the information available.

ANALYSES BASED ON DOLLARS

Some analyses that are presented to support delay conclusions are based on the relationship between dollars and time. There is a widely held, but mistaken, belief that the dollar value of the work performed is directly related to the time progress of the job. Here are some of the arguments:

Contractor: I was able to complete 90 percent of the work in ten months on the job. Then, because of the Owner delays in inspection and punch list, it took four months to complete the last 10 percent of the work.

Owner: During the period of the alleged delay to the Contractor, he was able to perform 25 percent of the dollar value of his work on the Project. The dollar value of work accomplished during that period reflected the same rate of progress as that before and after the delay. Therefore, we could not possibly have delayed the Contractor.

While both of these arguments may appear plausible and logical from an initial quick reading, there is no quantifiable linear relationship between time progress and dollars on any Project. The dollar value at each stage of the Project depends on the nature and cost assigned to the specific activities completed. For instance, some high-dollar-value items of work may be performed in a short period of time, whereas some low-dollar-value items of work may take more time. Finally, the dollar value method does not identify and track the activities on the critical path of the Project, which was identified earlier as the most reliable method for determining delays on the Project.

Again, using the bridge example, the abutment/approach activities may have a low dollar value compared with the main bridge work. Yet, the abutment and approach work items caused a delay to the Project. Unfortunately, some public Owners have structured their Contract documents to reflect the idea that time and dollar value have a linear relationship. For instance, some Owners grant time extensions based on the dollar value ratio of a change order to the original Contract amount. Here is an example:

- Original Contract amount: $100,000
- Dollar value of change: $10,000 (or 10 percent of original Contract amount)
- Original Contract duration: 100 days

Therefore, time extension granted is 10 days (10 percent of original Contract time).

In fact, an Owner could make a change that would increase the Contract amount by 10 percent but would not affect the duration of the Project. Likewise, the Owner could make a change with a minimal (direct) dollar impact that could significantly delay the Project.

S Curves

Some Owners, such as the Corps of Engineers, have fostered this time-dollar relationship. In the past, the Corps of Engineers required Contractors to provide "S" curves to show job progress. The Contractor provides a bar chart of

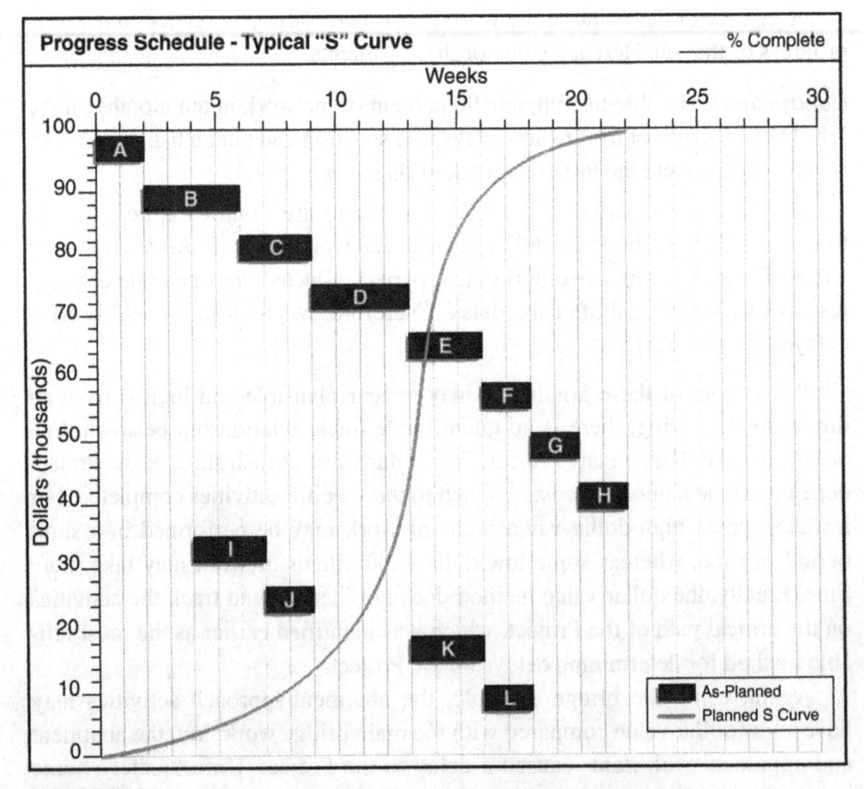

FIGURE 7.18

the major activities on the job and applies the dollar value from the schedule of values to the bar chart. By summing the dollars over time, an "S" curve is generated. Figure 7.18 shows an example of a typical "S" curve for a Project.

The Contractor must then submit an updated "S" curve with each monthly pay request. Unfortunately, the monthly submission also has an entry for "progress to date," recorded as a percentage (see Figure 7.19). This update has often been used to measure the Contractor's progress in time, which is very often misleading. "S" curves developed from a Contractor's total billing dollars do not measure the time progress of the Project but merely show the progress of billings over the course of time. There are several reasons why the authors do not recommend using an "S" curve for determining job progress. The original planned "S" curve might not be accurate due to "front end loading," or other factors. It should also be noted that the updated "S" curve information might be misleading because of payments for stored materials and equipment.

The Contractor could overbill or "front end load" its bid items to recover its costs as early as possible, and, therefore, reliance on the "S" curve could portray more "progress" than had actually occurred. Because they are unreliable, time-to-dollar relationships should not be used to establish delays on a Project.

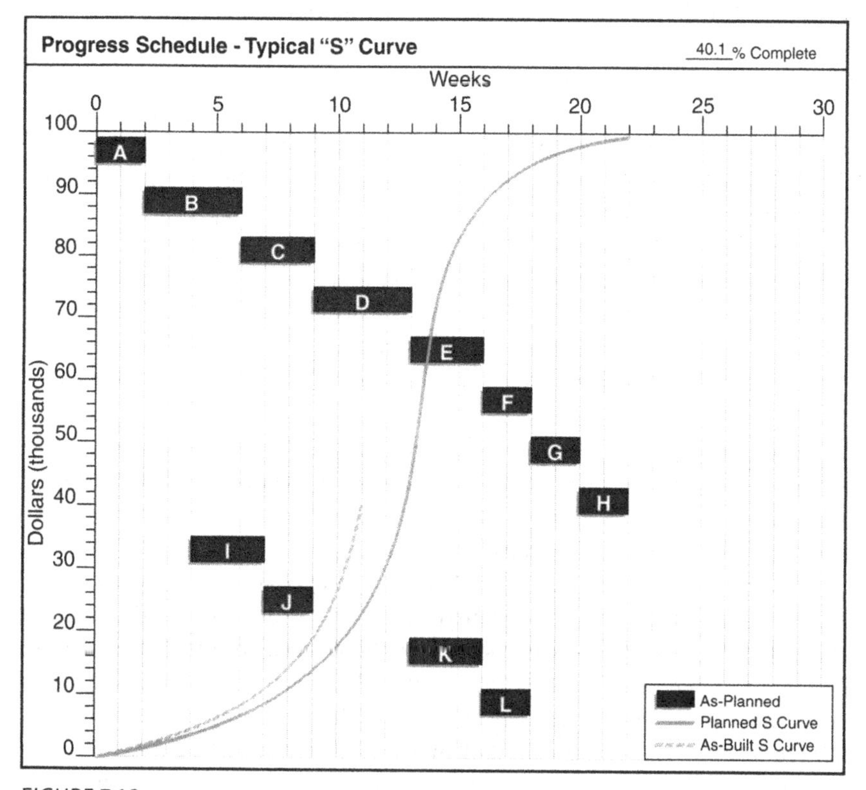

FIGURE 7.19

BUT FOR SCHEDULES, ANALYSES, AND ARGUMENTS

Another school of thought exists where analysts rely on the "but for" approach to establish and, more often, to refute the cause of delays. In a "but for" argument, the analyst takes the position that regardless of what occurred, there were other delays that would have had virtually the same effect or would have been responsible for the same amount of delay—for example, *but for* the Owner's delay to the shop drawing process, the Contractor would have been late anyway, since he had not mobilized his Subcontractor on time.

This approach completely denies the concept of critical path method scheduling. Using this approach normally results from an analyst's comparing the entire as-planned schedule with the overall as-built schedule. In this comparison, the analyst determines that several activities experienced delays. The analyst then identifies and argues that an Owner-caused delay is offset by an apparent Contractor-caused delay. This analysis fails to recognize the fact that the original Owner-caused delay provided the Contractor with float on other activities, which, therefore, no longer had to be accomplished in the originally scheduled time frame. For example, Figure 7.20 is a bar chart for a Project showing the comparison between the original as-planned schedule and the actual as-built diagram. The Contractor asserts that an activity was delayed because of a change by the Owner. The Owner responds that "but for" its change, the Contractor would have delayed the Project, and since the Contractor did not start the activity until much later, its delay was not related to the Owner's change. The real situation is that once the Contractor was delayed by the change orders, it chose not to proceed or pace its work, since the work could be done more efficiently by waiting until the change order work was accomplished.

Contractors sometimes use a "but for" analysis when there is no contemporaneous schedule to support their delay position. The Contractor may insert the Owner's delays into the original Project schedule and then argue that "but for" these delays, he could have finished the work much sooner. The difference between the actual completion date and the "but for" schedule is then the measure of the delays caused by the Owner.

When using the "but for" approach, the party making the argument may even go so far as to admit that it is responsible for *some* of the delay. Purportedly, this adds credibility to the analysis (i.e., "If I admit that some of the delays are my fault, then I am being fair in my evaluation").

The "but for" approach relates back to the concept of concurrency discussed earlier. In the discussion of concurrency, we also addressed the Primacy of Delay. Reflecting back on that concept, one should recognize that any analysis must walk through the Project in time, day by day, if possible, and in that manner determine the critical path as it may change. It must always be kept in mind that the schedule is not static; it is dynamic, changing over the life of the Project. What may be the measure of a delay one month may not be the accurate measure

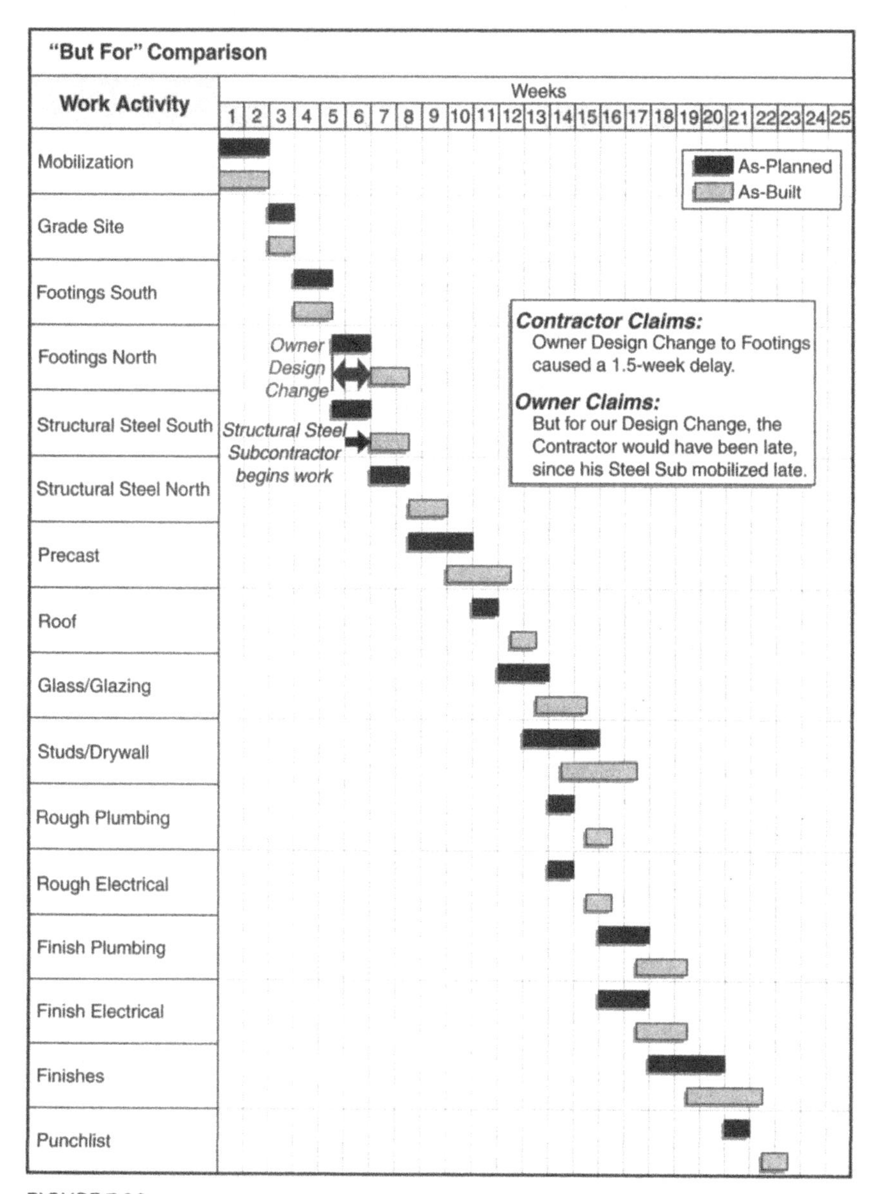

FIGURE 7.20

of a delay that occurs two months from that time. The schedule does not guarantee that the job will be built a certain way. It only provides a "road map" or plan for construction and for the durations of the activities. Therefore, any time an analyst "freezes" the schedule to measure delays, the analyst is asking for trouble. One cannot apply a static measure to a dynamic situation.

Chapter | eight

The Owner's Damages Due to Delay

LIQUIDATED DAMAGES

When a Project is delayed, the Owner, Contractor, or both may incur added costs. The determination of the amount of these costs is based on the results from the delay analysis and the determination of liability once the critical delays have been identified. This chapter addresses the types of damages that the Owner may experience if it is shown that the Contractor has caused the delay. In the broadest sense, the Owner's damages are either liquidated damages or actual damages. Both of these categories are discussed in this chapter.

The general concept for recovery of Owner's damages is similar to the concepts that apply to Contractors. Damages serve to place the Owner in the same position it would have been in had the Contract been performed as planned. Under these guidelines, an Owner's delay damages may include the loss of income, the cost of alternate facilities, increased financing costs, extended overhead costs, and lost profits. Because these costs are typically difficult to measure, the liquidated damages clause is often a preferred alternative to the reimbursement of actual damages.

Liquidated damages are predetermined prior to the execution of the Contract. The exact amount of the liquidated damages is specified in the Contract. A typical Contract clause that incorporates liquidated damages appears in Figure 8.1. Liquidated damages clauses can take many forms. The Owner should seek the assistance of qualified counsel in structuring the wording of the clause and should carefully compute the damages it may reasonably sustain if a delay occurs to all or a portion of the Project. One might ask why an Owner would specify a liquidated damages amount in advance as opposed to seeking recovery for actual damages if a delay occurs. The answer is that liquidated damages are desirable when it is difficult or impossible to accurately determine the actual damages that the Owner

Sample Liquidated Damages Clause

Should the contractor fail to complete the contract within the time allowed by the contract to include time extensions allowed by executed change orders, then for each calendar day of delay, the owner has the right to withhold the amount of $500 per calendar day as liquidated damages.

These liquidated damages are compensation to the owner for costs the owner may experience due to the contractor's delay and are not to be construed as penalties.

FIGURE 8.1

would incur in the event of a delay, particularly for public projects. Projects such as highways and transit systems have a value to the public, but delay damages for such often cannot be easily calculated. Therefore, the Owner cites a liquidated damages amount to cover the estimated damages that will be sustained if the Project is completed late. The Owner should check with counsel before calculating the amount of liquidated damages to ensure that all legalities are considered.

Many Owners believe that the inclusion of a liquidated damages clause is a deterrent to lateness. In other words, fear of having to pay liquidated damages "motivates" the Contractor to complete the Project on time. In reality, liquidated damages clauses are not intended to deter lateness, and they do not have that effect. The vast majority of Contractors are optimistic and bid on projects believing they can and will finish on time.

Generally, the inclusion of a liquidated damages clause does not affect Contractor bids. Contractors recognize that with or without a liquidated damages clause, they are still liable for actual damages should they finish late. Therefore, during Project performance, the liquidated damages clause may not be a motivating factor. In fact, if a Contractor falls behind schedule during the Project, the liquidated damages clause allows it to determine whether acceleration efforts would be cost effective. For example, if a Contractor is behind schedule by ten days on a Project and the liquidated damages are $300 per day, the potential exposure is $3,000. If the cost of accelerating the work to make up the ten days is $7,000, then the cost-effective decision is to finish late. The Contractor may still decide to accelerate the work because of other considerations.

An Owner should consider the following items when preparing an estimate of liquidated damages:

- Cost for Project inspection
- Costs for continued design services
- Costs for the Owner's staff
- Costs for maintaining current facilities
- Costs for additional rentals
- Costs for additional storage
- Lost revenues

- Costs to the public for not having the facility
- Additional moving expense
- Escalation costs
- Financing costs

While other costs may be included in the liquidated damages estimate, this list provides a general idea of the types of costs to consider.

Estimating Liquid Damages

A note of caution: Liquidated damages are specific for each Project. Unfortunately, Owners often use standard tables for liquidated damages that may not reasonably reflect the damages they will sustain if a delay occurs on their Project. For instance, many State Departments of Transportation have liquidated damages clauses similar to the one shown in Figure 8.2.

While the use of standard tables is convenient, the Owner should ensure that the amounts in the table are valid and indeed were based on a reasonable estimate. Fortunately, the amount stated in the tables is most often low. Nevertheless, it is important that Owners recognize that each Project is different and will have specific amounts for liquidated damages.

Owners must recognize that liquidated damages do not necessarily bear a direct relationship to the Contract amount. Two different projects of equal value can have very different potential damages. An Owner who uses a standard table to figurc liquidated damages may risk either understating the damages and thereby shortchanging herself or overstating the damages and becoming susceptible to a successful legal challenge by the Contractor.

Sample Liquidated Damages Clause

Should the contractor fail to complete the contract within the time allowed by the contract to include time extensions allowed by executed change orders, then for each calendar day of delay, the owner has the right to withhold the amount of $500 per calendar day as liquidated damages.

These liquidated damages are compensation to the owner for costs the owner may experience due to the contractor's delay and are not to be construed as penalties.

From More Than	To and Including	Daily Charge
$ 0	$ 100,000	$ 300
100,000	500,000	550
500,000	1,000,000	750
1,000,000	2,000,000	900
2,000,000	5,000,000	1,150
5,000,000	10,000,000	1,350
10,000,000	—	1,400

FIGURE 8.2

When Do Liquated Damages Begin and End?

The Contract should clearly specify when the assessment of liquidated damages will begin and end. Often, the Contract will state that the liquidated damages will apply to the time period from the specified Contract completion date until the completion of all work. Sometimes the Contract will tie the assessment of liquidated damages to the substantial completion date. Another common milestone for liquidated damages is the date when the Owner obtains beneficial use or occupancy of the Project.

Regardless of the time period specified, the Contract should be as clear as possible on this point, and the estimate used to determine the liquidated damages amount should be based on the same factors. For example, if the Owner assesses liquidated damages from the Contract completion date to the completion of all work, the Contractor may be able to successfully argue that the Project achieved substantial completion and the Owner gained use of the Project well before the date that all work was completed and that because the liquidated damage amount was based on the Owner not having use of the Project, the amount assessed was excessive and a penalty.

In another case, the Contract may refer to substantial completion without a clear definition of precisely what needs to be complete in order to achieve that milestone. Again, the parties may have an argument as to the proper period for assessing the liquidated damages.

Application to Project Milestones

Liquidated Damages clauses can also be written to apply to milestone dates or events during the Project. For instance, liquidated damages may be linked to the completion of work phases, such as building close-in date or completion dates for a section of the Project. The amounts of these milestone liquidated damages may be separate from the liquidated damages amount that applies to Project completion. For instance, a Project may involve the construction of several buildings. In the Contract, the Liquidated Damages clause may specify separate damages for the completion of each building, as well as a liquidated damages amount for the overall completion of the Project. A highway construction Project may specify liquidated damages for the completion of each bridge and for Project completion.

Hourly Fees

For some projects, liquidated damages are specified on an hourly basis. In certain critical highway projects, the Owner specifies hourly liquidated damages for failing to open portions of the roadway to traffic at set times for each day of the Project.

Graduated Damages

In some cases, liquidated damages may be graduated. For example, the liquidated damages may be $1,000 per day up to a certain date or for a defined number of days, and then may increase to $1,500 per day for delays beyond the date or in excess of the initial number of days. These graduated liquidated damages should reflect the Owner's increase in damages as the delay continues.

Bonus or Incentive Clauses

In the construction industry, it is often presumed that liquidated damages must also have a corresponding bonus or incentive. This is not true. There is no requirement that the Owner offer a bonus or incentive merely because the Contract includes a Liquidated Damages clause. The lack of a bonus or incentive does not justify a challenge to the Liquidated Damages clause.

The Owner may, in fact, include a Bonus or Incentive clause in the Contract for early completion. If a bonus or incentive is included, it does not have to match the amount of the liquidated damages. The bonus can be higher or lower, and can have limitations. For example, the bonus in a Contract could allow $1,000 per day for each day the Contractor finishes the Project earlier than the specified Contract completion date, up to a limit of $50,000. Alternatively, the Owner may allow a bonus for early completion that increases or decreases over time. For example, the Owner will offer a bonus of $1,000 per day for early completion up to 50 days, and for every day that the Project is finished early in excess of 50 days, the bonus may be increased to $1,500.

The bonus is computed from the Contract completion date. Therefore, if the Contract completion date is extended by a change order, the bonus is computed from the new Contract completion date. In some instances, the benefit that the Owner will realize from an early completion may evaporate after a certain calendar date. Therefore, this bonus clause may not apply if the completion date is changed by change order during the course of the Project. In such cases, the Owner should draft a clause that determines the bonus from a specified calendar date. In this case, regardless of the Contract completion date, no bonuses will be paid if the Contractor does not finish before the specified calendar date.

Such clauses can be difficult to write and should be drafted by qualified counsel. Furthermore, the Owner should make an extra effort to ensure that all bidders understand the intent of the bonus clause to minimize future disputes.

Enforceability

One of the Owner's major concerns when using a Liquidated Damages clause is whether or not it will be enforceable. A reasonable amount of case law exists, and, with proper guidance by counsel, the Owner should be able to structure a clause that will be upheld.

If a Contractor completes a Project late and is assessed liquidated damages by the Owner, it is possible that the assessment may be challenged. There are two basic approaches that a Contractor may use to overcome the assessment of the liquidated damages. First, he may attack the propriety of the assessment by disclaiming responsibility for the delay. Second, the Contractor may claim that the specified amount of the liquidated damages is excessive and, consequently, acts as more of a penalty than a damages compensation.

If a Contractor challenges responsibility for a given delay, it must show through a delay analysis that the delays to the Project were excusable delays and, therefore, warrant a time extension. If the delay analysis proves the assessment of the liquidated damages is inappropriate, the Contractor may be granted relief from the damages. Similarly, the Contractor may attempt to show only partial responsibility for delays to a Project, arguing that the Owner also caused some concurrent delays. As previously discussed, if it can be shown that delays were also caused by the Owner, then the Contractor may be granted relief from the assessment of some or all of the liquidated damages.

The second approach used to challenge liquidated damages is based on the magnitude of the damages specified. The Contractor may argue that the amount specified was excessive and was in effect a penalty rather than a statement of the Owner's loss. Some Owners may feel that it does not matter whether the amount reflects a penalty or a loss, since the damages were clearly specified in the Contract that bears the Contractor's signature.

However, in construction Contract law in the United States, penalties in a construction Contract are not enforceable. If it is found that the amount specified was too high, it may be judged as a penalty and not a liquidated damage. In such cases, the courts will not enforce the clause. For this reason, most knowledgeable attorneys carefully avoid the use of the word *penalty* anywhere in the Contract. Judges have been known to disallow clauses merely because the word was used in the Contract wording.

High Estimates

When the Contractor challenges the amount of liquidated damages, the Owner must substantiate the validity of the damages. This does not mean that the Owner must demonstrate that actual damages are comparable to the liquidated damages specified in the Contract. The issue that must be decided is whether or not the estimate for liquidated damages was reasonable at the time it was prepared. In other words, when the Contract was drafted, given what was known at that time, was the estimate a reasonable one? Clearly, it is in the Owner's best interest to maintain the documentation used to estimate the liquidated damages figure.

If it is determined that the Owner's estimate was not reasonable or did not reasonably approximate the liquidated damages amount specified, the clause may not be enforceable. For example, if the Owner's estimate showed potential damages of $4,500 per day, but the liquidated damages amount specified in the Contract was $10,000 per day, then the amount specified most likely would be construed as a penalty and, therefore, not be enforced.

Low Estimates

Most of the time, the amount of liquidated damages specified in a Contract is too low. The Owner's damages are often greater than the specified liquidated damages. Can the Owner recover its damages when they are greater than the specified liquidated damages? In most cases, the Owner is limited to the liquidated damages amount specified. There are very few exceptions where an Owner can recover more than the liquidated damages amount. The argument is that the Owner wrote the Contract and calculated the damages and is not entitled to collect more than the specified amount. Because it allows the Owner to recover his losses but protects the Contractor from excessive penalties, the Liquidated Damages clause is often referred to as the "Owner's Sword" and the "Contractor's Shield." If a Liquidated Damages clause is considered a penalty and becomes void, the Owner may sue for actual damages, which may exceed the amount under the Liquidated Damages clause.

Once liquidated damages are involved in litigation, there are further issues to be addressed. Even if a Contractor is successful in challenging the amount of the liquidated damages, it does not mean the Owner is not entitled to any damages. It simply means that the Owner must prove the actual damages. Just remember that a Liquidated Damages clause is often used because it was difficult to accurately calculate the actual damages. While the Owner has the opportunity to recover actual damages, it may be very difficult to substantiate that the Owner is entitled to damages above those specified in the Liquidated Damages clause.

ACTUAL DAMAGES

If the Owner does not include a Liquidated Damages clause in the Contract, or if the clause is deemed legally unenforceable, then the Owner may seek actual damages for a delay caused by a Contractor. When attempting to recover actual damages, the Owner has the legal burden to substantiate the damages. The Owner should seek qualified counsel to properly prepare the damage computations. In order to assess all the damages of a Contractor-caused delay, the Owner should document all damage items to the maximum extent possible. The items listed in Figure 8.3 should be considered during the process of calculating actual damages.

The last item of damages in Figure 8.3—lost revenues and profits—is often difficult to recover. The courts and boards look upon lost revenues and lost profits as being highly speculative and, therefore, not subject to exact quantification. However, this does not mean that Owners should not claim for lost revenues and lost profits. Instead, Owners should be realistic about their chances of recovery. Chance of recovery of lost revenues or profits may be enhanced, for instance, if the Owner is able to show a measurable difference in the production rates of the old facility and the new facility. This is possible primarily where the Project is a replacement facility, as opposed to a new product facility.

Figure 8.4 is an example of an Owner's calculation of actual damages sustained because of a delay caused by the Contractor. In general, actual damages are more difficult to recover, since by their nature they may not be accurately quantified. Consequently, the Owner must decide before the Project begins whether or not it is most advantageous to utilize a Liquidated Damages clause or to seek actual damages if a delay occurs. The less amenable the damages are to accurate calculation, the more reason to utilize a Liquidated Damages clause.

Considerations for Making Damage Calculations

The owner should have its designer **keep separate records** that reflect those costs for inspection, site visits, and so forth that are the result of the project having a longer duration.

The owner should assess if the delay caused an **increase in its moving costs**, such as escalation of moving costs, storage of items, temporary facilities, and so on.

If the owner had to pay **additional rent** to occupy its previous facilities longer than planned, those costs should be carefully documented.

If the owner had staff involved with the construction project, then any **extended staff costs** should be documented.

Depending on the nature of the project, the owner may have increased cost associated with **temporary lodging**—particularly with respect to housing developments. These temporary lodging costs should be documented.

If the project is financed in some fashion, then the owner will be exposed to **additional interest expense** relative to the cost of financing. These costs must be tracked.

The owner may face damages from **other follow-on contractors** who have been delayed by the delay caused by the first contractor. These damages would then be a part of the owner's actual delay damages.

If the project is a production facility, the owner may claim the **lost revenues from not being able to produce its product**.

FIGURE 8.3

Statement of Actual Damages—ABC Corporation

As noted by our expert report prepared by Trauner Consulting Services, Inc., the project was delayed 150 calendar days, attributable to the contractor. Based on a delay of 150 calendar days, the following damages are claimed by the ABC Corporation.

Additional Design Services
Based on the records of our design firm, the A/E made 22 more site visits than the number originally scheduled. The exact cost for these 22 site visits was $10,560.

Moving Costs
Because of the delay, ABC Corporation was forced to renegotiate its contract with its moving company. The renegotiated contract was $6,520 higher than the original moving contract.

Rental Costs
ABC Corporation was forced to stay in its previous facilities for an additional five months. The documented rental costs during this period were $617,000.

FIGURE 8.4

(Continued)

Staff Costs
ABC Corporation's in-house construction staff remained active on the project for an additional 150 days. The staff costs based on salaries and benefits were $61,845.

Temporary Lodging
ABC Corporation's plant start-up engineer moved to the project site based on a promised completion date (four months later than the original date) by the contractor. The contractor exceeded that date by 35 days. The lodging and subsistence paid to the start-up engineer during that 35-day period totaled $2,265.

Finance Cost
ABC Corporation financed this project at an interest rate of .5% over prime. Because of the delay, ABC Corporation recorded additional financing costs of $792,000.

Claims by Follow-On Contractor
ABC Corporation received a delay claim from its start-up contractor. The claim was settled for the amount of $29,000.

Lost Revenues
Because of the delay, ABC Corporation was unable to benefit from the increased production capability of the new facility. Based on sales records, the lost revenues total $2,675,000.

Total Damages due ABC Corporation: $4,174,190.

FIGURE 8.4—CONT'D

Chapter | nine

The Contractor's Damages Due to Delay

Even after entitlement to recover additional compensation has been established, the change or claim can remain unresolved due to problems associated with the calculation and presentation of damages. Problems may include overstated or incorrectly calculated claim amounts, claims for damages not adequately supported with appropriate documentation, claims for damages that either contradict or expand the terms of the Contract, and claims for damages that conflict or are inconsistent with the basis for recovery and the legal aspects underpinning the claim.

In the following sections we explore the proper use of several methods used to compile delay damages. Before we cover some of the specific damage components (material, equipment, and labor costs; labor and equipment inefficiency; etc.), it is important to better understand the basic principles underlying the recovery of damages. Merely asserting that one has been harmed will not suffice. The calculation of damages must be properly supported by the pertinent Contract provisions and the applicable facts supporting the claim (whether they come from the documents, cost records, testimony, etc.).

GENERAL GUIDELINES FOR THE PRESENTATION AND RECOVERY OF DAMAGES

Following are some basic guidelines for the Contractor in preparing its statement of damages in order to expedite the recovery of damages.

1. The initial step in formulating the damage calculation, whether caused by delays or other reasons, is to carefully review and closely follow the

Contract provisions. A typical agreement will require both contracting parties to fulfill specific requirements and that a certain measure of risk or loss can accrue to either party if the requirements are not satisfied. For example, a termination for convenience clause will likely provide rules for measuring damages, and the Contractor will be limited to the terms set forth in the Contract. It is essential that the damages claimed follow these specific rules, or recovery may be more difficult.

2. Avoid frivolous claim items and overstated claim amounts. Although some parties believe you must start high and negotiate down, frivolous claim items and inflated claim amounts cause a Contractor to lose credibility and ultimately delay the process of resolution. Contracts often now include clauses that classify overstated claim amounts as false claims, with harsh consequences. A well-documented damage claim should satisfy any audit requirements of the Contract.
3. Be prepared to negotiate and compromise. Just about any unresolved dispute involves shades of gray and is not a case of black and white, where one party is absolutely right and the other party is totally wrong. Consequently, settlement is seldom reached unless each party is willing and able to compromise.
4. Carefully evaluate your opponent's ability to settle. Many public agencies are not in a position to negotiate due to the need to justify any settlement amount to others, such as taxpayers or other agencies. Generally, these types of claims will require thorough documentation to facilitate the approval process.
5. Support damages based on the verifiable facts, and not one's expectations anticipated prior to the award of the Contract. If a Project incurs a loss, it is the reasons for the loss that are being claimed with appropriate support and measurement, not the bottom-line loss.
6. Summarize your damages within a format that will allow subsequent updates or revisions, especially if the quantitative measure is not fully available until a future date.
7. Avoid duplication of claimed costs and calculation or posting errors. Cross-check and double-check the calculation as it evolves over time.
8. Remember the burden of proof is borne by the party submitting the claim. The claimant must produce facts to establish that the damages were incurred as a result of actions of the other party, along with the measure of the damages.
9. There is always an obligation to mitigate damages. If certain damages could have been mitigated, these costs are unlikely to be recoverable.
10. Generally a claimant's actual costs are considered reasonable when measuring damages, unless proven differently. However, actual costs do not automatically prove that damages were incurred as a result of the action of the other party.
11. The evidence used in supporting damages must be admissible under the jurisdictional rules governing the case.

12. The evidence supporting your damages is generally more persuasive if it is prepared contemporaneously with the actual progress of the Project, as compared to documents prepared after the fact.
13. Remember to establish a cause-and-effect relationship to the extent possible for each component of damages being claimed. You need to show how you were damaged as a result of the other parties' actions or inactions.
14. When feasible, submit your claim for damages using the suggested format of the Owner. This process will lessen the areas of disagreement with the Owner about the format and structure of the claim and keep the discussion focused on the content of the claim.

Overview of the Cost Presentation

A Contractor's claim for damages resulting from delay will need a carefully planned program of cost recovery, including well-documented cost support for each component of the claim. Overall, the calculation of damages is based on several interrelated factors. The initial guidelines of the claim would be based on the rules outlined within the Contract agreement in conjunction with the applicable laws governing the Contract. The parties to the Contract are also an important consideration. Many public projects do not lend themselves to negotiated settlements and may require a lawsuit to pursue the claim. Before measuring damages, the Contractor must carefully review all applicable facts supporting the claim in order to establish that it has been harmed and suffered a loss based on the harm. The calculation of damages must be based on the facts and merits of the situation and not on unrealistic demands or expectations. To the extent possible, damages should be based on a cause-and-effect relationship with the change for which damages are sought. Although a Contractor may have suffered a loss, that does not always support a valid claim for damages. Too often, the damage calculations are not realistic or adequately supported, and a settlement becomes remote. A well-presented claim will lend itself to subsequent revisions if deemed necessary. The more supporting documentation accompanying the claim, the more likely a settlement may be negotiated.

Keep in mind that it is a Contractor's responsibility to mitigate damages whenever practical in order to not create damages greater than those reasonably justified under the circumstances. When in a delay situation, carefully document and review any and all efforts to mitigate damages when applicable. It is generally recommended to identify these efforts in correspondence to the Owner. Correspondence addressing the need to maintain equipment on site, reduce or not reduce manpower, and so forth will serve to strengthen the Contractor's claim.

The Contractor has to prove its claim and related damages. The Contractor's actual costs are generally presumed reasonable unless the Owner can prove the costs are unreasonable. Once liability for delay has been established, the resulting damages do not have to be proven to an exact measure, but amounts should be reasonable and adequately supported.

Although this section pertains to delay damages and actual costs, the reader should be aware that other methods of recovery may be applicable under varying circumstances. These methods are usually specified by the Contract, and legal counsel should be used to pursue items such as wrongful termination, total cost claims, modified total cost claims, quantum meruit, and jury verdict methods.

TYPES OF DELAY DAMAGES

Delay damages are summarized in Table 9.1, although not all may be recoverable. A more thorough review of these costs follows. This and the next few chapters deal separately with various kinds of damages to which Contractors may be entitled for excusable, compensable delays.

The next few sections deal separately with various kinds of delay-related damages that Contractors may be entitled to recover as the result of an excusable, compensable delay. Not all categories will be addressed in every detail, as the possibilities and permutations are virtually unlimited. Rather, an example of the general methodology is presented so the reader may apply the same principles to her or his specific situation.

Table 9.1

Extended and Increased Field Costs	These damages include the additional labor, material, and equipment costs resulting from project delays. These costs are typically quantified and supported by actual measurements of increased units and/or rates.
Home Office Overhead (Addressed in Chapter 10)	When a project is delayed, the Contractor may experience unabsorbed and unrecovered overhead costs. Generally, the Contactor's home office costs that support all projects are allocated to the delayed project and recovered based on a daily rate applied to compensable days of delay.
Inefficiency or Lost Productivity Costs (Addressed in Chapter 11)	Depending on the nature of the delay, a contractor may also experience some measure of inefficiency. The resulting increased labor and equipment costs may be included within the damage claim.
Acceleration Costs (Addressed in Chapter 12)	When the labor force is accelerated to mitigate delays, the resulting increased labor and equipment costs may also be included within the damage claim.
Other Categories of Delay Damages (Addressed in Chapter 13)	A delay claim may include other related damages such as Noncritical Delays, Legal and Consulting Costs, Lost Profits, Lost Opportunity Costs/Bonding Impairment, Constructive Acceleration, Interest, and Other Impacts.

Extended Field Costs

One type of additional cost that a Contractor may claim is extended field costs. For example, when a Project experiences a delay, the Contractor's field staff and field equipment may be on site longer than originally scheduled. The discussion in this chapter addresses the major categories of extended field labor and equipment costs. This chapter explains how to calculate these damages to clearly illustrate the additional cost to a Contractor due to a delay and the nature of its delay-related damages.

Depending on the Project-specific circumstances, when a Project encounters a delay, the Contractor would typically retain its supervisory team at the job site. Personnel such as the Project Manager, Project engineer, superintendent, assistant superintendent, and administrative support positions are in this category. These personnel represent a direct labor cost to the contracting firm.

To calculate the delay-related direct labor cost to maintain these people on site, add the daily cost of each staff member's salary, including burden, and then multiply that sum by the number of days of excusable, compensable delay. Make sure that if the delay is expressed in calendar days that the costs are also expressed in cost per calendar day.

(Aggregate of each daily salary + burden) × Number of days of compensable delay = Damages for extended field labor (supervisory personnel)

While this calculation is straightforward, it is important to note that the analyst also needs to address the propriety of the Contractor's claim to recover extended field costs for supervisory personnel.

Propriety of Claiming Extended Field Supervisory Damages

In order to address the propriety of claiming damages for extended field supervisory labor, the analyst should start with a review of the Contractor's normal accounting procedures for these costs. For example, if the Contractor routinely charges the Project directly for the Project Manager, then claiming damages for him would be appropriate but subject to the Contract requirements. However, if the Contractor routinely charges the Project Manager to home office overhead, then claiming this cost as a direct field cost may require further investigation. The Owner would likely argue that the Project Manager's salary is an overhead item when resolving the delay damages. The same principle applies to the salary of any other field supervisory personnel extended on the Project because of the delay.

Union Supervisory Personnel

On certain projects, labor is provided through signatory agreements that require additional nonworking personnel to be at the job site while work is being performed. Some examples include a master mechanic, the shop steward, or certain levels of foreman. If there is a delay, these personnel also represent an extended labor cost. The Contractor can provide copies of the union requirements to substantiate these costs to the Owner's analyst.

Idle Labor

Another category of delay-related labor costs that might arise is idle labor. If a Project is delayed or suspended, the Contractor's workers may be at the Project site but may be unable to productively work. For potential recovery of this type of cost, the Project daily reports need to show that the Contractor's workers were on the site but were unable to perform their work. The Owner would appropriately question why the Contractor was unable or unwilling to shift those workers to other tasks or other jobs or lay them off. The Contractor should be prepared with Project daily reports to substantiate an idle labor claim. If the Owner does not retain such daily reports, there may be no basis for a challenge to the Contractor's documented claim. For this reason, both the Owner and the Contractor should diligently maintain documentation of labor activity on each day throughout the Project. It should be noted that the Contractor has an obligation under most contracts to mitigate the damages when a delay occurs; therefore, if possible, the Contractor needs to look at shifting its labor to other work during a delay.

ESCALATION OF LABOR COSTS

Some delays can cause the Contractor to experience escalation of its labor costs. This may occur if the delay shifts labor into a more expensive time period than that originally scheduled. This situation is sometimes referred to as extended labor, but it should more appropriately be considered labor cost escalation. The following example illustrates the difference between idle and escalated labor costs.

EXAMPLE 9-1

A Project is scheduled to start on a specific date and finish 300 calendar days later. The Contractor is using union labor contracted under a collective bargaining agreement that specifies pay increases at certain intervals. In this example, the pay increase for common laborers is to occur on day 150. The rate before the wage increase is $25 per hour; it escalates to $27 per hour after the increase. In the third month of the Project, the Owner causes a ten-day delay to the Project. The Owner accepts responsibility for the delay, so there is no question of liability or the magnitude of the delay. The Project is extended by change order an additional ten days, and the Contractor must present the labor costs associated with the delay (see Figure 9.1).

The Contractor's records show that during the delay, there were 200 labor hours of idle time for common laborers. To figure delay costs, the Contractor multiplies the number of hours by the wage rate:

$$200 \text{ hours} \times \$25.00 \text{ per hour} = \$5{,}000.00$$

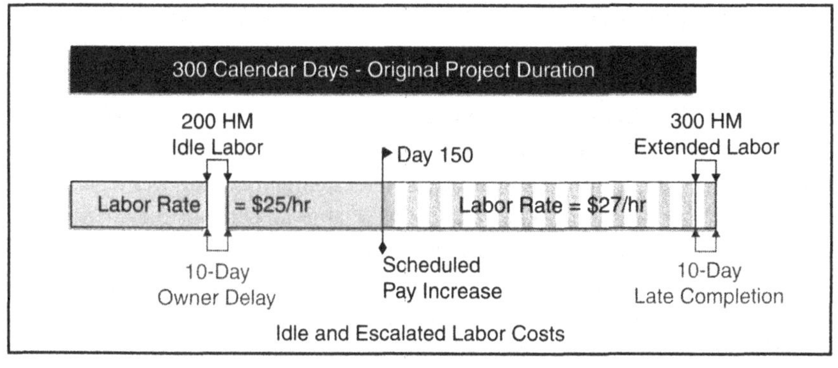

FIGURE 9.1

The Contractor's records also show 300 labor hours for common laborers during the last ten days on the job, which occurred in a time frame later than that originally planned. The Contractor seeks compensation for the additional $2 per hour ($27.00 – $25.00) for the 300 labor hours.

$$\$2/\text{hour} \times 300 \text{ labor hours} = \$600$$

The Contractor presented a total claim of $5,600, with $5,000 for idle labor and $600 for extended labor.

The Contractor's approach, however, is incorrect. To correctly determine the damages, the analyst must examine each component of the claimed additional costs separately. In that case, the idle labor request of $5,000 was correct. Of course, the records provided by the Contractor must validate the idle labor hours and substantiate that it was not possible to reassign the idle workers into other productive work activities. The request for $600 for the extended labor was incorrect. Had there been no delay, the 300 labor hours worked at the end of the Project would have been worked only ten days before the date they were actually worked. Therefore, these hours still would have been worked at a rate of $27 per hour. There was no impact to these last 300 hours. Instead, the Contractor should be looking at the ten-day time period immediately following the wage increase. The labor hours in this time frame were shifted from the less expensive time period into the more expensive time period. The Contractor's certified payrolls showed that 400 labor hours were worked during the ten-day period immediately after the wage increase. Therefore, the Contractor would be entitled to an additional $800 for escalated cost of its labor due to the delay.

$$\$2/\text{hour} \times 400 \text{ labor hours} = \$800$$

Thus, the Contractor's damages were actually $5,800 for the direct labor costs caused by the delay. The Contractor would naturally add the appropriate burden to this amount to calculate the total additional labor costs attributable to the delay.

This simple example illustrates that the analyst should not estimate damages without first clearly understanding the nature of the delays. Hasty conclusions or conclusions based on unverified assumptions often produce erroneous cost estimates. A more realistic and complex Project example might involve several trades under different collective bargaining agreements, each with varying wage increases at different dates. A wage increase might extend over a period of more than a year for some trades. The updated labor agreements may make new and different requirements for apprentice or supervisory labor. The Contractor or Owner must assess how the shift in the work performed by the labor force increased labor costs or exposed the Contractor to escalated wage rates. The example just given is further simplified by the fact that no additional work was added by the delay. It also assumes that once the delay ended, the work was performed in the same sequence as originally planned.

What happens when a delay alters the work sequence or when a delay combines with other changes to affect the actual labor distribution on the job? In these situations, the Contractor must substantiate the original planned distribution of labor compared with the actual distribution. The following example illustrates this approach.

EXAMPLE 9-2

A Project is scheduled for a duration of 300 calendar days. During the Project, the Contractor maintains a resource-loaded CPM schedule to show the distribution of manpower planned over time. Figure 9.2 shows the distribution of planned carpenter hours for the Project taken from the original schedule. Figure 9.3 shows the distribution of actual carpenter hours on the Project, based on the delay caused by the Owner.

As can be seen from the figures, the wage increase for the carpenters took effect on day 150. In the original schedule, the Contractor planned to use 720 carpenter labor hours after day 150, at the higher labor rate of $36 per hour. In the actual distribution of labor, the Contractor used 860 carpenter labor hours during the higher wage rate period. Therefore, 140 labor hours (860 – 720) were

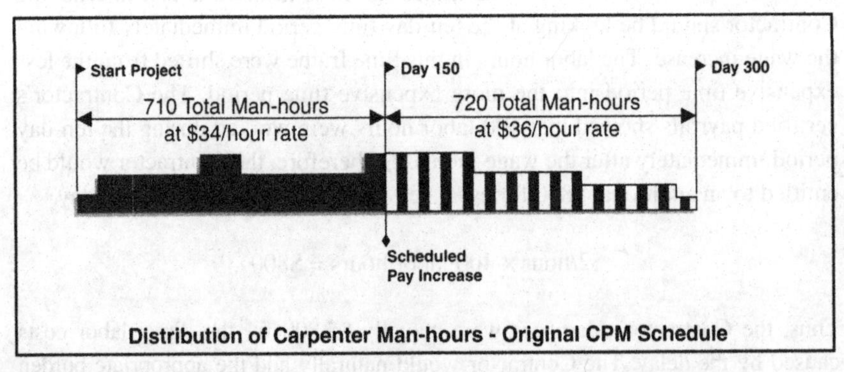

FIGURE 9.2

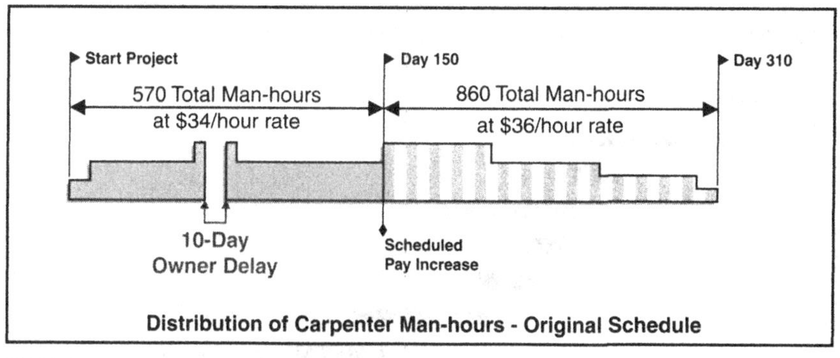

FIGURE 9.3

shifted into the period of the increased labor rate. The Contractor should calculate $280 ($2/hour × 140 hours) for the escalated labor cost of the carpenters.

860 actual hours – 720 planned hours = 140 labor hours

140 hours × $2/hour wage increase = $280.00

In this example, the actual distribution of carpenter labor was caused solely by the Owner's delay. In a different situation, if the Owner can demonstrate that this was not the case, then the delay damages would not include the escalated labor calculation.

Of course, many projects do not have a resource-loaded CPM schedule to allow a reasonably precise comparison between the planned and the actual distribution of labor. Without the CPM schedule, the analyst can estimate the planned labor distribution from the Project bar chart. The analyst can compare the labor hours actually expended on the job (taken from the Project daily reports) to the planned labor hour distribution. The following example illustrates this procedure.

EXAMPLE 9-3

A Contractor plans to perform the work in accordance with the bar chart shown in Figure 9.4. No labor hours are shown in the bar chart. Due to an Owner-caused delay, the Contractor's work is performed later than originally planned. As a result, the Contractor experiences escalated labor cost on the Project, after a rate increase (from $35 per hour to $36 per hour) went into effect on day 200 of the Project.

Figure 9.4 is the Contractor's bar chart for the work as originally planned. Figure 9.5 shows the actual bar chart for the Project, reflecting the delays. It also shows the actual carpenter labor hours worked on each activity. These labor hours were taken from the contemporaneously recorded Project daily reports.

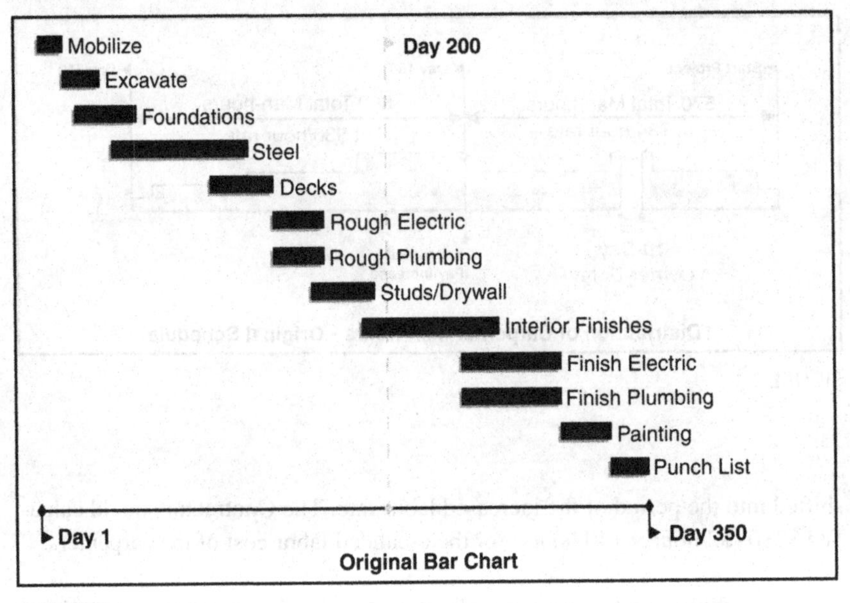

FIGURE 9.4

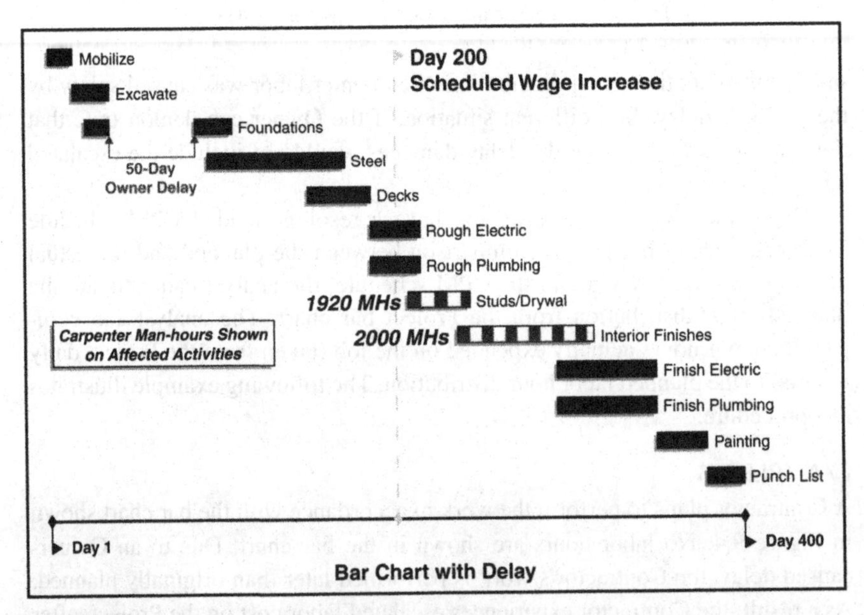

FIGURE 9.5

In order to calculate the cost escalation, the Contractor uses its original bar chart (Figure 9.4) and applies the actual carpenter labor hours to each respective task. This is shown in Figure 9.6. A comparison of Figures 9.5 and 9.6 shows that there were 2,740 carpenter labor hours shifted into the more expensive time frame. Therefore, the Contractor's cost escalation for this trade is $2,740.

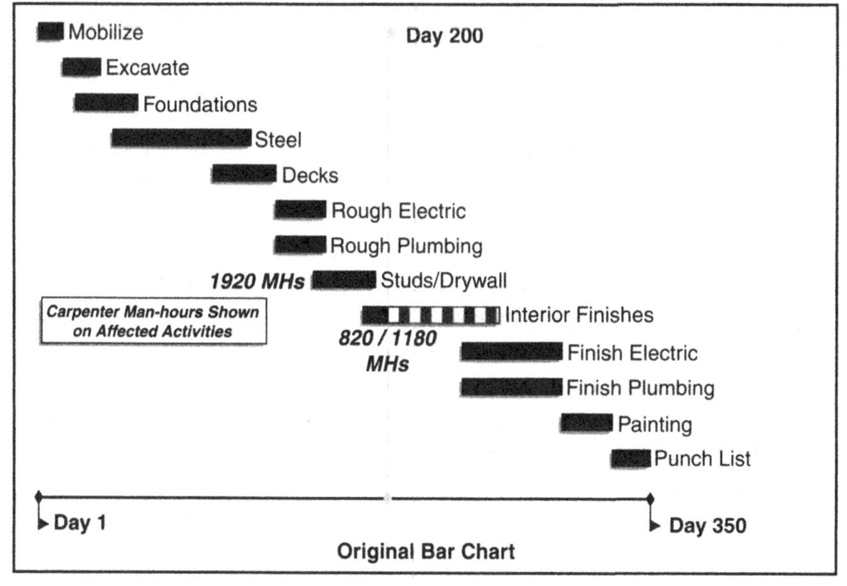

FIGURE 9.6

$$\$1/\text{hour} \times (1{,}920 \text{ hours} + 820 \text{ hours}) = \$2{,}740$$

It should be noted that this conclusion assumes the delay only shifts the work into the more costly time period and that no labor inefficiency resulted.

EQUIPMENT COSTS

A Contractor's equipment costs on a Project can also be affected by a delay. For example, delays may cause the equipment to be idle. The amount of damages that can be claimed for idle equipment depends on the Contract provisions. In some cases, the Contract may address idle or standby equipment and allow only a reduced rate or no compensation at all. If the Contract is silent on this issue, the Contractor would likely claim damages at the full rates of the equipment during the idle time. As in the case of idle labor, the Contractor would need to identify specific equipment on the Project in the daily reports to show exactly when this equipment was idle and the duration of the idle time.

The Owner will generally question whether the equipment was idle because of the delay. It is often worthwhile for the Owner to verify the Contractor's use of equipment prior to and immediately following the period of the delay. If the equipment was idle both before and after the delay period, then the Owner may appropriately question whether the Contractor actually sustained any damages from the delay.

As with labor costs, a Contractor's equipment costs may also be subject to escalation. If a Contractor is using rental equipment, the delay could shift the Contractor's equipment into a time period with a higher rental rate. This situation can and should be documented with invoices.

Other costs can be associated with equipment in the case of a delay. For example, if a significant delay occurs to a portion of a Project, specialized equipment originally planned for the work may no longer be available. The Contractor's own paving machine may have been assigned to a Project for a particular period of time, but it is no longer available after the delay period. The Contractor is forced to replace the owned equipment with more expensive rental equipment in order to meet the Project's revised schedule. In another case, the Contractor originally planned to perform mass excavation work using scrapers. Because of the delay, the Contractor now finds it must use loaders and trucks to excavate the material. The damages to the Contractor would be measured by comparing the cost of the two methods, the originally planned versus the actual. In some respects, this type of cost borders on lost productivity, which will be addressed later in this book. In the purest sense, however, it does represent the delay cost of the equipment. The following example illustrates both cost escalation and inefficiency in equipment use caused by a delay.

EXAMPLE 9-4

An Owner is constructing a waste water treatment plant addition. The Project site is an inactive landfill. The excavation Subcontractor is required to excavate approximately 200,000 cubic yards of material and dispose of it at a nearby active landfill. Prior to beginning excavation, the Owner is advised that the site may contain hazardous waste materials. Consequently, the Owner stops the Contractor's excavation activity in order to perform toxicity testing. The testing takes three months to complete, at which time the excavation Subcontractor is allowed to again proceed with the work as originally specified. Because of the delay, however, the Subcontractor experiences changes in its equipment cost and availability.

Originally, the Subcontractor planned to excavate 170,000 cubic yards with scrapers and the remaining 30,000 cubic yards with a hydraulic excavator and a 20-ton dump truck. Due to the delay, the scrapers are no longer available because they have been committed to another Project. Consequently, the Subcontractor must now perform all of the excavation work without the use of scrapers. The Subcontractor experiences two problems because of the delay. First, the excavators it intended to rent have increased in cost. Second, the unit cost for the excavation work is higher using the dump trucks than it would have been using the scrapers. Based on these factors, it calculates the following increased costs.

1. Escalation of rental equipment cost: Cat hydraulic excavator—quoted rate: \$2,800/week (quote attached), new rate: \$2,900/week (invoice attached). Difference of \$100/week. Planned equipment time was 10 weeks (2 excavators for 5 weeks). Extra cost: \$100/week × 10 weeks = \$1,000.

2. Increased excavation cost: Excavation with scrapers: \$4.50/cubic yard (calculations attached). Excavation with dump truck: \$7.20/ cubic yard (calculations attached). Difference of \$2.70/cubic yard. Extra cost: \$2.70/cubic yard × 170,000 cubic yard = \$459,000.
3. Total cost from delay: \$460,000 = \$459,000 + \$1,000.

Of course, the Contractor must always show that it complied with its obligation to mitigate damages. In this case, the Subcontractor may need to demonstrate that the premium to rent scrapers over and above the cost of the planned scrapers was greater than the damages that resulted from the change in work methods.

MATERIAL COSTS

The most common added material cost caused by a delay is price escalation. Because of the delay, the Contractor is forced to buy materials to be incorporated into the work in a period when the price has increased.

The calculation of escalation cost for materials can be performed in the same manner as that for the escalation of labor cost. A comparison must be made between the quantity of each material that would have been purchased in the original schedule and the quantity of each material purchased later under the delayed schedule. The following example illustrates this method.

EXAMPLE 9-5

A Project was originally scheduled to take 300 calendar days. The Project experienced a 90-day, Owner-caused delay. The delay occurred at the beginning of the Project and forced the Contractor to pay more for concrete. The Contractor can demonstrate it had a purchase order for concrete at \$80 per cubic yard from the beginning of the Project until day 100. After day 100, by the terms of the purchase order, the cost of concrete increased to \$85 per cubic yard. Due to the delay, the Contractor purchased more concrete material at \$85 per cubic yard than it originally planned.

A review of the schedule shows that the Contractor planned to have placed concrete on six of eight floors before the price increase took effect. A quantity takeoff establishes this as 2,000 cubic yards of concrete. In the actual performance of the work, the Contractor placed only 1,200 cubic yards of concrete before the price increase took effect. The material cost increase is directly attributable to the excusable delay. Therefore, the associated damages will be at least \$4,000 (\$5 per cubic yard × 800 cubic yards) due to the escalated price of concrete.

2,000 cubic yards planned – 1,200 cubic yards actual = 800 cubic yards

Additional cost of \$5/yard × 800 yards = \$4,000 damages due to delay

STORAGE COSTS

Delays may also affect a Contractor's material costs because of storage costs. The Contractor may be forced to store materials either on or off the site as a result of a delay. The Contractor must support claims for these costs with invoices. In some cases, the Contractor may be able to purchase materials before they are required for installation and store them, as this may be less expensive than the escalated cost of the materials purchased at a later date. The Contractor should notify the Owner in advance of this course of action.

The following example shows a Contractor's damage calculations, including the items described in this chapter. It is intended to help present a clear picture of the procedures to use for calculating delay damages for an excusable delay.

EXAMPLE 9-6

A Project has a Contract duration of 400 calendar days. The Contractor estimates the construction schedule to last the full 400 calendar days, as shown in Figure 9.7. The Project has a delay of 200 calendar days and finishes on day 600 (see Figure 9.8). Of the 200 days of delay, it was determined that the Owner caused 150 days of delay, and the Contractor was responsible for the remaining 50 days. The Owner has issued a two-part change order for its delays. Part I issues a time extension of 150 days. The Owner is now in negotiation with the Contractor over the compensation for the delay.

Figure 9.9, a summary of delays to the schedule, shows that the Owner delayed the start of the Project by suspending the work for 50 days. The next

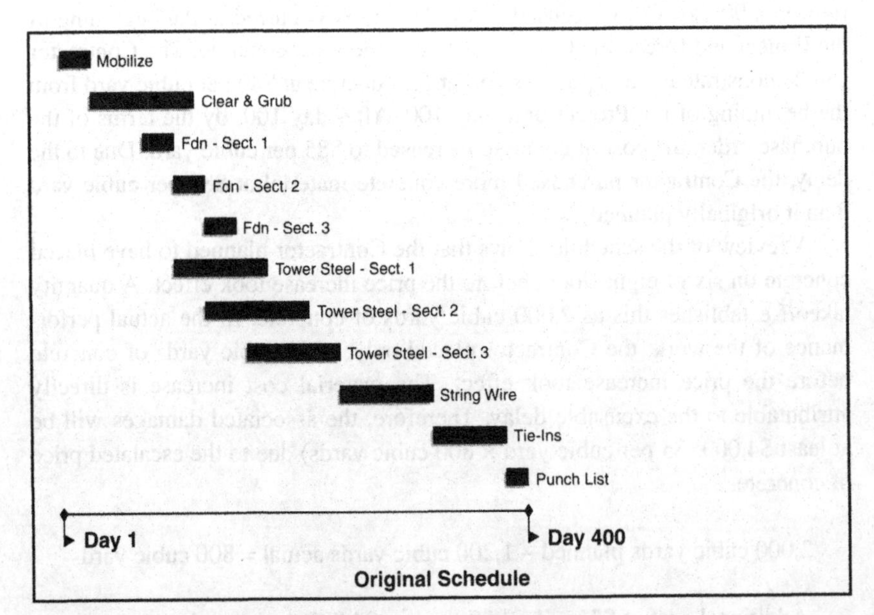

FIGURE 9.7

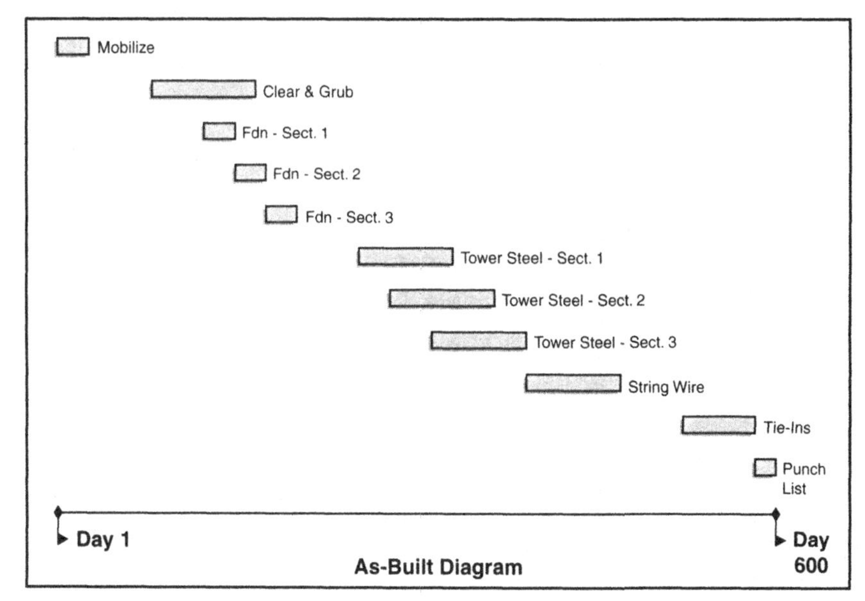

FIGURE 9.8

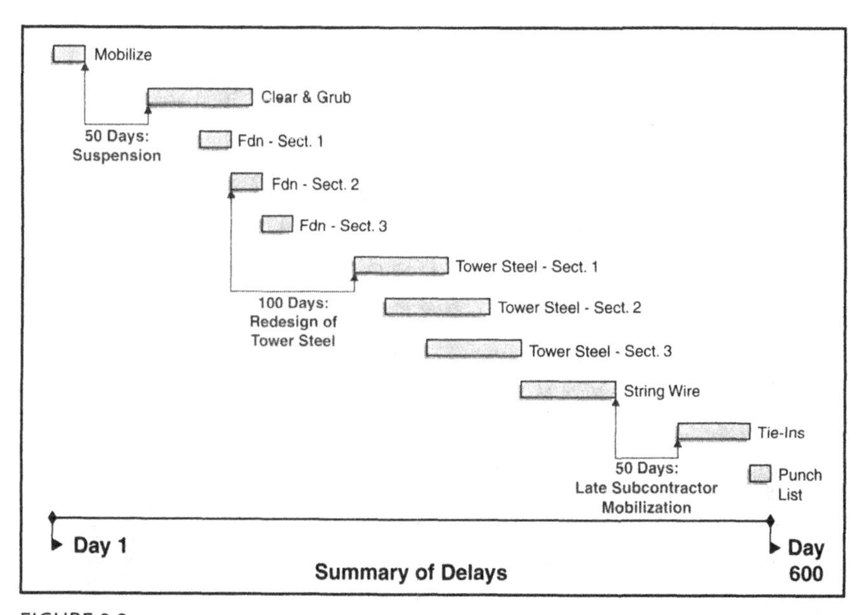

FIGURE 9.9

delay was also caused by the Owner, who, by requesting design revisions, delayed the Project an additional 100 days. The final delay was 50 days, attributable to the Contractor for failing to mobilize a Subcontractor to perform work. The presentation shown in Figure 9.10 was made by the Contractor to support the additional compensation it requested.

Request for Equitable Adjustment

Ace Construction Company hereby requests the following compensation for the 150 days of delay caused by factors beyond our control. In Part 1 of Change Order No. 7, the owner has granted a time extension of 150 calendar days. Because of this 150-day delay, Ace Construction accrued the following extra costs.

Extended Field Overhead

Labor				
Project Manager	150 days @ $600/day for 25% of time	=	$22,500	*Note: Certified payrolls attached.*
Superintendent	150 days @ $520/day	=	$78,000	*Project Manager is a direct job charge.*
Clerk	150 days @ $200/day	=	$30,000	
Master Mechanic	150 days @ $480/day	=	$72,000	
Subtotal Labor			$202,500	
Equipment				
Superintendent's truck	150 days @ $25/day	=	$3,750	*Note: Invoices for equipment attached.*
Portable toilet	5 mos. @ $150/month	=	$750	*Superintendent's truck costed by*
Project trailer	5 mos. @ $160/month	=	$800	*normal company accounting practices.*
Tool trailer	5 mos. @ $120/month	=	$600	
Subtotal Equipment			$5,900	
General Conditions				
Utilities	Electricity: 5 mos. @ $150/month	=	$750	
	Water/Sewer: 5 mos. @ $100/month	=	$500	
Dumpsters	2 dumpsters: 5 mos. @ $150 ea/month	=	$1,500	
Copy machine	5 mos. @ $120/month	=	$600	
Office supplies	5 mos. @ $80/month	=	$400	
Scheduling updates	5 mos. @ $1,000/month	=	$5,000	
Subtotal General Conditions			$8,750	
Total Extended Field Overhead			**$217,150**	

FIGURE 9.10

Construction Equipment Costs				
Extended Equipment				
100-ton crane	3 extra mos. @ $6,000/month	=	$18,000	Note: Rental invoices attached.
	2 mos. @ increased rental rate of			Company-owned equipment costed
Escalation of Equipment	$400/month	=	$800	in accordance with normal company
D-7 dozer	327 hours @ $70/hour	=	$22,890	accounting practices.
Idle Equipment	240 hours @ $75/hour	=	$18,000	
Scrapers				
D-7 dozer				
Total Construction Equipment Costs			**$59,690**	
Labor Costs				
Idle Labor				
Carpenters	725 hours @ $37/hour	=	$26,825	Note: Applicable timesheets and
Laborers	816 hours @ $25.50/hour	=	$20,808	certified payrolls attached.
Ironworkers	420 hours @ $37/hour	=	$15,540	Union agreement and calculation of
Subtotal Idle Labor		=	$63,173	rates to include burden and overhead
Escalation of Labor	escalation: 700 hours @ $1.60/hour	=	$1,120	attached.
Carpenters	escalation: 250 hours @ $1.50/hour (1st increase)	=	$375	
Laborers	escalation: 320 hours @ $1.20/hour (2nd increase)	=	$384	
Ironworkers	escalation: 650 hours @ $1.50/hour		$975	
Subtotal Labor Escalation			$2,854	
Total Labor Costs			$66,027	
Material Costs				
Escalation of concrete	1,200 yards @ $3.00/yard		$3,600	
Storage of rebar	Supplier Charge		$650	
Total Material Costs			**$4,250**	
Grand Total All Costs			**$347,117**	

FIGURE 9.10—CONT'D

OTHER DELAY COSTS

Delays on a construction Project may cause a number of different increased costs to the Contractor, such as the following:

- Temporary utility and facility costs
- Extended warranties
- Maintaining and protecting work during delays
- Inefficiencies
- Increased bond costs

Normally, all of these items fit into one of the general cost categories, such as field overhead, inefficiency, and so forth. The analyst structuring the claim must carefully assess all of the additional costs to the Project occasioned by the delay. In this way, items such as these will not be overlooked.

Chapter | ten

Home Office Overhead

WHAT IS HOME OFFICE OVERHEAD?

In the event of an Owner-caused delay, a Contractor may seek to recover delay damages associated with home office overhead. Despite being a commonly sought element of delay damages, home office overhead remains a contentious issue, due in part to the lack of a universally accepted method of calculating the associated damages. Adding to the potential for conflict is the question of what constitutes *home office overhead*, a term that can take on different meanings given the financial structure of the Contractor's organization and the accounting principles employed. Normally, however, home office overhead consists of the Contractor's fixed costs of operating its principal or home office. It is in the home office that executive and administrative functions are performed on behalf of the Contractor's entire organization. Examples of such costs are shown in Figure 10.1.

Specifically excluded from this definition of home office overhead are the direct costs of labor, equipment, and materials expended to manage, administer, and construct a specific Project. The cost of providing a job site trailer, for example, is not a home office cost as it is incurred specifically to support a particular Project.

The auditing or accounting standards employed by the Owner can further restrict the definition of home office overhead. For example, under contracts governed by Federal Acquisition Regulation (FAR) cost principles, certain costs, such as those expended on marketing and entertainment, may not be recovered as home office overhead. Under FAR, the Contractor could only recover such costs through profit markup.

Regardless of the exact nature of its home office overhead, a Contractor must pay for these costs in some manner through the projects it performs as a constructor. Normally, the Contractor includes home office overhead costs in

Rent
Utilities
Furnishings
Office equipment
Executive staff
Support and clerical staff not assigned to the field
Estimators and schedulers not assigned to the field
Mortgage costs
Real estate taxes
Non-project-related bond or insurance expenses
Depreciation of equipment and other assets
Office supplies (paper, staples, etc.)
Advertising
Marketing
Interest
Accounting and data processing
Professional fees and registrations

FIGURE 10.1

some part of the bid price for each Project. Usually, the Contractor calculates the final bid price by adding a percentage for markup to the direct cost bid amount. The exact markup depends on the amount of home office overhead costs the Contractor incurs in a given period, usually one year. The number of projects the Contractor has under construction at any one time also affects the markup. For instance, if a Contractor works on only one Project at a time, 100 percent of the home office costs for the period of construction would be added to the direct Project cost. As the number of projects increases, the percentage allocated to each job would be reduced, as shown in Figure 10.2. The allocation of home office overhead costs to individual projects is typically a function of direct labor costs or total Project revenues and billings, rather than a fixed dollar markup for each Project.

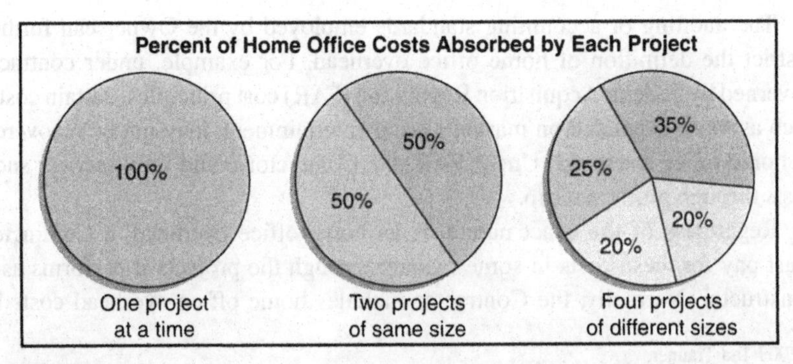

FIGURE 10.2

EFFECTS OF DELAYS ON HOME OFFICE COSTS

As just mentioned, a Contractor typically includes in its bid price for a particular Project a percentage markup through which it will recover some portion of its home office overhead. If this Project were to then experience a delay, Project revenues may be earned over a longer period of time, disrupting the basis under which the Contractor originally allocated its home office overhead costs. This effect can be best demonstrated through the use of examples.

One Project at a Time

For the Contractor who performs one Project at a time, the effect of a delay is relatively straightforward. In the example shown in Figure 10.3, the Contractor did not receive additional compensation for extra work on change orders to offset the increase in home office costs to be covered by this Project. If, however, the one-month delay was the result of extra work with direct costs of only $10,000, and the overhead markup allowed by the Contract was 10 percent, the Contractor would receive $1,000 for overhead as compensation for the extra work. If, as in the previous example, actual home office overhead costs were $4,000 per month, the Contractor would be $3,000 short in recovering home office costs using the overhead markup allowed in the Contract. The difference between delays caused by a suspension and those that result from extra work are explained later in this chapter.

Multiple Projects

In the example in Figure 10.4, the Contractor has multiple ongoing projects. Clearly, as the number of projects increases, the calculations become more difficult, as is apparent from the numerous articles written on this subject.

A contractor has home office costs of $48,000 per year and has a one-year contract for $1,000,000. The project experiences a one-month delay, during which time home office costs of $4,000 accrue. If the delay were the result of a compensable suspension of work, the contract amount of $1,000,000 would remain the same, but the contractor would now have to cover a total home office cost of $52,000 over the term of the contract, an increase of $4,000.

Contract amount: $1,000,000
Yearly home office costs: $48,000
New home office expense, including extending the project one month: $52,000
Increase in home office costs to be covered by this project: $52,000 — $48,000 = $4,000

Therefore, the contractor may be entitled to receive $4,000 in damages to cover home office costs for the delay.

FIGURE 10.3

> This contractor has home office overhead costs of $1,000,000 per year. The contractor normally has four projects in progress at any one time, for a total yearly volume of $20,000,000. The contractor allocates home office costs by adding a 5 percent markup to each bid. Each of the four projects is valued at $5,000,000. What would happen if one project were to experience a delay?
>
> Originally, each project absorbed $250,000 of home office costs, or about $21,000 per month. If one project experiences a one-month delay, then the contractor would receive no additional revenue to cover home office costs from that project and would therefore suffer a net loss in recovered home office overhead costs of $21,000. If the delay were attributable to extra work, the contractor might receive compensation from the markup on a change order. However, in the absence of extra work, the project experiencing the delay would underabsorb its share of the home office costs, leaving the other projects to overabsorb these costs.

FIGURE 10.4

In characterizing home office overhead damages, many authors choose to distinguish between the terms *extended home office overhead* and *unabsorbed home office overhead*. Some distinguish the terms on the basis of whether a Project was formally suspended (extended home office overhead) or only partially or informally suspended (unabsorbed home office overhead). The calculation of damages for each may differ, but in practice, most courts and boards have not always maintained this clear distinction in terms. Regardless of the nature of the delay, the analyst should establish the damages the Contractor actually experiences. The remainder of this chapter presents some methods for calculating delay damages associated with home office overhead costs.

EICHLEAY FORMULA

The difficulty in precisely establishing home office overhead damages has led to the development of formulaic approaches that approximate the effect of an extended performance period on the Contractor's home office overhead costs. One such method is the Eichleay Formula. The formula originated from a decision by the Armed Services Board of Contract Appeals in 1960 (Eichleay Corporation, ASBCA 5183, 60-2 BCA 2688). In its appeal before the Board, the Eichleay Corporation proposed a formula for calculating its damages for home office overhead during a delay. The Board accepted this formula, which has since become known as the Eichleay Formula, as a reasonable method of calculating such damages.

The Eichleay Formula is a simple three-step formula. First, the total Contract billings are divided by the total company billings for the Contract period, as shown in Figure 10.5. The quotient is then multiplied by the total home office overhead costs during the Contract period to produce the allocable overhead.

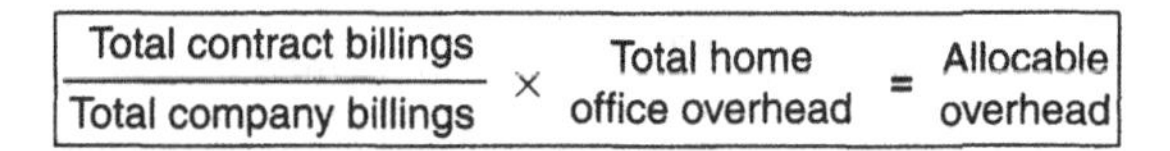

FIGURE 10.5

$$\frac{\text{Allocable overhead}}{\text{Number of days of contract performance (including delay)}} = \text{Daily allocable overhead rate}$$

FIGURE 10.6

$$\text{Daily allocable overhead rate} \times \text{Number of days of compensable delay} = \text{Home office overhead damages}$$

FIGURE 10.7

Next, the allocable overhead is divided by the number of days of Contract performance, including the delay. This produces the daily allocable overhead rate (Figure 10.6). Finally, in Figure 10.7, the daily allocable overhead rate is multiplied by the number of days of compensable delay to produce the home office overhead damages.

A short example illustrates application of the formula. A company has total billings of $50,000,000 during the Contract period. The Contract billings total $5,000,000. Total home office overhead is $1,000,000 during the Contract period. The Project duration totals 200 days, and compensable delays total 30 days. The first calculation to produce the allocable overhead of $100,000 (Figure 10.8) is as follows:

$$\frac{\text{(Contract billings)}}{\text{Total contract billings}} \times \text{(Home office overhead)} = \text{(Allocable overhead)}$$

$$\frac{\$5{,}000{,}000}{\$50{,}000{,}000} \times \$1{,}000{,}000 = \$100{,}000$$

FIGURE 10.8

The next calculation produces the daily allocable overhead rate (Figure 10.9). And finally, in Figure 10.10, the daily allocable overhead rate is multiplied by the number of days of compensable delay to yield home office overhead damages of $15,000.

Problems with the Eichleay Formula

While the Eichleay Formula is simple to apply, one might reasonably question its accuracy. The Eichleay Formula is an estimated allocation and may, therefore, be somewhat inaccurate, yielding results that are either too high or too low.

$$\frac{\text{(Allocable overhead)}}{\text{No. of days of contract performance}} = \text{Daily allocable overhead rate}$$

$$\frac{\$100{,}000}{200 \text{ days}} = \$500/\text{day}$$

FIGURE 10.9

Daily allocable overhead	×	Number of days	=	Home office overhead damages
$500/day	×	30 days	=	$15,000

FIGURE 10.10

Despite this limitation, the federal courts and boards have, for the most part, accepted the formula as a reasonable approximation of the damages sustained. With less case law to draw upon, state courts have been less receptive to outright acceptance of the formula.

Debate over the use of the formula has therefore led to refinements regarding its application. For example, in Excavation-Construction, Inc., ENG BCA 3851 (1984), the Engineering Board of Contract Appeals (ENG BCA) recognized the use of the Eichleay Formula to determine the cost not only of a suspension of work but also of a delay caused by extra work. The board's opinion is noted in Figure 10.11. The board did, however, subtract from the Eichleay calculation the amount of overhead that was already being paid by virtue of the markup on the change order.

Figure 10.12 shows the Armed Services Board of Contract Appeals' response in George E. Jensen Contractor, Inc., ASBCA no. 29772 (1984). The most common argument against the use of the Eichleay Formula is that the Contractor

"The Board believes in general that an Eichleay-type approach is the preferred way to determine home office overhead in a suspension situation and that a markup of direct costs is the preferred way to determine home office overhead for a change. However, no automatic approach can be applied in avoidance of a careful scrutiny of the facts. In this appeal, the changed work on the retaining walls is only part of the reason for extension of the contract period. Much of the delay and disruption occurred because the change was not timely made. The parties have agreed that the net effect was to extend the period necessary for performance by 99 days. Measurement of the effect on home office overhead by the costs alone is likely in these circumstances to understate the amount to which E-C is entitled. Therefore, the Board considers this appeal to be a proper one for application of the Eichleay Formula."

FIGURE 10.11

"Finally, the Government argues that the home office or extended overhead costs are fixed costs that would have been incurred even if there had been no delay. Its argument continues that to allow relief by utilizing the Eichleay Formula would permit recovery of overhead costs much greater than the direct costs incurred during the periods of delay. This argument misses the point. Home office expenses are indirect costs usually allocated to all of a contractor's contracts based on each contractor's incurred direct costs. When a government-initiated delay causes a contractor's direct costs to decline greatly, that contract does not receive its fair share of the fixed home office expenses. The Eichleay Formula is one method approved by boards and courts over a long period of time that corrects this distortion in the allocation of these indirect expenses."

FIGURE 10.12

receives compensation of home office overhead by virtue of the markup on a change. The obvious problem with this argument is that a Contractor likewise receives this same markup whether or not the change causes a delay.

Unless the markup clearly contains an allocation for home office overhead, the argument that Eichleay should not be used may not be valid. Figure 10.13 shows how the ASBCA addressed the point in Shirley Contracting Corporation, ASBCA No. 29848 (1984).

These examples are not to suggest that application of the Eichleay Formula is automatic. Like all claims for delay damages, to recover home office overhead costs, the Contractor must first establish that the delay was indeed compensable. This typically entails establishing that the delay was the Owner's fault or responsibility and could not have been reasonably anticipated or otherwise mitigated by the Contractor. In addition, many courts that have accepted the Eichleay Formula have attached other prerequisites to its application. These can include the Contractor establishing (1) an Owner-imposed suspension of critical work, (2) an Owner requirement that the Contractor stand-by during the associated delay, and (3) proof that while standing-by, the Contractor was unable to take on additional work.

When to Apply the Eichleay Formula

By now the reader has probably recognized that the Eichleay Formula is a calculation applied at the end of the Project after all the work and delays are

"We find no support for this position. The Corps' Area Engineer testified that he did not know the composition of the 15 percent allowed but nonetheless approved it as a matter of course. He also admitted that the 15 percent overhead was allowed even 'on modifications that involved no delay at all.'"

FIGURE 10.13

completed. If the parties attempt to resolve the question of home office overhead during the Project, some form of a modified Eichleay Formula may be appropriate. One approach is to apply the Eichleay Formula from the beginning of the Project up to the point of negotiations. Thus, the total Contract billings, the total company billings, and the total number of days from the start of the Project up to the approximate date of the calculation are used.

When using the Eichleay Formula in federal government cases, some costs (such as advertising, entertainment, interest, etc.) that are normally considered home office overhead are not allowable. The government will normally disallow these costs during its audit procedures. In its Contract, the federal government has the right to audit a Contractor's records. For delay claims in excess of $100,000, the government normally will perform an audit. The various federal agencies have different internal guidelines mandating the dollar level at which an audit is required.

The use of the Eichleay Formula in other jurisdictions (such as at the state and municipal level) may or may not allow such costs. The Contractor, on the other hand, might maintain that the Owner cannot reduce the allowable costs unless specific case law so dictates. The nonfederal Owner should encourage that use of the Eichleay Formula follow federal guidelines, as its origins are in federal cases.

Home office overhead costs in a delay situation can represent a significant percentage of the overall delay damages. Owners should carefully consider this fact in drafting their construction contracts. Some Owners prevent problems in this area by defining in the Contract the allowable damages for delays, including a computation for home office overhead, if any.

CANADIAN METHOD

An alternative method of calculating home office overhead costs for a delay, used extensively in Canada, is known as the Canadian Method. The Canadian Method uses the Contractor's actual markup for overhead in its calculation. This markup is based on either the Project bid documents or an audit of the Contractor's records. An audit would reveal the historical percentage markup for home office overhead applied to each Project. Figure 10.14 shows that the percentage markup is multiplied by the original Contract amount and then divided by the original number of days in the Contract. This yields a daily overhead rate based on the amount the Contractor bid. This rate is then applied to the number of days of compensable delay, as illustrated in Figure 10.15.

$$\frac{\text{Percentage markup} \times \text{Original contract sum}}{\text{Original number of days in the contract}} = \text{Daily overhead rate}$$

FIGURE 10.14

Daily overhead	×	Number of days of compensable delay	=	Compensation for home office overhead

FIGURE 10.15

Example of the Canadian Method Used to Estimate Overhead Cost Associated with a 50-Day Delay	
Project bid	= \$5,000,000
Contract duration	= 500 days
Overhead from bid papers	= 10%
Compensable delay	= 50 days
Daily Rate = $\frac{10\% \times \$5{,}000{,}000}{500 \text{ days}}$	= \$1,000/day
Damages = \$1,000/day × 50 days	= \$50,000

FIGURE 10.16

If the delay is significant (a couple of years, for example), then the daily rate may be escalated to account for an inflationary increase in overhead over time. For example, say a Project is bid for \$5,000,000, and the contracted duration is 500 days. Based on the original bid documents, the overhead markup is 10 percent. Multiply the Contract amount, \$5,000,000, by the percent overhead, and divide by the number of days to determine the daily overhead rate, \$1,000/day, as shown in Figure 10.16. The daily overhead rate of \$1,000/day is then multiplied by 50 days to determine the overhead for the delay period, \$50,000.

Because typically no consideration is given to unallowable costs, the Canadian method is simpler to apply than the Eichleay Formula. However, despite this simplicity, it has not been widely used in the United States.

A variation of the Canadian Method, known as the Hudson Method, has been used in Great Britain. In this method, the percentage markup portion of the formula includes a profit allocation. As with any method, the Contractor must demonstrate that the underlying markup and cost assumptions are reasonable before recovery under these alternate methods will be allowed.

CALCULATION USING ACTUAL RECORDS

Some Owners are reluctant to include home office overhead costs in compensation for delays. This is particularly true when these damages are based on a formula (Eichleay or Canadian) that provides only an approximation of damages. The Contractor could strengthen its argument by maintaining accurate records in the home office that would support its specific claim for damages.

For example, say a Contractor has a home office staff of 20 people. The staff includes Project Managers, estimators, schedulers, clerical workers, and accountants. The company requires that all employees maintain accurate time sheets

Calculation Based on Actual Records

A project is suspended for 50 calendar days. The home office staff consists of eight people, including the CEO who does not maintain a time sheet. Staff time for this project during the delay was as follows

Position	Project Hours	Total Hours
Secretary	25	288
Estimator	24	288
Accountant	96	288
Project Manager	288	288
Project Manager	0	288
General	40	288
Superintendent	10	288
Clerk	543	2016

543 project hours/2016 total hours = 26.9%

Home office costs during suspension = \$51,000

26.9% × \$51,000 = \$13,719

FIGURE 10.17

by activity and by Project. This documentation may be useful in supporting the Contractor's request for home office overhead costs when a delay occurs. In this approach, the Contractor can use the records to determine the staff's percentage effort expended, either throughout the Project or during the specific delay period, if appropriate. The time sheets show the hours the home office staff expended on the delayed Project. This percentage can be applied to the other fixed home office costs to apportion those to the delayed Project. Alternatively, the computation in Figure 10.17 could also be performed on a salary basis to generate the percentage of salary expended on the delayed Project, as opposed to time.

NET PRESENT VALUE ANALYSIS

In addition to the home office overhead damages that may result from a delay or suspension to a Project, a Contractor may also incur additional indirect damages, such as disruptions to its cash flow. Such indirect damages can be addressed using the Net Present Value Analysis (NPVA), as illustrated in the following example.

A Contractor has five projects, A through E, each with a planned duration of one year. The value of each Contract is \$1,000,000. The Contractor's home office overhead costs are \$100,000 per year. The Contractor's direct expenses are expected to be 93 percent of revenues. The Contractor receives notice to proceed on all five projects, but shortly after starting the work, Project C is halted for a year. Given this information, the Contractor can formulate a budget and actual income statement for the period, as shown in Figure 10.18.

Contractor's Budgeted and Actual Income Statements for Year 1

	Budget Income Statement	Actual Income Statement	Variance
Revenues	$5,000,000	$4,000,000	
Costs	4,650,000	3,720,000	
Gross Profit	350,000	280,000	
Home Office Overhead	100,000	100,000	
Income	$250,000	$180,000	$70,000 (unfavorable)

Net Percent Value Analysis Example

FIGURE 10.18

Contractor's Budgeted and Actual Income Statements for Year 2

	Budget Income Statement	Actual Income Statement	Variance
Revenues	$5,000,000	$6,000,000	
Costs	4,650,000	5,580,000	
Gross Profit	350,000	420,000	
Home Office Overhead	100,000	100,000	
Income	$250,000	$320,000	$70,000 (favorable)

Net Percent Value Analysis Example

FIGURE 10.19

As shown, the delay of Project C results in an unfavorable variance of $70,000 in the income from operations for the period. The next year, the Contractor wins five more contracts (F through J), each valued at $1,000,000. The Contractor's cost and overhead data remain the same as in year 1. However, the suspension of Project C is lifted, and the Contractor completes that Project in year 2, along with contracts F through J. The budgeted and actual income statements for year 2 are shown in Figure 10.19. In contrast to year 1, in which the Contractor showed an unfavorable variance of $70,000 due to Project C's underabsorption, in year 2 the contributions from Project C create a favorable balance because of Project C's overabsorption in that year. As a result, the Contractor's cash flow (in year 1) is adversely affected by the delay. The cost of this effect is addressed in the next section.

Discounted Cash Flow

The NPVA not only considers the amount of cash that flows in and out of a Project but also the time value of cash flows. A dollar received sometime in the future is not worth as much as the dollar received and available for reinvestment today. The NPVA can measure the indirect effect of delays, such as cash flow problems. In this way, NPVA is also known as the Discounted Cash Flow Method.

The only difficulty in using NPVA is choosing the appropriate rate for discounting the cash flow. The most objective rate to use in performing the analysis is the Contractor's cost of capital. For publicly held corporations, this is the Weighted Cost of Capital (WCC). The WCC is a weighted average of the component costs of debt, preferred stock, and common equity. The WCC is determined by using the Capital Asset Pricing Model (CAPM). CAPM is a formula that considers (1) the firm's rate of return, (2) the risk-free rate of return, and (3) the firm's "beta" value (a measure of a firm's inherent risk).

The Cost of Capital

The CAPM yields a WCC value for Contractors that is slightly higher than the prime interest rate. Because the WCC calculation is based partly on equity markets, it is not the most accurate discount rate for privately held Contractors to use for short-term borrowing. This rate is determined by a lending institution's loan officer instead of the capital markets. As with the WCC rates, the private Contractor's short-term borrowing rate will be slightly higher than the prime interest rate.

Example Estimate for Indirect Effect of Delay

In this section, the Net Present Value Method will be used to evaluate the indirect effect of a delay or suspension on the Contractor. Using the previous example, recall that the Contractor has an unfavorable variance of $70,000 in year 1 because of Project C's underabsorption. In contrast, the contribution from Project C in the following year creates a favorable balance in year 2. To simplify the example, revenues and expenses are assumed to be realized and paid in 12 equal monthly segments, as shown in Figure 10.20. The example further assumes that the Contractor's cost of capital is 12 percent.

To determine the indirect damages associated with disruption to the Contractor's cash flow, it is necessary to first discount back to the present the budgeted and actual cash flows for years 1 and 2. Steps 1 through 3 following

Net Cash Flows

Year 1			
Time Period	Budgeted	Actual	Variances
Each Month	$20,833	$15,000	–$5,833
Year 2			
Time Period	Budgeted	Actual	Variances
Each Month	$20,833	$26,667	+$5,833

Indirect Impact on the Contractor Using Net Present Value Analysis

FIGURE 10.20

describe this discounting process. Note: Texts dealing with construction engineering economics provide tables for the time value of money for various periods and different interest rates.

STEP 1

Determine the present value of budgeted cash flows. The year 1 and year 2 projects, as budgeted, forecast a monthly net cash flow to the beginning of the year that yields a present value of $234,477.50. Discounting the second-year cash flows back to the beginning of year 2, and then discounting that sum as a lump sum back to the beginning of year 1, yields a present value of $209,364.96. The sum of these two present values, $443,842.46, is the Net Present Value of the budgeted or as-planned cash flow for years 1 and 2. Calculations are shown in Figure 10.21.

STEP 2

Calculate the Net Present Value (NPV) of actual cash flow. As shown earlier, because of the delay to Project C, the Contractor's monthly net cash flows were reduced in year 1, but they increase in year 2. The same procedure is followed in this step as in Step 1. First, year 1 actual monthly net cash flows are discounted to the present at a 12 percent cost of capital. Next, year 2 cash flows are twice discounted as in Step 1 to arrive at a present value for the year 2 actual monthly net cash flows. Calculations are shown in Figure 10.22.

STEP 3

Determine the indirect effect of disrupted cash flows. This step involves simply subtracting the NPV of the actual cash flow from the NPV of the budgeted cash flows to determine the effect on the Contractor's cash flow, as shown in Figure 10.23.

Estimating Contractor's Indirect Impact Using Net Present Value Analysis
Step 1
Determine Present Value Budgeted Cash Flow
Year 1 Cash Flow
PV^1 = $20,833 X PVIFA
PV^1 = $20,833 X 11.2551
PV^1 = $234,477.50
Year 2 Cash Flow
PV^2(a) = $20,833 X PVIFA
PV^2(a) = $20,833 X 11.2551
PV^2(a) = $234,477.50
PV^2(b) = PV^2(a) X PVIF
PV^2(b) = $234,477.50 X .8929
PV^2(b) = $209,364.96
Net Present Value (as budgeted) = PV^1 + PV^2(b)
NPV = $234,477.50 + $209,364.96
NPV = $443,842.46 (as budgeted)

FIGURE 10.21

Estimating Contractor's Indirect Impact Using Net Present Value Analysis
Step 2
Calculate NPV of Actual Cash Flow
Year 1 Cash Flow
$PV^1 = \$15{,}000 \times PVIFA$
$PV^1 = \$15{,}000 \times 11.2551$
$PV^1 = \$168{,}826.50$
Year 2 Cash Flow
$PV^2(a) = \$26{,}667 \times PVIFA$
$PV^2(a) = \$26{,}667 \times 11.2551$
$PV^2(b) = PV^2(a) \times PVIFA$
$PV^2(b) = \$300{,}139.75 \times .8929$
$PV^2(b) = \$267{,}994.78$
Net Present Value (actual) = $PV^1 + PV^2$
NPV = \$168,826.50 + \$267,994.78
NPV = \$436,821.28 (actual)

FIGURE 10.22

Estimating Contractor's Indirect Impact Using Net Present Value Analysis
Step 3
Determine Indirect Impact to the Contractor Resulting from Disputed Cash Flows
Indirect Impact = NPV Budgeted – NPV Actual
= \$443,842.46 – \$436,821.28
= \$7,021.18

FIGURE 10.23

STEP 4

Determine the current value of the indirect costs. The indirect cash flow impact calculated in steps 1 through 3, \$7,021.18, represents the cost to the Contractor of disrupting the cash flow. This impact is valued in pre–year 1 dollars. The final step of this process involves calculating the current value of the pre–year 1 dollars, as shown in Figure 10.24. In this example, settlement is made at the end of year 2.

Estimating Contractor's Indirect Impact Using Net Present Value Analysis
Step 4
Determine Current Value of Indirect Impact
Indirect Impact per Year 1 Dollars = \$7,021.18
(from Steps 1–3)
Current Value = \$7,021.18 × CVIF (1.12 × 1.12)
= \$7,021.18 × 1.2544
= \$8,807.37

FIGURE 10.24

The resulting $8,807.37 represents the indirect cost to the Contractor for the Project suspension in year 1. Note that this figure cannot be calculated if the delay is analyzed using a pure accounting approach that does not consider the time value of money.

Calculating Indirect Damages Using Cost-Loaded Schedules

Of course, the preceding example was quite simple. The technique, however, is equally applicable to more complex situations. In some cases, a Contractor can project a cash flow using a computerized, cost-loaded, CPM schedule that becomes the as-planned part of the analysis. The pay requisitions that the Contractor submits serve as the as-built cash flow data. The analyst can calculate variances using these two pieces of data. The analyst would then have to evaluate these variances in light of the as-planned versus as-built analysis. The schedule analysis will show the liability for variances. From this information, the financial (direct cost) extent of the liability for schedule variances can be analyzed. The final step in this process is the calculation of the indirect affect using Present Value Analysis, as shown earlier. The following example illustrates this approach.

A Contractor has an as-planned cost-loaded CPM schedule for a Project. Based on this schedule, the following cash flow is projected:

Year 1: $1,360,000
Year 2: $2,540,000
Year 3: $ 950,000

Because of delays and changes by the Owner, the Project is performed later than planned and in a different sequence. The actual cash flow based on the pay requests shows the following:

Year 1: $ 900,000
Year 2: $1,000,000
Year 3: $1,950,000

Using the same method as shown earlier, the variance in years 1, 2, and 3 can be calculated and the NPVA performed to establish the costs incurred by the Contractor.

It should be recognized that the effect would only occur in the areas of overhead and profit. Costs for labor and materials, although shifted in time, should not result in damages to the Contractor, in that the Contractor merely passes along these dollars and does not have beneficial use of them. The NPVA process can also be used in situations where a Contractor's progress is not delayed across the board but rather only some work activities are halted. Once again, the schedule analysis, along with the Project documents, form the basis for assessing liability for the delays. Delays that extend throughout the Project duration but are not compensated for by change orders are evaluated on the basis of their effect on the Project cash flow. Of course, there is a difference between a partial delay and a complete suspension of work. The effect on cash flow may be analyzed either way. It may be necessary to apply the present value analysis to cash flow from different pay items to calculate the indirect effect of delays. The procedure for the line-by-line analysis for a partial delay remains the same as for a full suspension of work. However, this analysis is more complex.

Summary

Regardless of the method used to calculate home office overhead damages resulting from delay, it is necessary to demonstrate two fundamental points. The first is that the Contractor actually incurred a damage related to its home office overhead costs as a result of the actions or inactions of the Owner. Generally, that means that the Contractor's period of performance was extended with no commensurate increase in revenues from which home office overhead costs for the extended period were recovered. In some venues, this may be considered to have occurred only in a pure suspension period. In other venues, the Contractor may be able to demonstrate that a large delay during which only minimal change order work occurred still resulted in a damage or loss.

The second point is that the method chosen to calculate the home office overhead damages results in a reasonable estimate of the damages incurred. The most appropriate method will be a function of a variety of factors including the type of businesses the Contractor's company is engaged in and the historical trends of the company's revenues and costs. As with any damage being claimed, the Contractor has the burden of demonstrating that it actually incurred a damage of the degree being claimed.

Chapter | eleven

Inefficiency Caused by Delay

In addition to the delays and damages presented in previous chapters, there are other delay-related effects that may occur. High on this list is a decrease in the Contractor's efficiency caused by delays. The delay may either directly cause the inefficiency or be caused by the inefficiency.

WHAT IS INEFFICIENCY?

Perhaps the best way to define *inefficiency* is to start with a definition of *efficiency*. Efficiency is a measure of units of work performed per units of resources consumed to perform that work. Inefficiency (also referred to as loss of efficiency or lost productivity) is a relative measurement. An operation is inefficient when it consumes more units of resources to perform a unit of work than should have been consumed or than were consumed by the same type of activity performed at another time.

This chapter is not intended to explain every type of inefficiency or to present techniques for measuring productivity. Rather, it will show how a delay can affect the productivity of the job and discuss how that reduction in productivity can be measured, as long as accurate, contemporaneous records are maintained.

WAYS THAT DELAY CAN LEAD TO INEFFICIENCIES

It is not the intention of the authors to attempt to present every possible way that a delay can contribute to a loss of efficiency. Rather, some of the more common instances of delays contributing to inefficiency will be addressed so that the reader will have a better understanding of the relationship and be alert to it for specific Project situations. The following are examples of how delays can lead to inefficiencies.

Shifts in the Construction Season

A delay to a Project can shift work originally scheduled for one season into a different season. For example, work scheduled for late summer and early fall may be pushed into the winter months by a delay. The effect of the delay on the Contractor's efficiency depends on the type of work. Several examples follow.

EXAMPLE #1

A Contractor plans to complete all concrete operations before the winter season. A delay forces the Contractor to continue concrete work through the winter months in a cold-weather environment. As a consequence, the concrete crews do not work as efficiently as they would under ideal conditions, and the Contractor experiences an increase in the unit cost for placing concrete. The Contractor is also forced to change the concrete mix design to include accelerators, which further increases the unit cost, in this case, for materials. Finally, the Contractor must use winter concrete placing techniques, including extra winter protection and steam curing in some instances. All of these extra items were the direct result of the shift in season caused by the initial delay to the Project.

EXAMPLE #2

A highway Contractor plans to complete all paving operations before the winter, which marks the seasonal shutdown of local asphalt plants. The Project is initially delayed, and as a result, the Contractor cannot finish paving before the winter begins. Because the asphalt plants shut down and the Owner's specifications do not allow paving from November 1 to April 1, the initial delay is compounded by the winter shutdown period. The Contractor must now finish the work during the next season. In this case, there may not be a direct inefficiency in the Contractor's labor or equipment productivity, but the Contractor experiences additional demobilization and remobilization costs. It is also possible that some loss of efficiency may result because new workers may have to be trained and, initially, may have a less productive period before reaching peak productivity levels.

EXAMPLE #3

A Contractor performing a heavy-earth-moving operation is delayed so work that was to be performed during a relatively dry season is forced into a wetter season. The earth-moving operation is adversely affected by muddy conditions in the cut and fill areas. In wet weather, the overall productivity of cubic yards per crew/equipment day is reduced.

EXAMPLE #4

An HVAC Contractor is scheduled to install the heating system in a building to be operational by March. Because of some initial delays, the work is resequenced and the HVAC Contractor must accelerate the work to ready the system for operation by the end of November. Since the Project is in a cold-weather climate and the subsequent crews will be working inside by December 1, the building will now be a heated structure in which to work, which would not have been the case according to the original schedule. The result should be an increase in productivity.

Numerous other scenarios could develop from the shift of work from one season to another. The important issue for the delay analyst is to assess whether a delay caused the operations to shift into another season and if that shift had any effect on productivity. When work is shifted into adverse seasonal conditions, the analyst should evaluate the work that was shifted from that season into more favorable conditions.

Availability of Resources

At times, delays can affect the availability of resources in the areas of manpower, subcontracts, or equipment. The following examples illustrate the effects of unavailable resources.

EXAMPLE #1

A General Contractor plans to complete a Project in April. Because of a delay, the work extends into the summer. However, because the construction workload in that location is at its peak during the summer, there is less available labor from which to draw. Therefore, the Contractor may not be able to obtain enough labor to finish the work by the revised schedule for completion. This is particularly true of weather-related work such as exterior painting, site work, and landscaping. The inability to complete the work according to the original schedule may be a compounding delay and not have a component of inefficiency. Conversely, the Contractor may hire less-experienced crews or "travelers" and have either reduced efficiency for a portion of the work or a higher unit cost for the work.

EXAMPLE #2

An earth-moving Contractor plans to excavate several hundred thousand cubic yards of material using scrapers. The Project is delayed at the beginning. By the time it gets under way, the scrapers are committed to another Project and are no longer available. Consequently, the Contractor must either rent equipment at a higher cost than his owned equipment or use loaders and dump trucks to move the material. The productivity resulting from the use of loaders and dump trucks is significantly lower than that originally planned based on the use of scrapers, and therefore the operation is more costly.

EXAMPLE #3

A Contractor constructing a bridge must schedule a portion of the work during a specific interval because of the availability of certain equipment—for instance, the use of a snooper crane. Because of a delay to the Project, the work shifts and the equipment is no longer available. The Contractor must now perform the work using a new method, thereby increasing Project cost. When a delay occurs, the analyst must look closely at exactly what the effects are on resources, such as equipment and manpower, and how to quantify those effects.

Manpower Levels and Distribution

Certain types of delays affect the level of manpower and its distribution on a Project. These changes may occur in the form of additional manpower, erratic staffing, or variations in preferred/optimum crew size. Any of these situations may affect the level of efficiency of the work.

Additional Manpower

Delays to specific activities may force the Contractor to work on more activities than planned at one time and to increase the levels of manpower significantly for a specific trade. Depending on the union rules, additional manpower may also require more foremen or master mechanics.

Also, as the Contractor increases the crew size, it is not uncommon for the added personnel to be less productive than the original crew. Contractors often say that as they draw more personnel from the union hall, they see a decline in the level of productivity.

Erratic Staffing

In the face of a delay, a Contractor may staff a Project erratically in order to address specific needs as they arise. Theoretically, a Contractor would like to staff a Project in a bell curve fashion: starting with a small crew, building up to optimum size, and then tapering down toward the end of the Project.

Constant fluctuations in the size of the crew on the site are not desirable. However, the Contractor may in some circumstances be forced to man the Project erratically in order to achieve schedule goals. In such situations, there may be a measurable reduction in efficiency.

To demonstrate the negative effect of a forced change in labor distribution, the Contractor would be well advised to plot the original schedule to graphically portray the planned distribution of labor and then plot the actual distribution of labor caused by the delay and compare the two.

Preferred/Optimum Crew Size

Another factor that should be considered is preferred/optimum crew size. For example, a finish Contractor has a standing force of eight carpenters employed through the year. Because the crew works together throughout the year, they have established a smooth and efficient routine. If a delay now causes that Contractor to accelerate his work and increase his staff above his optimal crew, there can be some measured loss of efficiency as the original crew assimilates the new personnel and brings them "up to speed."

Sequencing of Work

Delays to critical and noncritical activities can also force a Contractor to resequence the work. The resequencing itself is not a problem, but its effects may reduce the Contractor's productivity in a number of ways. The Contractor's crew may be hampered in their work by the presence of another trade, or the crew may be obstructed by material stockpiled in the work area. With such interferences, workers may experience some reduction in productivity.

QUANTIFYING INEFFICIENCY

There are many ways in which a Contractor's work can be affected because of changes to the work schedule. The delays may cause these problems directly or indirectly. The delays may be to critical or noncritical items. The Contractor must be able to measure and demonstrate how the delays adversely affected the workers' productivity if it is to be compensated for the additional costs. There are several methods for quantifying productivity loss. The delay analyst should be aware of each of these options. The following list ranks the different methods by their accuracy in measuring losses in productivity:

1. Compare unimpacted work with impacted work.
2. Compare similar work on other projects with the impacted work on the Project in question.
3. Use statistically developed models.
4. Use expert testimony.
5. Refer to industry published studies.
6. Use the total cost.

Compare Unimpacted with Impacted Work

The impacted versus unimpacted method, usually referred to as a measured mile, is the preferred method to measure losses in productivity. The Contractor must show a comparison between unimpacted and impacted work. For example, if a Contractor's work is shifted into a cold-weather season, it should compare the work during the cold-weather season with productivity during the more favorable weather. Of course, the comparison must be made on the same type of work.

Here is a more specific example. A Contractor plans to set reinforcing steel during the summer. A delay pushes this activity into the winter months. The Contractor's records show that during the favorable weather, the work crews were able to set two tons per crew-day. During the less favorable weather, however, the same crews are able to set only 1.5 tons per crew-day. Thus, the loss of productivity was 25 percent.

To measure productivity in this manner, all information must be recorded in a form that can be converted into productivity units. The other methods mentioned become less accurate as one progresses down the list. Total cost is the least desirable. The reasons for this can be seen in the following example.

Total Cost Method

In the total cost method, a Contractor argues that it estimated a certain cost for its work. Because of the delay and the subsequent inefficiency of a shift in work seasons, the actual cost was higher. Therefore, the Contractor claims the difference in damages. This method is carried out as follows:

Actual cost of paving operation: $1,975,000
Estimated cost of paving operation: $1,250,000
Damages claimed because of inefficiency: $725,000

This method assumes that the Contractor's estimate was accurate. It also assumes that the Contractor in no way contributed to the reduced efficiency and that all additional costs are solely attributable to the delays cited. All of these assumptions may be challenged.

This chapter is not intended to be a treatise on the subject of inefficiency or on the techniques for measuring productivity. Rather, the intent is to point out that a delay may adversely affect productivity on the Project. Also, it must be recognized that detailed, accurate, and contemporaneous information must be maintained in order to measure productivity impacts associated with a delay.

QUANTIFYING THE COSTS OF INEFFICIENCY

The costs associated with inefficiency are direct costs. Since we are discussing delays as the catalyst for the inefficiency, all indirect costs should be associated with the period of delay. Therefore, the costs associated with inefficiency will be directly related to costs of labor, equipment, or materials. As such, if the analyst can reasonably measure the magnitude of the loss of efficiency, the cost calculations are straightforward.

Chapter | twelve

Acceleration

WHAT IS ACCELERATION?

The *Merriam-Webster, On-line Dictionary* defines *accelerate* as "1: to bring about at an earlier time; 2: to cause to move faster; 3a: to hasten the progress or development of." All three definitions convey the idea of completing work in less time. These definitions certainly apply to the progress of a construction Project relative to the scheduled completion date. A construction Project is accelerated when the Contractor must complete its original scope of work in less time. However, a construction Project may also experience acceleration when the scheduled completion date is unchanged. Two examples are performing additional work on the critical path of the Project within the same Contract performance period and performing noncritical items of work in less time than planned.

The following example illustrates acceleration of noncritical work. The example also makes the important point that acceleration and increased costs resulting from acceleration are not limited to work on the critical path.

EXAMPLE 12-1

The Contractor's schedule of May 1, 2007, shows a noncritical path of work through the construction of a bridge pier in a river that is scheduled to finish on September 25, 2007. For environmental reasons, the Contract prohibits work within the river from October 1, 2007, to April 30, 2008. After starting pier work in May 2007, a differing site condition suspended work on temporary cofferdams for ten calendar days. Although the suspension was not a critical delay, the delay would push the scheduled completion of the pier until after September 30, 2007. The delay increases the risk that the cofferdams would have to be repaired due to damage over the winter. To eliminate the risk of damage, the Owner may choose to direct the Contractor to accelerate the pier work by five calendar days to finish the river pier before October 1, 2007.

Many people consider acceleration the result of a change. While it is possible that acceleration may result from a change, a better view of acceleration is that it is a Contract change. Like any change to a construction Contract, the Contractor must show that the acceleration on a Project was a change in accordance with the Contract clauses. After establishing that acceleration was a change, the Contractor must also show that the acceleration had some definable effect and resulted in additional costs. The following two examples illustrate the concept of change, effect, and damages.

EXAMPLE 12-2

A Contractor planned to perform two five-day activities sequentially that required different crews and equipment because the work for each activity was in the same location. To recover a two-day delay caused by the Owner, the Contractor actually performed the two activities sequentially but with 12-hour workdays, thus accelerating the work. The Contractor worked each activity four days with 12-hour workdays for a total of eight days, recovering the two-day delay.

The scope of work for the two activities was unchanged, and there was no delay to the Project. The acceleration was a change because the Contractor changed its plan and performed the two activities by utilizing longer workdays. The effect of the acceleration was that each activity required more man-hours than planned (48 man-hours versus 40 man-hours) and premium rates. The damage would be the premium cost and any measurable inefficiency of the increased daily man-hours for labor.

EXAMPLE 12-3

A Contractor planned to perform two five-day activities that required different crews and equipment sequentially. To recover a five-day delay caused by the Owner, the Contractor actually performed the two activities at the same time, thus accelerating the work. The Contractor worked on both activities for five days, recovering the five-day delay.

The scope of work for the two activities was unchanged, and there was no delay to the Project. The acceleration was a change because the Contractor changed its plan and performed the two activities at the same time. The Contractor was able to complete each activity in the planned duration without additional labor and equipment costs. The acceleration did not affect the Contractor's work, did not increase the Contractor's resources, and did not incur any increased costs for the Contractor.

WHY IS A PROJECT ACCELERATED?

As indicated previously, a Project is accelerated when there is a need for the Contractor to complete some portion of the work in less time. The most common reasons a Project is accelerated relate to money. This includes saving money by avoiding delay damages or reducing overhead costs. It also includes making more money by allowing an earlier income from the facility or by freeing the Contractor to begin other work. Sometimes, acceleration is required to meet some other need, such as the early use of the facility or a commitment to a user. A Project is accelerated when it is necessary for the Project to complete more quickly than it would otherwise. Most of the time, projects are accelerated because they are behind the required Project completion date.

CONSTRUCTIVE ACCELERATION

While Owners should grant time when it is due, sometimes an Owner will not accept or resolve a legitimate time extension request. If the Contractor is due an extension to the Contract time but is not provided one and later accelerates its work in order to finish in the time provided, the Contractor may have been constructively accelerated. Constructive acceleration, similar to a constructive change, is subtle and less readily recognized by an Owner. Let's look at two similar situations to help distinguish between directed acceleration and constructive acceleration.

EXAMPLE 12-4

The Contractor experiences an excusable 20-day delay associated with revised sitework drawings. The Contractor properly notifies the Owner and requests a time extension, which the Owner grants. Later, the Owner directs the Contractor to accelerate the Project by 20 days. The Contractor accelerates the work, makes up the 20 days of delay, and finishes the Project on time. This is directed acceleration. The Owner is likely to recognize and assume liability for the costs associated with the directed acceleration.

Say we have the same excusable 20-day delay, but the Owner refuses to grant the requested time extension. In addition, the Owner requires the Contractor to finish within the original Contract time. The Contractor accelerates the work despite protesting that it is due the time extension, makes up the 20 days of delay, and finishes the Project on time. In this instance the Contractor may have been constructively accelerated. Often Owners do not recognize and assume liability for the costs associated with constructive acceleration.

While the exact requirements may vary by jurisdiction for constructive acceleration, the Contractor will typically be required to show the following:

1. It was entitled to an excusable extension of time.
2. It properly requested a time extension in accordance with the Contract.
3. It had its time extension request denied or ignored.
4. It was directed by the Owner to finish in accordance with the schedule excluding the excusable delay period.
5. It accelerated the work on the Project.
6. It incurred additional costs in accelerating the work.

If all of these requirements have been met, the Contractor may recover for the additional costs incurred. It should be noted that it is not necessary for the Contractor to complete the Project by the earlier date required by the Owner. As long as the preceding requirements are present, then constructive acceleration may exist.

HOW IS A PROJECT ACCELERATED?

In accelerating a Project, the critical work must be performed more quickly, and/or the sequences must be changed to allow more of the critical work to occur at the same time. There are several potential ways to accomplish this: change the sequence of activities in the schedule, increase manpower, add equipment, change the materials used, change the method of construction, or improve productivity. We say "potential" because any effort to accelerate may not necessarily be effective. Different methods of acceleration can be combined. For instance, you can increase both equipment and manpower to accelerate a Project.

A common method to accelerate a Project is through the addition of more manpower. This comes in many forms, including increasing the daily work hours, increasing the number of workdays, adding additional shifts, increasing the number of workers, or any combination of increased hours, days, shifts, or workers. The hope is that if manpower is increased, the rate of work in place would increase as well. However, while the rate of work in place may increase with increased resources, the relationship may not necessarily be linear. In other words, if you double resources, you may not double output. This occurs because the productivity of the work is often affected by the addition of manpower to a Project. While productivity might improve, more commonly the addition of manpower will result in a decrease in productivity. Lower productivity may occur for a variety of reasons; some examples include less available workspace, overlapping of trades, worker exhaustion, learning curves for new employees, or lack of supervision or introduction of new supervision. If productivity is lowered due to the addition of more manpower, it can limit acceleration efforts. It is also important that the addition of more manpower improves the critical work. If the materials are not being distributed on site because the material lift is at maximum capacity, then increasing the workforce will have little chance of accelerating that work. When attempting to accelerate through increased manpower, it is important to monitor productivity and make sure the critical work will benefit from increased manpower.

The addition of more equipment is another way to accelerate. More equipment, assuming the crews are available, should result in increased production. Time savings from equipment will be limited to the work activities affected by the equipment, and more equipment will often necessitate additional crews. So having an extra backhoe helps to accelerate if the work affected by the backhoe is critical and there are enough workers available to support it. When using equipment to accelerate, recognize that only certain areas of work may be affected and that when additional equipment is used, additional manpower may also be required.

Materials can be changed to accelerate the work. One common example is the use of accelerators in concrete or high-early-strength concrete. Other examples might include using expansion anchors in place of epoxy anchors, ordering stock marble instead of custom, using laminate floors instead of marble. The use of different materials either accelerates the work by being easier to incorporate into the Project or by being more readily available. Much like the previous example, the use of different materials will only affect certain work activities on the Project and may require specialized training or different skills than currently available on the Project.

The method of construction can be changed to accelerate the work. Some examples may include changing from welded to mechanical connections, the addition of more false work to limit or remove an interim phase, changing an application method from rolled to sprayed, or using precast concrete members instead of cast in place. Changing the method of construction to accelerate is effective only to the extent that the new method of construction affects the critical path of work and takes less time than what was originally planned.

A Project can be accelerated through improved productivity. Though not exhaustive, productivity can be improved through better equipment and tools, optimizing crew sizes, incentives, better supervisors, training, proper work sequencing, a specialized workforce, clear direction, and good planning. And as with every method used to accelerate, this will only be successful to the extent it affects the critical work.

Though not necessarily a way to accelerate in and of itself, planning and managing acceleration is an essential part of the process. While it is easy to resequence three areas of work from occurring sequentially to occurring concurrently, a time savings will not occur unless the work is critical and you can actually staff all three areas effectively. A successful acceleration effort will identify the critical path of work and focus on selectively applying resources toward shortening the critical path's duration.

QUANTIFYING THE TIME SAVINGS ASSOCIATED WITH ACCELERATION

Often, managers attempt to measure the time savings gained through acceleration by comparing the Project's original completion date to the actual completion date. Unfortunately, if you just measure the difference in the end dates, not only do you have to wait until the end of the Project, but you are likely to end up with the wrong answer. If a Project is accelerated to make up for existing delays or if delays or improvements are experienced after the acceleration effort is put into place, then the difference in the end dates would be a compilation of the acceleration effort, Project delays, and perhaps even some Project improvements. This would not be an accurate measure of the time savings associated with acceleration.

In order to isolate the time savings associated with acceleration, you need to determine the difference in the duration of the critical path activities before and after the acceleration. Since the schedule is the tool used to identify the critical path, it is the schedule that is used to quantify the time savings associated with accelerating. In practice, the schedule is updated and then revised based on the acceleration plan. The difference in length of the critical path activities being accelerated prior to and after the schedule revisions is the time savings associated with acceleration.

By using the schedule in this manner, changes in the critical path due to acceleration will be identified. Once a path is accelerated to the point where the critical path shifts, activities on the new critical path will also have to be accelerated. An example of this is provided later in this chapter.

QUANTIFYING THE COSTS OF ACCELERATION

The cost of acceleration is the difference between what it would have cost to do the work as originally planned versus what it will cost to do the work in the accelerated time frame. After acceleration has occurred, costs can be evaluated using actual data, or, prior to the acceleration, costs can be put together based on detailed estimates. In assessing acceleration costs, there are several categories that need to be evaluated, including additional material costs, labor premiums, inefficiency, additional equipment costs, and other miscellaneous expenses.

Additional material costs are simply the difference in the cost of the materials that would have been required to execute the work prior to the acceleration plan versus the cost of the materials needed to perform the work after the acceleration plan is put in place. If an additive is used in order to accelerate the concrete cure time, then the additive is an additional material cost. If the plan is changed to include temporary falsework, the falsework is an additional material cost. Any cost for more or better material that is a direct result of the acceleration plan is an additional material cost.

Labor premiums are additional costs associated with the manpower needed to accelerate. Overtime and holiday pay are examples of labor premiums. Others include having to use higher-paid employees or higher-paid Subcontractors. On some projects, a large-scale acceleration effort can affect the prevailing rate of local labor as the demand for workers exceeds the supply. When the average cost of an hour of labor is increased as a result of the acceleration effort, the difference in the old and new hourly rate is part of the labor premium.

One thing Owners should be cautious of when accepting an overtime premium is the flat 50 percent markup on the accepted or average burdened labor rate provided to them previously on the Project. While it is true that an employee will typically receive time-and-a-half for overtime, the increase is applied to base salary and may not affect the overhead and benefits package that are included in the burdened rate originally provided.

Let's start with an example of labor premium costs that will be expanded to include other acceleration costs through this section. A framing Subcontractor originally planned to frame the second floor with two four-man crews working eight hours a day for ten workdays. Instead, the Subcontractor will increase to three four-man crews working 12 hours per day and complete the work in five days. Prior to acceleration, the average burdened labor rate was $15 per hour, and after acceleration the average burdened labor rate was expected to be $20 per hour. In addition, a third supervisor was required. Thus, the cost of each hour of labor was expected to increase $5 per hour due to the acceleration, and a third supervisor will be needed for a week.

In many cases, as you increase your manpower, the efficiency of the work suffers. In the previous example, the Contractor had originally planned to frame the second floor with two four-man crews working eight hours a day for ten workdays (640 man-hours) but will instead increase to three four-man crews working 12 hours per day to complete the work in five days (720 man-hours). As a result, the same work is expected to take 80 man-hours longer to perform. This 80 hours is inefficient time associated with the acceleration.

Another common cost of acceleration is additional equipment. Again, using the previous example, to support the additional crew, the framing Contractor had to purchase two new nail guns and rent a third compressor. The additional rental cost of a compressor and any associated delivery charge may be a legitimate acceleration cost. The nail guns are questionable, since the Contractor may keep them and gain the full benefits of their use over time. On the other hand, this Contractor may have no need for new nail guns and no desire to run three crews on future projects. In that case, the nail guns may also be a legitimate cost of the acceleration.

Miscellaneous costs may include the costs of using express mail, the housing of additional staff and labor, the administrative costs of planning the acceleration and revising the schedule (if allowed by Contract), markups for profit and overhead, additional cleaning costs, running additional temporary power, evening meals, and other miscellaneous expenses that would not have been incurred if the work had not been accelerated. Sometimes, additional supervisors are captured as a labor expense, and sometimes they are identified as an overhead item.

One area often overlooked, is the savings associated with acceleration. The acceleration effort will decrease the amount of time required to complete the work, and as a result, any time-related costs, as discussed earlier, should also be decreased. Continuing our previous example, the Contractor may be able to return the two compressors it originally rented a week (five workdays) earlier. Likewise, the supervisors, who are salaried employees, should also be finished with this job a week earlier than originally planned. Using the information from the previous examples, the acceleration costs for the framing Contractor to accelerate its work from ten workdays of duration to five workdays are estimated as shown in Figure 12.1. Thus, the total cost to accelerate the framing work by five days was $5,500 dollars.

Managing Acceleration

Before embarking on an acceleration of a Project, the manager needs to carefully plan what tasks will be accelerated and how the acceleration will be effected. All too often when a decision is made to accelerate a Project, the response is to go to overtime or increased workweeks for every facet of the Project. This may be an unnecessary waste of effort and money. The focus on any acceleration is to shorten the time on the critical path at the lowest cost.

Item	Cost
Labor premium	
$5/hour for 640 hours	$3,200
Additional supervisor for 1 week	$1,465
Inefficiency	
$20/hour for 80 hours	$1,600
Additional equipment	
One week air compressor rental	$116
Two nail guns	$246
Miscellaneous	
Evening meals, 5 days @ $50/day	$250
Savings	
Supervisor A 1 week	$(946)
Supervisor B 1 week	$(916)
note that the original supervisors were less costly than the late addition	
Air compressor rental (2 @ 1 week each)	$(232)
Subtotal	$4,783
Profit and overhead @ 15%	$717
Total	$5,500

FIGURE 12.1

One of the preferred approaches to determine the most cost-effective way to accelerate work is to use a cost slope calculation. The basic equation is:

$$\text{Cost Slope} = \frac{\text{Crash Costs} - \text{Original Costs}}{\text{Original Time} - \text{Crash Time}}$$

The Crash Time is the absolute fastest time that the activity can be performed. The Crash Cost is the cost of performing the activity within the Crash Time. The Original Costs and Original Time are those that existed prior to any consideration for acceleration. Using the previous example:

$$\text{Cost Slope} = \frac{\$5{,}500}{10\text{ WDs} - 5\text{ WDs}} = \$1{,}100/\text{day}$$

The Cost Slope then tells the manager the incremental costs per day for any specific activity to shorten that activity by one day. The Cost Slope of the individual work activities and the Project schedule can be used to plan the Project's acceleration. This is demonstrated in the next example.

EXAMPLE 12-5: USING THE SCHEDULE TO ACCELERATE INTELLIGENTLY

The Project in Figure 12.2 has nine remaining work activities. The critical path is highlighted in red. The Project is currently expected to finish in 32 days. The Owner has directed the Contractor to accelerate the work so the

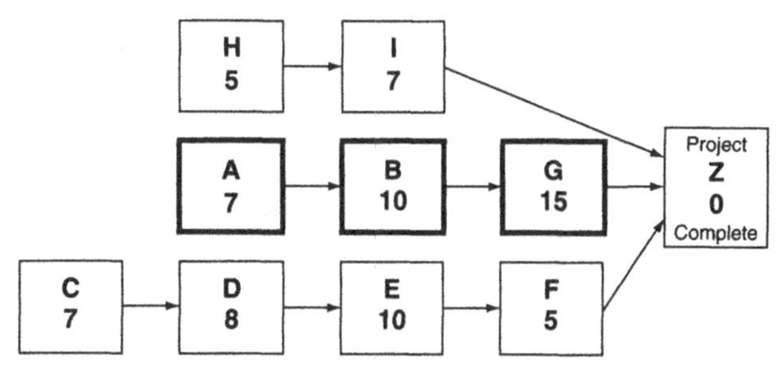

FIGURE 12.2

Activity	A	B	C	D	E	F	G	H	I
Cost Per Day	$5K	$10K	$6K	$500	$5K	$2K	$1K	$500	$2K
# Days Possible	2	2	2	2	2	1	3	2	2

FIGURE 12.3

Project finishes in 25 days. Therefore, the Contractor must shorten the overall duration by seven days. Figure 12.3 shows the daily incremental cost (cost slope) and the maximum number of days that each of the work activities can be reasonably accelerated.

In looking at the chart, the most affordable work activity to accelerate along the critical path would be activity G. We can accelerate this by two days before the critical path shifts to include another path of activities. After accelerating activity G by two days, our schedule would look like Figure 12.4.

Thus, after the second day of acceleration, it will be necessary to accelerate both the path that begins with Activity A and the path that begins with Activity C in order to improve the completion date. The activities accelerated and the cost to do so are summarized in Figure 12.5. As you can see, the total cost to accelerate the Project by seven days is $46,000.

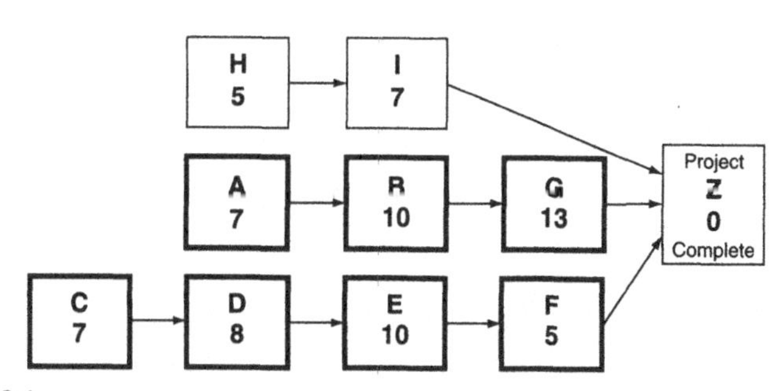

FIGURE 12.4

From Day	To Day	Activity(ies) Accelerated	Cost
32	31	G	$1K
31	30	G	$1K
30	29	G & D	$1.5K
29	28	A & D	$5.5K
28	27	A & F	$7K
27	26	B & E	$15K
26	25	B & E	$15K
		Total	**$46K**

FIGURE 12.5

It is worth noting in this example a few characteristics common to projects that are attempting to accelerate. When accelerating intelligently, each day may become more expensive than the preceding day. Accelerating some areas of work will be a waste of money (activities H and I in the previous example). Sometimes the work that is currently ongoing should be left alone in favor of accelerating future less-costly work (activity C in the previous example). By applying a reasoned plan to acceleration, the Project will avoid unnecessary expenses and wasted effort.

Chapter | thirteen

Other Categories of Delay Damages

As the reader may now understand, the calculation of delay damages is as much an art as a science. The appropriate damage calculations are Project specific and situation specific. No book can address every combination and permutation. This chapter addresses certain points that, although not obvious, are important in defining damages related to delays. Included are some situations that frequently occur but are not well understood.

DAMAGES ASSOCIATED WITH NONCRITICAL DELAYS

Thus far, this book has focused on delays to the critical path or delays that are associated with a delay to the overall Project. That does not mean, however, that delays to noncritical activities may not also cause damages. Activities not on the critical path can be delayed and can have damages without ultimately affecting the completion date. The following example illustrates this concept.

EXAMPLE 13-1

A Contractor has a Contract for the construction of a hospital complex. The complex consists of three buildings. Two buildings (A and B) already exist and must be renovated. The third (C) is new construction. The Contractor schedules the Project using the critical path method. The schedule shows that the critical path of the overall Project is controlled by the construction of the new building, C. The other two buildings need only be completed within that overall duration and have 12 months of float (see Figure 13.1).

Early in the Project, the Owner discovers asbestos in buildings A and B. The Contractor cannot proceed until the Project Architects develop a method for the safe removal of the asbestos. The hold on the two buildings remains in effect for ten months, at which time an acceptable method is devised and the

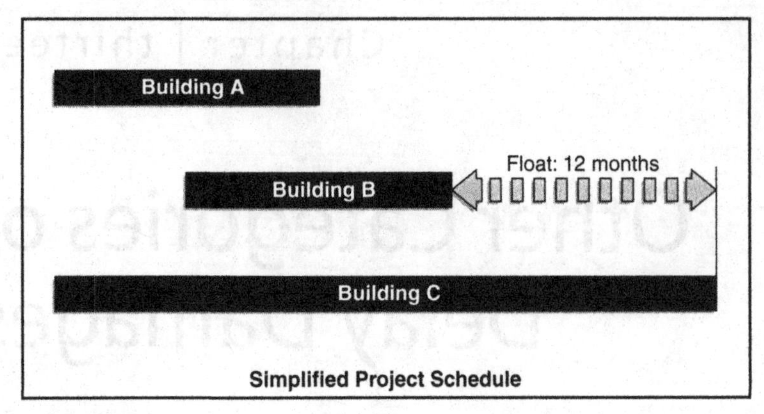

FIGURE 13.1

Contractor is released to continue work. In the interim, the Contractor works on the construction of building C. The Owner and Contractor meet to negotiate a change order for the delay and the extra work associated with removing the asbestos in buildings A and B. The direct cost of the asbestos removal is a straightforward calculation. The Owner and Contractor can agree on the work involved and the cost of the work. The Contractor, however, requests additional compensation for the delay to buildings A and B. The Owner argues that the overall Project was not delayed and, therefore, the Contractor is not due any delay damages or any extra cost associated with the delay.

Did the Contractor experience any damages from the delayed start of these two buildings? To answer this question, the Contractor must establish the effect of the delay and the corresponding damages. The Contractor explains that the work sequence in the original schedule would have been more efficient and that the delay affects (1) the escalation of labor, (2) the additional supervision, and (3) the reduced efficiency.

Escalation of Labor

The Contractor had originally planned to work each trade through the three buildings in a consecutive sequence. For instance, the drywall crew was to work building A first, then move to building B, and then move to building C. This same sequence was planned for electrical, HVAC, plumbing, millwork, flooring, and painting.

Because of the delay, the Contractor now must work several crews in all three buildings at the same time to meet the Project completion date, instead of moving the same crew from building to building. As a result, the distribution of labor shifts to a later time frame. The Contractor can document a wage increase of $1.25 per hour for the sheetrockers. The shift in labor is plotted on a graph as shown in Figure 13.2. The graph shows that 1,560 man-hours have been shifted to a later time frame.

For sheetrock, the Contractor claims damages of $1,950 (an increase of $1.25 per hour for labor for 1,560 man-hours). The Contractor performs the

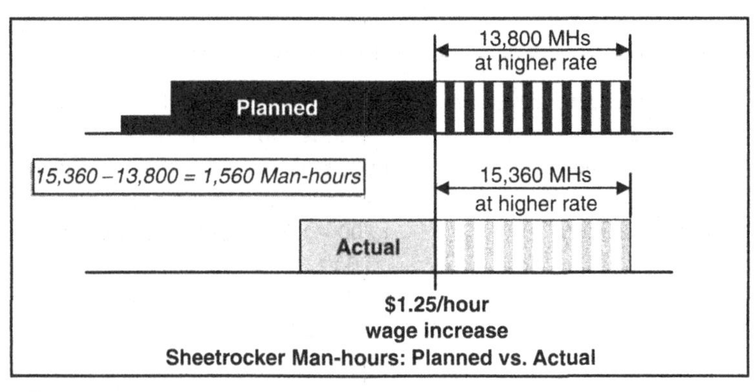

FIGURE 13.2

same analysis for each trade affected by the change in sequence. These calculations would be similar to that shown for the sheetrockers in Figure 13.2.

Additional Supervision

To allow for the inclusion of supervision costs, the Contractor explains that when crew sizes increase, additional nonworking foremen must be added. The Contractor calculates additional supervision for the affected trades and summarizes the claim as shown in Figure 13.3.

Reduced Efficiency

Finally, the Contractor argues that if it had been able to use the same crew throughout all the buildings, it would have absorbed the initial mobilization and learning curve to become most productive. Because each crew had to go through the mobilization/learning curve, the result is a lower productivity level than that originally planned. The Contractor argues that reduced productivity represents an increase in cost.

Additional Supervision
Trade: *Sheetrockers*
Crews Planned: 2
Duration Planned: 23 months
Thus, 1 nonworking foreman for 23 months

Crews Actual: 4
Actual Duration: 15 months
Thus, 2 nonworking foreman for 15 months each

Additional Supervision
30 man months – 23 man months = 7 man months
Damages 7 man months × $2,950/month = $20,650

FIGURE 13.3

Loss of Efficiency	
Demonstrated productivity by benchmark crew	1,200 sf per crew-day (based on daily reports and labor tickets)
Productivity of additional crews	1,000 sf per crew-day (based on daily reports and labor tickets)
Loss of efficiency	1,200 – 1,000 = 200 sf per crew-day 200/1,200 = 16.7%
Additional crews gross costs	$9,600 per month × 28 months = $268,800
Damages	16.7% × $268,800 = $44,890

FIGURE 13.4

Unfortunately, productivity is a very difficult item to document. In order to raise the Owner's level of confidence, the Contractor has used the best crew for each trade as a benchmark to measure the learning curve/startup effect and the difference in productivity caused by added crews. Figure 13.4 is an example the Contractor developed for the drywall crews.

Equipment

The preceding example shows how a delay to a noncritical activity can significantly affect the cost of the work without delaying the overall Project. The cost of work may also go up due to a change that affects equipment usage but does not involve a critical path activity. For example, if a delay to a noncritical activity forces the Contractor to mobilize an additional piece of equipment, the Contractor can claim compensation. Only the mobilization and demobilization costs would be a valid damage, since the cost of using that equipment (once mobilized) should be included in the Contract sum.

A Final Word on Noncritical Delays

Many Owners argue that the concept of damages associated with noncritical delays is really an argument of "who owns the float." Most contracts do not address who "owns" the float or the consequences if float is used by the Owner. In reality, the Contract owns the float, and either party may use it as long as it does not financially affect the other party. Some Owners claim they own the float and, if they use it, the Contractor will receive no additional compensation. This is a form of a no-damages-for-delay clause. The Owner should hire qualified counsel to review such a clause before inserting it into the Contract.

CONSULTING AND LEGAL COSTS

In general, the costs for attorneys and consultants are not recoverable in a claim situation. This does not mean that they will not be collected in a settlement but that they are not usually awarded in litigation or arbitration. The party asserting the claim should include legal/consultant costs as valid elements of its claim, recognizing the limited chance of recovery.

Although these costs are typically disallowed in litigation, the Contract may allow reasonable costs of experts expended to support a change order under the changes clause. In this case, the consulting costs would typically be connected to the administration, monitoring, or completion of extra work. The recovery of legal fees is generally precluded, but the Contract or legal statutes may provide for recovery.

LOST PROFITS/OPPORTUNITY COSTS

Contractors and Owners may each seek to recover damages associated with lost profits and lost opportunities. In general, these costs are extremely difficult to recover due to their remote and speculative nature. In order to achieve recovery, the Contractor or Owner must prove that the lost profits or lost opportunity costs are directly due to the Owner- or Contractor-caused delay. A well-supported claim that can clearly establish an economic loss (typical of a loss in bonding capacity) and prove that anticipated profits were not speculative may be able to recover these damages. Generally the period subsequent to delay would be limited.

Absent a Liquidated Damages clause in the Contract, an Owner may assert that a Contractor's delay caused it to lose profit. In that scenario, the Owner may argue that the facility could have produced a certain amount of revenue had it been completed on time. Most courts have held, however, that lost profits are highly speculative and subject to many diverse factors. Accordingly, courts have demonstrated a reluctance to award lost profit damages to Owners. The astute Owner will carefully consider this fact while drafting the Contract with the help of qualified counsel. If an Owner faces significant damages resulting from lost profits, the Owner should routinely consider including a Liquidated Damages clause in the Contract rather than relying on recovery of these damages through litigation.

A Contractor's chance for success in receiving compensation for its lost profits is no better than the Owner's. Some Contractors claim damages to other projects resulting from the delays to another specific Project. For example, a Contractor may claim that because of a delay to Project #1, it was unable to utilize equipment on Project #2, which began on the scheduled completion date of Project #1. As a result, it was necessary to rent other equipment at an increased cost. While this argument may seem viable, these types of costs are usually not recoverable because these damages were not reasonably contemplated by the Owner as a consequence of the delays caused. Therefore, they are typically not awarded. The Contractor is more likely to receive compensation for using rented equipment if it is employed on Project #1 and instead uses its owned equipment on Project #2.

Interest

A Contractor may incur interest or financing costs as a result of borrowing funds to finance the construction costs, including the damages identified in the claim. Although these interest costs may be a real cost or damage resulting from delays,

Contractors often are not successful in their recovery, as interest claims may be barred by the Contract or by statute. When allowable, interest may be claimed either on the overall value of the claim or as a component of the claim that represents the recovery of the cost of borrowed monies used to fund the work.

The interest claimed on the value of the claim may face several obstacles in recovery. Generally, interest on an unresolved claim represents prejudgment interest and is often excluded as an allowable damage either by the Contract or by statute. Contract agreements often preclude such interest charges but typically will allow interest to be paid once a claim amount has been successfully litigated or resolved. Generally, a negotiated settlement to a claim excludes interest. If a claim is litigated, interest may also be regulated by State and Federal laws, and interest rates may be set by statute.

When interest is included as a specific cost of the work, or the cost of funds borrowed to perform the work, the chances of recovery are increased. The Contractor must support his claim by identifying increased borrowing, and the measure of interest cost is based on the actual financing cost incurred. If such interest is measured as an actual cost, recovery is usually allowable, providing the Contractor has supplied adequate proof.

When measuring interest, the period of interest may vary based on the nature of the claim. The start date may be the date a payment became due or the date a claim was filed. Ending dates may be the date of recording the judgment or the date of actual payment. Other factors to consider include the type of interest allowable (compound or simple interest).

Many interest claims are not recoverable because the Contractor has not provided adequate support or has taken shortcuts in perfecting the claim. A successful interest claim must adhere to the Contract agreement and be adequately supported.

Chapter | fourteen

Determining Responsibility for Delay

Once the analyst determines the activities that were delayed, the magnitude of the delays, and the general nature of the delays (extended duration, late start, etc.), he or she can then assess the responsibility for each delay.

CONTRACT REQUIREMENTS

When determining who was responsible for delays, the analyst must first refer to the Contract. Generally, the analyst will look most closely at the General Conditions and Special Provisions of the Contract documents, as well as the Agreement or related documents and any referenced documents such as Standard Provisions that may not physically be included with the Contract documents. With respect to delays, the analyst will focus on clauses that generally address delays and time extensions, such as the changes clause, differing site conditions clause, time extensions clause, scheduling clause, clauses related to claims, and any exculpatory clauses such as a no-damage-for-delay clause. Based on these clauses, the analyst can determine what events allow for a time extension and the procedures to be followed to substantiate the justification for the time and how to determine the amount of time.

For the most part, delays to a Project that are the responsibility of the Owner are caused by some form of change, either directed or constructive. It could be a change in the design, an error or omission in the Contract documents, a differing site condition, failure to make approvals on time, failure to respond to requested information required to progress the work, or even stop work orders. Basically, some event in accordance with the Contract will give rise to the right to a time extension. The question whether a delay is compensable or not is also determined by the wording of the Contract. If it is determined that a delay was the result of an Owner-caused change, typically it would be excusable and compensable.

If a change occurs, the Contractor, in a timely manner, should document it in writing. The Contract may specify a set number of days that claims for additional time and compensation must be submitted to the Owner or its representative. The clause may further state that if the required claim information is not submitted within the time specified, then the Contractor forfeits the right to recover any additional time or money. Not only may the Contract specify a time limit for the filing of claims, but it may also specify exactly the information that the Contractor must submit when asserting a claim. Such information would include a clear statement of what is being claimed; an explanation of why the claim item differs from that already required in the Contract; references to the specific Contract clauses that apply; an explanation of the cause (or responsibility) for the claim; a clear definition of the specific effects associated with the claim (i.e., extra work, overtime, delay, etc.); and a detailed breakdown of the damages or extra costs with supporting information.

GATHERING THE FACTS

In addition to what is referenced in the clauses of the Contract, the process of assigning responsibility for a delay requires a thorough review of all the Project documentation to find out what factors influenced the performance of the specific activity. The facts required to assign responsibility will often come from prebid estimates and drawings; design drawings; preliminary schedules, updates, and revisions; change orders (COs); daily reports or diaries; general correspondence and e-mails; requests for information (RFIs); submittals; meeting minutes; photos and videos; and cost reporting data. For example, the Project daily reports may show that the Contractor experienced an equipment breakdown on two separate occasions, which caused the interruptions to a particular activity. Similarly, the analyst may discover that the late start of another activity was caused by the Project Designer's slow return of shop drawings.

For a specific delay, the analyst may find that no documentation exists that reasonably explains why an activity had an extended duration. Without documentation, it may be reasonable to conclude that the extended duration was the Contractor's fault. Either the scheduled duration was too optimistic or inadequate resources were applied to the task to accomplish the work in the time scheduled. Unfortunately, Project documentation seldom exists that will answer all the questions regarding the cause of delays for every activity.

Once the pertinent Project documents explaining a Project delay have been identified, it would be good practice to make copies of those documents and place them in chronological order for evaluating responsibility, asserting a claim, and quick reference during later review. The documents may be filed in chronological order as a reference to the delaying time period or the delaying issue. It should be kept in mind that when presenting a notice of delay or delay claim, the pertinent documents should be presented as exhibits.

EVALUATING RESPONSIBILITY

When evaluating the responsibility for a delay, it is necessary to review the Contract and other available Project documents just noted. These documents will enable the analyst to make a determination of the cause for the delay and who was responsible for that cause. As stated earlier, the delay analysis accounts for minor errors in the schedule. Essentially, this means that even if the delay analysis indicates that an activity in the schedule is delayed by a certain number of days, it may be the result of an issue not identified or logically related in the schedule. For example, the delay analysis may determine that the *Foundations at Building A* started ten days late. A review of all the available Project documents indicated that the equipment needed to excavate the footings was unable to mobilize to that portion of the scheduled work due to the ongoing work of the *Excavation and Placement of 48" RCP at Main Entrance*. Further review of the Project documents indicated that the Owner issued a change order for an additional 500 feet of 48-inch reinforced concrete pipe along the main entrance roadway. Even though these two activities were not linked logically in the schedule and were initially unrelated in the field, a review of the Project documents indicated that the Contractor could not start the *Foundations at Building A* until the installation of the 48-inch RCP was complete, which would then provide access for the equipment needed to install the Building A foundations.

WEATHER DELAYS

It is common for time to be "of the essence" in a construction Contract. It is also common for "Contract time" or the Contract completion date to be terms defined by the Contract. In addition, given the Contract's role in defining the sharing of risk, especially the considerable risk associated with weather, it is common for the Contract to specifically address how weather-related time extensions will be determined and administered.

It is important to know if, how, and in what form the Contract grants additional time for weather. The Contract language for granting extra time for weather delays can vary from allowing recovery of one workday for every workday lost to allowing no time at all. Typically weather-related time extension provisions only grant extra Contract time for unusually severe weather. When that is the case, the Contractor is required to consider and account for anticipated weather in its bid and work durations.

The ability to identify the number of days that the actual precipitation exceeded the monthly average or the number of days that the temperature fell below the norm is only the beginning. The Contractor must first demonstrate that the adverse weather experienced at the Project site limited its ability to perform work on those weather-affected days.

The second element the Contractor must demonstrate is that the unusually severe weather delayed the Project. This requires the Contractor to establish that the weather-affected work was on the Project's critical path, because only delays to the critical path can delay the Project. In most instances, the Contractor will need to rely on the Project schedule—hopefully a critical path method (CPM) network schedule—that allows the critical path to be identified easily to determine if the weather-affected work was, in fact, on the critical path and that the adverse weather resulted in a delay to the Project.

Let's reflect on each of the areas mentioned and gain a very simple perspective. We have the following elements that must be understood:

1. What constitutes weather that would allow for a time extension?
2. How is a time extension for weather effected?
3. If the Project experiences weather severe enough to allow a time extension, is that all that is required for time to be granted?

First, we need to review the Contract to determine exactly how weather problems are considered. As noted, the Contract may state that no time extensions will be granted regardless of the severity of the weather. In this case, the Contractor had best perform a very conservative assessment of the weather in the particular area of the Project and adjust its estimate accordingly. Admittedly, most contracts allow some relief for weather problems. The exact wording of the Contract must be clear and must be understood. Perhaps the most common approach to weather is for the Contract to allow excusable but noncompensable time extensions for weather-related delays. Normally, the Contract wording will refer to "unusually severe weather." A well-written Contract should further define what "unusually severe weather" means. It may clarify this as weather beyond that ordinarily anticipated in that location based on the National Oceanic and Atmospheric Administration (NOAA) historical records. If this type of wording exists, all parties to the Contract should refer to this standard in the resolution of weather-related time extension requests. Some agencies go as far as checking the weather experienced at the end of each month, and if a defined number of days exceeded the norm (as defined by the Contract), the agency will unilaterally grant weather-related time extensions each month. The more normal practice is for the Contractor to request weather-related time extensions.

Second, though the Project may experience weather of a nature that would qualify for a time extension, the mechanism for effecting a time extension must be spelled out in the Contract and must be followed. For example, as noted previously, normal practice requires the Contractor to request a time extension for weather-related problems. If the Contractor fails to make this request, it is possible that no time will be granted. Therefore, all parties need to understand the process for a weather-related time extension and must follow it as prescribed.

Third, merely because a Project experiences weather that would qualify for a time extension does not necessarily mean that one is due or should be granted. For example, if the Project is a midrise building and the building is completely enclosed and "water-tight," the occurrence of a significant rainfall may have no effect on the Contractor's ability to perform finish work inside the structure. Consequently, no time extension may be due. Conversely, if the Project is a heavy civil job involving earthwork, and significant rainfall occurs, it is possible that three days of severe rain may actually delay the Project for more than just three days. Additional time may be required to clean up the site and dry the soil such that work may continue in a productive fashion. One must also verify that the specific work that is adversely affected by the weather is on the critical path of the Project and will delay the Project completion date. If a Project experiences unusually severe weather that would qualify for a time extension, but the work affected had ample float with respect to the critical path, no time extension may be warranted.

In the resolution of questions concerning weather-related delays, the analysis must assess the basic requirements of the Contract, the validity of the weather occurrence with respect to the Contract requirements, the specific activities that are affected, and the Project critical path at the time of the event.

Chapter | fifteen

Risk Management

Construction is a business fraught with risk. One of the greatest areas of risk is controlling time and the cost of time. All parties can exercise better risk management in this area. By recognizing and planning, risks can be minimized and controlled. This chapter addresses some risk management considerations for each of the parties to the construction Project.

All parties in a construction Project must be keenly aware of the importance of good Project documentation. The analysis procedures presented in this book cannot be performed if good documentation does not exist. At the top of the list are daily reports. Also important are photographs/videos taken periodically over the course of the Project. Finally, detailed documentation on costs must be maintained and must be amenable to segregation to discrete issues.

OWNER'S CONSIDERATIONS

The Owner's considerations for risk start at the Project inception. First, the Owner must consider external constraints concerning time. Must the facility be completed to meet a critical production date? Must the Project finish by a certain date for political reasons? These factors influence the way in which the Owner pursues the Project. For example, certain time requirements and other factors may indicate a fast-track approach to the Project. The Owner must consider the realities of finishing the Project within the required time frame. Merely because external considerations require that a Project be completed by a certain date does not mean that the Project can, in fact, be completed by that date. The Owner should consult with knowledgeable advisors to determine a reasonable duration to specify in the Contract documents. Bidders should themselves thoroughly investigate the time constraints, but the Owner should point out any special considerations up front. For example, if the required duration can only be achieved by an accelerated effort such as multiple shifts and seven-day workweeks, the potential need for these elements should be stated in the Contract or at least discussed during the prebid meeting. The Owner is far better off alerting bidders to an urgent situation.

When the Contract is being drawn up, the Owner should also decide whether to include a Liquidated Damages clause. At this time, the Owner should carefully consider the potential damages if the Project is delayed. When writing a Liquidated Damages clause, the Owner should determine not only the amount of potential damages but also whether to include milestone liquidated damages.

In drafting the Contract, the Owner should also consider establishing time limits for the filing of claims by the Contractor. In other words, the Contract should clearly specify that claims for additional compensation must be submitted to the Owner or his representative within a set number of days after the event that gave rise to the claim. A period of between 30 and 90 days is a common time frame for the submission of claims. The clause should further state that if the required claim information is not submitted within the time specified, then the Contractor forfeits the right to recover any additional compensation. While this is a useful provision, the enforceability of such a clause may depend on the case law in your specific jurisdiction.

Not only should the Contract specify a time limit for the filing of claims, but it also should specify exactly the information that the Contractor must submit when asserting a claim, such as the following:

- A clear narrative of what the claim is with references to attached documents.
- An explanation of why the claim item differs from that already required by the Contract.
- References to the specific Contract clauses that apply.
- An explanation of the cause (or liability) for the claim.
- A clear definition of the specific impacts associated with the claim (i.e., extra work, overtime, delay, etc.).
- A detailed breakdown of the damages or extra costs with supporting information.

A sample clause addressing claims is shown in Figure 15.1.

ADMINISTRATIVE PROCESS

RESIDENT ENGINEER. When disputes and disagreements arising out of or relating to the contract of any work performed pursuant to the Contract, including additional work required in a Change Order or written or oral order or direction, instruction, interpretation, or determination by the Resident Engineer occur, the Contractor shall immediately give a signed written notice of intent to file a construction claim to the Resident Engineer. If such notification is not given and the Resident Engineer is not afforded the opportunity by the Contractor to examine the site of work or is kept from keeping a strict account of actual costs incurred to perform the disputed work or is not afforded the opportunity to review the Contractor's project records, then the Contractor shall waive all his rights to pursue the claim under the Contract.

FIGURE 15.1 105.17 Claims for Adjustments and Disputes

(Continued)

Unrelated claim issue will be processed as separate claims and therefore must be submitted as separate claims.

The Contractor shall supplement the written notice of claim within 15 calendar days of filing the notice of intent to file a construction claim with a written statement providing the following:

1. The date of the claim.
2. The nature and circumstance that caused the claim.
3. The Contract provisions that support the claim.
4. The estimated dollar cost, if any, of the claim and how that estimate was determined.
5. An analysis of the schedule showing any schedule change, disruption, and any adjustment of Contract time.

If the claim is continuing, the Contractor shall supplement the information required above in a timely manner.

The Contractor shall provide to the Resident Engineer full and final documentation to support the claim no later than 60 calendar days following the date the claim has been fully matured. A claim fully matures when all the direct damages (money and/or time) resulting from the claim issue can be reasonably quantified. The possibility of impact damages should not delay the submittal of full and final documentation of claims for direct damages.

The full documentation of the claim, as presented in the administrative process shall, at a minimum, contain the following elements.

1. A detailed factual narration of events that details the nature and circumstances that cause the claim. This detailed narration of events shall include, but is not limited to, providing all necessary dates, locations, and items of work affected by the claim.
2. The specific provisions of the Contract or laws which support the claim and a statement of the reasons why such provisions support the claim.
3. The identification and copies of all documents and the substance of any oral communications that support the claim. Manuals that are standard to the industry may be included by reference.
4. If an adjustment for the performance time of the Contract is sought:
   a. The specific days and dates for which it is sought.
   b. The specific reasons the Contractor believes a time adjustment should be granted.
   c. The specific provisions of the Contract under which additional time is sought.
   d. The Contractor's detailed schedule analysis to demonstrate the justification for a time adjustment.
5. If additional monetary compensation is sought, the exact amount sought and a breakdown of that amount into the following categories:
   a. Labor. Listing of individuals, classification, hours worked, and so on
   b. Materials. Invoices, purchase orders, and so on

FIGURE 15.1—CONT'D

c. Equipment listing detailed description (make, model, and serial number), hours of use, and dates of use. Equipment rates shall be at the applicable Blue Book Rate, which was in effect when the work was performed, as defined in Subsection 109.03.
d. Job site overhead.
e. Home office overhead (general and administrative).
f. Other categories

6. The above data shall be accompanied by a notarized statement from the Contractor containing the following certification:
Under penalty of law for perjury or falsification, the undersigned,

____
(Name)

____
(Title)

____
(Company)
hereby certifies that the claim is made in good faith; that the supporting data are accurate and complete to the best of my knowledge and belief; that the amount requested accurately reflects the Contract adjustment for which the Contractor believes the Department of Transportation is liable; and that I am duly authorized to certify the claim on behalf of the Contractor.

____
(Dated)
Subscribed and sworn before me this_____ day of
__________, 20_____.
Notary Seal
My commission expires: _______________
All pertinent information, references, arguments, and data to support the claim shall be included.

FIGURE 15.1—CONT'D

Scheduling Clauses

To control the Project's duration, the Owner must have a viable schedule. To ensure that the desired schedule is used, the Contract must specify the requirements for both a viable schedule and periodic updates. Depending on the Owner's degree of participation, different scheduling requirements may be dictated by the Contract. Figure 15.2 is a sample scheduling specification that clearly requires the construction Contractor to perform all the mechanics of the schedule production process. Contractors are not always well versed in CPM scheduling. Some may provide the absolute minimum documentation to meet the Contract requirements but never really use the schedule as a management

100.1 Scheduling Terms

1. **Activity.** A discrete, identifiable task that takes time, has a definable start and stop date, and can be used to plan, schedule, and monitor a project.
2. **Activity, Controlling.** The first incomplete activity on the critical path. *(May also be referred to as the controlling operation.)*
3. **Activity, Critical.** Any activity on the critical path.
4. **Activity ID.** A unique, alphanumeric, identification code assigned to an activity. *(It is recommended that owners and contractors in a particular industry or region adopt a standard activity numbering system to facilitate the integration of schedules across projects. This system could be tied to the standard specification formats adopted by certain industries, such as the format used by the Construction Specification Institute, the AASHTO Guide Specification, or other similar model or guide specification systems. It could also be tied to a standard work breakdown structure for work typical to the industry. The standard activity numbering system should be set forth in a Scheduling Manual that has been ratified by owners, contractors, subcontractors, and suppliers and then referenced as the standard for activity numbering in the scheduling specification.)*
5. **Activity Network Diagram.** *(Also called a pure-logic diagram or network diagram.)* A graphic representation of a CPM schedule that shows the relationships among activities.
6. **Bar Chart.** *(Also called a Gantt chart.)* A time-scaled, graphic representation of a schedule without relationships.
7. **Calendar Day.** A day on the calendar; beginning and ending at midnight.
8. **Completion Date, Contract.** The original date specified in the contract for completion of the project or a revised date resulting from owner-executed time extensions. The contract may also specify completion dates for interim milestones, phases, or other portions of the project.
9. **Completion Date, Scheduled.** The project completion date forecasted by the schedule. The schedule may also forecast completion dates for interim milestones, phases, or other portions of the project.
10. **Constraints, Activity.** A restriction imposed on the start or finish dates of an activity that modifies or overrides the schedule's logic relationships.
11. **Critical Path.** The **Longest Path.**
12. **Data Date.** The first day in the Initial or Baseline Schedule and the first day for performance of the work remaining in the Monthly Schedule Update or Revised Schedule. *(May also be defined as the date from which a schedule is calculated.)*

FIGURE 15.2 Sample Design-Bid-Build Scheduling Specification

(Continued)

13. **Duration, Original.** The estimated time, expressed in workdays, needed to perform an activity.
14. **Duration, Remaining.** The estimated time, expressed in workdays, needed to complete an activity.
15. **Float, Total.** The number of workdays between an activity's early and late dates. An activity's Total Float can be calculated as its late start date minus its early start date or its late finish date minus its early finish date. (If the term "float" is referenced elsewhere in the Contract Documents, then it will be necessary to reconcile the definition of "float" with this definition of "Total Float." For example, it is common to address the ownership of float in the contract. A typical provision might be written as follows: "Float is a shared commodity owned by the project and available for use by the project participants." When float is used in this context, the definition of the term "float" needs to be more general than the definition of "Total Float" provided here. An example definition of "float" might be written as follows: "Float is defined as the difference between when an activity can start or finish and must start or finish so as not to delay the scheduled project completion date or an intermediate milestone date.")
16. **Holidays.** Holidays observed are: *(This list of holidays is typical for many public construction projects. Please revise this list to coordinate with the holiday schedule adopted by your industry.)*
    1st day in January (New Year's Day)
    3rd Monday in January (Martin Luther King Jr. Day)
    3rd Monday in February (Presidents' Day)
    Last Monday in May (Memorial Day)
    4th day in July (Independence Day)
    1st Monday in September (Labor Day)
    11th day in November (Veterans Day)
    4th Thursday in November (Thanksgiving Day)
    25th day in December (Christmas Day)

    For holidays that fall on a Saturday, both the Saturday and the preceding Friday are considered to be holidays. For holidays that fall on a Sunday, both the Sunday and the following Monday are considered to be holidays.

17. **Longest Path.** The longest path of work in the network.
18. **Milestone.** An activity with no duration; typically used to represent the beginning or end of the project or its interim phases or stages.
19. **Narrative Report.** A descriptive report submitted with each schedule. The required contents of this report are set forth in this specification.

FIGURE 15.2—CONT'D

20. **Open End.** The condition that exists when an activity has either no predecessor or no successor, or when an activity's only predecessor relationship is a finish-to-finish relationship or only successor relationship is a start-to-start relationship.
21. **Predecessor.** An activity that is defined by schedule logic to precede another activity. A predecessor may control the start or finish date of its successor.
22. **Relationship.** The interdependence among activities. Relationships link an activity to its predecessors and successors. *(A schedule's relationships are sometimes referred to as the logic of the schedule. Examples of relationships are: finish-to-start, start-to-start, and finish-to-finish.)*
23. **Schedule.** Activities organized by relationships to depict the plan for execution of a project.
24. **Schedule, Initial.** The schedule showing the original plan for the first 60 calendar days of work.
25. **Schedule, Baseline.** The accepted schedule showing the original plan to complete the entire project. *(Sometimes known as the as-planned schedule.)*
26. **Schedule, Monthly Update.** A schedule produced by incorporating the project's actual progress *(sometimes known as as-built information or data)* over a routine interval, usually monthly, into the Baseline Schedule or the latest Monthly Update Schedule.
27. **Schedule, Revised.** A schedule prepared and submitted by the contractor that includes a significant modification to the schedule logic or durations, usually for the purpose of depicting a significant change in the contractor's plan.
28. **Schedule, Final.** The last schedule update containing actual start and finish dates for every activity.
29. **Successor.** An activity that is defined by schedule logic to succeed another activity. The start or finish date of a successor may be controlled by its predecessor.

100.2 Administrative Requirements

1. **General Requirements.** Plan and schedule the project and report progress to the owner. Provide a schedule using the critical path method (CPM). The owner's acceptance of any schedule, whether initial, baseline, update, or revised, does not modify the contract or constitute endorsement or validation by the owner of the contractor's logic, activity durations, or assumptions in creating the schedule. By accepting the schedule, the owner does not guarantee that the project can be performed or completed as scheduled. If the contractor or the owner discovers errors after the schedule has been accepted, correct the error in the next schedule submission.

FIGURE 15.2—CONT'D

2. Required Schedules.

2.1 Initial Schedule. The owner will use the initial schedule to monitor progress until the baseline schedule is accepted. Prepare and submit a schedule for the first 60 calendar days of work in accordance with subsections 100.3.3.1 and 100.3.3.2, plus a summary bar chart schedule for the balance of the project. Activity durations on the summary chart may exceed 15 working days.

At least ten calendar days before the preconstruction meeting, submit the initial schedule to the owner. Ensure that the schedule shows milestone and completion dates no later than the specified contract milestone and completion dates.

The owner will review the initial schedule at the preconstruction meeting. At this meeting, be prepared to generally discuss the proposed schedule for the entire project, not just the 60-day period covered by the initial schedule. If deviations to the staging, phasing, or sequencing required by the contract documents are proposed, be prepared to discuss these deviations.

Within five calendar days of the first project meeting, the owner will respond by accepting the initial schedule, rejecting the schedule and identifying the reason for rejection, or by asking for more information. Address the reasons for rejection or provide the information requested and resubmit the revised initial schedule no more than five calendar days after the owner's response. The owner may withhold 25% of each progress payment until the contractor submits and the owner accepts the initial schedule.

2.2 Baseline Schedule. No more than 30 calendar days after approval of the initial schedule, prepare and submit a baseline schedule to the owner for review in accordance with the requirements of subsections 100.3.3.1 and 100.3.3.2.

Within ten calendar days of receipt of the baseline schedule, the owner will respond by accepting the baseline schedule, rejecting the schedule and identifying the reason for rejection, or by asking for more information. Address the reasons for rejection or provide the information requested and resubmit the revised baseline schedule no more than ten calendar days after the owner's response. The owner may withhold 25% of each progress payment until the contractor submits and the owner accepts the baseline schedule.

2.3 Monthly Schedule Update. Prepare and submit a monthly schedule update to the owner that depicts the status of the project on the first workday of each month in accordance with the requirements of subsections 100.3.3.1 and 100.3.3.2. The update will reflect a new data date, work performed up to, but not including,

FIGURE 15.2—CONT'D

the new data date, and the plan for completing the project. Submit the schedule update by the first Monday of the following month.

Within ten calendar days of receipt of the schedule update, the owner will respond by accepting the schedule update, rejecting the schedule update and identifying the reason for rejection, or by asking for more information. Address the reasons for rejection or provide the information requested and resubmit the revised schedule update no more than ten calendar days after the owner's response. The owner may withhold 25% of each progress payment until the contractor submits and the owner accepts the schedule update.

2.4 Revised Schedule. The owner has the right to request a revised schedule. Circumstances leading to such a request include, but are not limited to:

2.4.1 A forecasted delay to scheduled interim or completion dates of more than 15 calendar days

2.4.2 A significant difference between the actual sequence or duration of work and that depicted in the schedule

Prepare and submit the revised schedule in accordance with the requirements of subsections 100.3.3.1 and 100.3.3.2 no more than ten calendar days after the owner's request.

Within ten calendar days of receipt, the owner will respond by accepting the revised schedule, rejecting the schedule and identifying the reasons for rejection, or by requesting more information. Address the reasons for rejection or submit the information requested no more than ten calendar days after the owner's request. The owner may withhold 25% of each progress payment until the contractor submits and the owner accepts the revised schedule.

2.5 Final Schedule. Within 30 calendar days of final acceptance of the project, submit a final schedule with actual start and finish dates for each activity. Include with the submission a certification signed by the principal of the firm stating:

> *"To the best of my knowledge, the enclosed final schedule reflects the actual start and finish dates of the activities contained herein."*

100.3 Technical Requirements

1. **Software Compatibility Requirements.** The owner uses Primavera Version 6.0, or the most current version, to schedule and monitor its construction program. Prepare and maintain the schedule using one of the following software options:

1.1 Primavera Version 6.0 and My Primavera, in which case the schedule is prepared and maintained on the owner's database.

FIGURE 15.2—CONT'D

1.2 Primavera Version 6.0, in which case the schedule is prepared on a separate database and maintained through file submission as described in subsection 100.3.3.2.

1.3 Primavera Contractor, in which case the schedule is prepared on a separate database and maintained through file submission as described in subsection 100.3.3.2.

1.4 Any other software that is compatible with Primavera Version 6.0, in which case the schedule is prepared on a separate database and maintained through file submission as described in subsection 100.3.3.2.

2. Schedule Requirements. Provide a schedule that meets the following requirements:

2.1 Calculate the schedule using the Retained Logic scheduling option unless written authorization is obtained from the owner to use the Progress Override scheduling option.

2.2 Do not use the following types of logic relationships:

2.2.1 Negative lags.

2.2.2 Lags in excess of ten workdays.

2.2.3 Start-to-finish relationships.

2.2.4 Open ends. Only the first activity will have no predecessor and only the last activity will have no successor.

2.2.5 Constraints. The contractor may use a limited number of the following constraints with the owner's written authorization: start on or before, start on or after, finish on or before, finish on or after, as late as possible.

2.2.6 Manually modified dates. The contractor may manually modify dates only with the owner's written authorization.

2.2.7 The contractor may use lags with finish-to-start relationships with the owner's written authorization prior to use.

2.3 Includes the following work activities, as applicable:

2.3.1 Work to be performed by the contractor, subcontractors, and suppliers.

2.3.2 Work to be performed by the owner, other contractors, and third parties such as government agencies and authorities, permitting authorities, utilities, or other entities required for completion of the project.

2.3.3 The project start date, scheduled completion date, and other contractually mandated milestones, start or finish dates for phases, or site access or availability dates.

2.3.4 Submittal, review, and approval activities when applicable, including time periods for the owner's approval as specified in the contract documents. *(A specific contract reference is preferred here.)*

FIGURE 15.2—CONT'D

- 2.3.5 Fabrication, delivery, installation, testing, and similar activities for materials, plants, and equipment.
- 2.3.6 Sampling and testing activities.
- 2.3.7 Settlement or surcharge periods.
- 2.3.8 Cure periods.
- 2.3.9 Utility notification and relocation.
- 2.3.10 Installation, erection and removal, and similar activities related to temporary systems or structures such as temporary electrical systems or shoring.
- 2.3.11 Punch list, substantial completion, final cleanup, and similar activities.
- 2.3.12 Required acceptance testing, inspections, or similar activities.
- 2.3.13 Durations for receipt of permits or acquisition of rights of way.

2.4 Define the following attributes for each activity in the schedule:

- 2.4.1 A unique alphanumeric Activity ID as specified in the owner's Scheduling Manual.
- 2.4.2 A unique descriptive name, using such attributes as work type and location to distinguish activities as specified in the owner's Scheduling Manual. *(This should be coordinated with a standard work breakdown structure and follow the guidelines established in the Scheduling Manual.)*
- 2.4.3 A duration stated in workdays of no more than 15 workdays, unless a longer duration is requested by the contractor and approved by the owner.
- 2.4.4 Uses codes for responsibility, phasing, and staging as specified in the Scheduling Manual.

3. Schedule Submission Requirements

3.1 Preparing Schedule on Owner's Database. If the schedule is prepared using My Primavera in the owner's database, then for each schedule submission, submit the following items:

- 3.1.1 A transmittal letter to the owner identifying which schedule in the database is being submitted for review.
- 3.1.2 A narrative report.

3.2 Preparing Schedule on Separate Database. If the schedule is prepared using Primavera Version 5.0, Primavera Contractor, or some other software compatible with Primavera Version 6.0, then, for each schedule submission, submit the following items:

- 3.2.1 A transmittal letter.
- 3.2.2 A narrative report.

FIGURE 15.2—CONT'D

3.2.3 A Primavera Version 6.0 compatible electronic file of the schedule on a computer disc (CD).

3.2.4 The critical path in bar chart format (Longest Path sort).

3.2.5 Work paths with total float values within 20 workdays of the critical path's total float value in bar chart format. For example, if the critical path has a total float value of zero, then show all of the work paths with total float values of 20 or less.

3.2.6 An activity network diagram plotted in color, on E-size paper, with each sheet of the plot including a title, match data for diagram correlation, a page number, and a legend. The activity network diagram should only be submitted with schedules with revised relationships or activity durations.

3.2.7 A Predecessor/Successor report with the following items for each activity:

3.2.7.1 Activity ID and description
3.2.7.2 Original duration
3.2.7.3 Remaining duration
3.2.7.4 Calendar ID
3.2.7.5 Predecessors and successors
3.2.7.6 Early start date
3.2.7.7 Early finish date
3.2.7.8 Late start date
3.2.7.9 Late finish date
3.2.7.10 Total float
3.2.7.11 Relationship type
3.2.7.12 Lags
3.2.7.13 Constraints

3.3 Narrative Reports for the Initial and Baseline Schedule. For each submission of the initial and baseline schedule, provide a narrative report that includes the following information:

3.3.1 Explanation of the overall plan to complete the project, including where the work will begin and the how the work and crews will flow through the project.

3.3.2 Use and application of the workdays per week, number of shifts per day, number of hours per shift, holidays observed, and how the schedule accommodates adverse weather days for each month or activity.

3.3.3 If the project is a multiyear project, then identify the work to be completed in each construction season.

3.3.4 A description of problems or issues anticipated.

3.3.5 A description of anticipated delays, including:

3.3.5.1 Identification of the delayed activity by activity ID and description
3.3.5.2 Type of delay

FIGURE 15.2—CONT'D

3.3.5.3 Cause of the delay
3.3.5.4 Effect of the delay on other activities, milestones, and completion dates
3.3.5.5 Identification of the actions needed to avoid or mitigate the delay

3.3.6 A description of the critical path.

3.3.7 A description of work paths with total float values within 20 workdays of the critical path's total float value. For example, if the critical path has a total float value of zero, then describe all of the work paths with total float values of 20 or less.

3.3.8 A statement identifying constraints and an explanation of the reason for and purpose of each constraint.

3.3.9 A statement describing the status of required permits.

3.3.10 A statement describing the reason for the use of each lag.

3.3.11 Identification of number, size, and composition of work crews.

3.3.12 Identification of major pieces of equipment.

3.3.13 Identification of proposed productions rate for major operations.

3.4 Narrative Reports for the Monthly Schedule Update and Revised Schedule. For each submission of the monthly schedule update and revised schedule, provide a narrative report that includes the following information:

3.4.1 A description of the status of the scheduled completion date; focus on any differences from the last submission.

3.4.2 A statement explaining why the scheduled completion date is forecast to occur before or after the contract completion date. An explanation stating why any of the contract milestone dates are forecasted to occur late.

3.4.3 A description of the work performed since the last schedule update.

3.4.4 A description of unusual labor, shift, equipment, or material conditions or restrictions encountered or anticipated.

3.4.5 A description of the problems encountered or anticipated since the last schedule submission.

3.4.6 A statement that identifies and describes any current and anticipated delays. A discussion of delays in the narrative report does not constitute notice and does not replace the need for the contractor to provide notice as required by the contract. Include the following:

3.4.6.1 Identification of the delayed activity by activity ID and description
3.4.6.2 Type of delay
3.4.6.3 Cause of the delay

FIGURE 15.2—CONT'D

3.4.6.4 Effect of the delay on other activities, milestones, and completion dates

3.4.6.5 Identification of the actions needed to avoid or mitigate the delay

3.4.7 A description of the critical path.

3.4.8 A description of changes in the critical path and schedule completion date (for the project or its milestones) from the last schedule submission.

3.4.9 Descriptions of the status of work paths that have total float values within 20 workdays of the critical path identified in the previous schedule submission.

3.4.10 Descriptions of work paths with total float values within 20 workdays of the critical path's total float value. For example, if the critical path has a total float value of negative 25, then show all of the near critical paths with total float values of negative five or less.

3.4.11 Changes to planned labor, equipment, or productivity identified in the Initial or Baseline Schedule.

3.4.12 A statement or Claim Digger (or equivalent) report that identifies the changes made between the previous schedule submission and the current proposed schedule, including, but not limited to:

3.4.12.1 Data date
3.4.12.2 Completion date
3.4.12.3 Activity code assignments
3.4.12.4 Scheduling options
3.4.12.5 Activity descriptions
3.4.12.6 Added activities
3.4.12.7 Deleted activities
3.4.12.8 Added activity relationships
3.4.12.9 Deleted activity relationships
3.4.12.10 Activity original durations
3.4.12.11 Activity remaining durations
3.4.12.12 Activity actual start and finishes
3.4.12.13 Percent complete
3.4.12.14 Constraints
3.4.12.15 Activity resources
3.4.12.16 Activity costs
3.4.12.17 Activity coding

3.4.13 A statement providing status of pending items, including, but not limited to:

3.4.13.1 Permits
3.4.13.2 Change orders
3.4.13.3 Time extension requests

FIGURE 15.2—CONT'D

tool. If that potential exists, the Owner may consider including in the Contract a specification such as that shown in Figure 15.3.

In both the scheduling clauses, there may or may not be a requirement for manpower, equipment, and cost loading. These requirements may be useful tools, both for the Contractor and the Owner. While it appears worthwhile to have a manpower, equipment, and cost-loaded schedule, the Owner

SECTION 01310
NETWORK ANALYSIS SYSTEM
PART 1-GENERAL

1.1 DESCRIPTION:

A. The Contractor shall develop a Network Analysis System (NAS) plan and schedule demonstrating fulfillment of the contract requirements, shall keep the network up-to-date in accordance with the requirements of this section and shall utilize the plan for scheduling, coordinating and monitoring work under this Contract (including all activities of subcontractors, equipment vendors and suppliers). Conventional Critical Path Method (CPM) Precedence Diagramming Method (PDM) technique will be utilized to satisfy both time and cost applications. All schedule data and reports required under this specification section shall be based upon regular total float, not relative total float schedules.

1.2 CONTRACTOR'S REPRESENTATIVE:

A. The Contractor shall designate an authorized representative in the firm who will be responsible for the preparation of the network diagram.

B. The Contractor's representative shall have direct project control and complete authority to act on behalf of the Contractor in fulfilling the requirements of this specification section and such authority shall not be interrupted throughout the duration of the project.

1.3 CONTRACTOR'S CONSULTANT:

A. To prepare the network diagram and diskette(s) that reflect the Contractor's project plan, the Contractor shall engage an independent CPM consultant who is skilled in the time and cost application of scheduling using network techniques for construction projects, the cost of which is included in the Contractor's bid. This consultant shall not have any financial or business ties to the Contractor, and shall not be an affiliate or subsidiary company of the Contractor, and shall not be employed by an affiliate or subsidiary company of the Contractor.

FIGURE 15.3

(Continued)

B. Prior to engaging a consultant and within ten calendar days after award of the contract, the Contractor shall submit to the Contracting Officer:
   1. The name and address of the proposed consultant.
   2. Sufficient information to show that the proposed consultant has the qualifications to meet the requirements specified in the preceding paragraph.
   3. A list of prior construction projects along with selected PDM network diagram samples on current projects that the proposed consultant has performed complete project scheduling services. These network diagram samples must show complete project planning for a project of similar size and scope as covered under this contract.

C. The Contracting Officer has the right to approve or disapprove employment of the proposed consultant, and will notify the Contractor of its decision within seven calendar days from receipt of information. In case of disapproval, the Contractor shall resubmit another consultant within ten calendar days for renewed consideration. The Contractor must have their CPM Consultant approved prior to submitting any diagram.

FIGURE 15.3—CONT'D

should carefully consider if the benefits are worth the effort. Clearly, having resource and cost loading in a schedule allows for more precise tracking of progress and a means for more objective payment, but these enhancements complicate the schedule and can lead to confusion and other problems. It may be just as beneficial for the Owner to specify a basic CPM schedule with no resource loading but with the ability to require the Contractor to submit such information for specific activities at the request of the Owner. The scheduling requirement should also specify schedule updates during the Project and daily progress reporting of activities as defined in the CPM schedule.

Scheduling clauses should include the requirement to provide a baseline schedule before work will commence. The baseline schedule should be due within a prescribed period of time, and there should be a specific time period for Owner review. These requirements often appear in contracts, but enforcement varies greatly. As with any provisions, unenforced scheduling provisions accomplish nothing. An Owner could consider allowing the Contractor to submit a basic schedule to get the Project started but require more detail within a prescribed time frame. The same should be true of updates: There should be specific requirements about when to submit them, and the submission should be tied to progress payments.

Change Orders

During the course of the Project, the Owner must carefully monitor the management of change orders. Every change order has two parts: time and money.

Every change order should state whether or not additional time is warranted. Obviously, this task is far easier if an up-to-date CPM schedule is maintained throughout the Project.

Delay Damages Clauses

Another risk management consideration for the Owner is the area of delay damages. The Owner can insert a no-damage-for-delay clause in the Contract, thus attempting to shift the burden of risk for delays to the Contractor. However, the use of this type of exculpatory language may increase the amount bid by a Contractor and is still no guarantee that a dispute over delays will be prevented. The Owner should research the use of a no-damage-for-delay clause with qualified counsel before including it in the Contract. An alternative approach is to specify limits to what types of damages are allowable in the event of a delay. Some government agencies use this approach.

CONSTRUCTION MANAGER'S CONSIDERATIONS

Some projects include a construction manager, typically abbreviated by the letters "CM." The CM can be hired by the Owner as its construction representative, in which case the CM is responsible for representing the Owner and protecting the best interests of the Owner. This arrangement is typically called "agency CM."

In other cases, the CM may have a financial interest in the Project and may be performing the Project at a preestablished maximum cost, often called a "guaranteed maximum price," or GMP. The CM may be working in a GMP arrangement with some sharing of the savings below the GMP. These types of arrangements are often called "CM at Risk."

Other variations in the relationships between the Owner and the CM or the CM and the Contractors exist, but the discussion of these relationships is beyond the scope of this book. More details about these arrangements can be found at the website of the Construction Management Association of America at cmaanet.org, or the Design Build Institute of America at DBIA.org.

Regardless of the arrangement, all of the CM's considerations stem from the fact that the CM's role is unique. The CM is expected to be the Project expert in construction management and all its many facets, including the often difficult task of management of time and delays.

It is not uncommon for a Project to get behind schedule, and it may be difficult to recognize and enact ways to make up time. In terms of comparing time management and budget management, it may be simple to solve a budget crisis by securing additional funding. However, it is impossible to create more time, especially on a Project that has a fixed deadline.

CM and the Project Timetable

The CM's considerations regarding time begin at the inception of the Project during the planning phase. The CM must ensure that the overall Project schedule includes adequate time for all parties to perform their work, including time for the exchange of Project performance information between the Owner and Designer during the design phase; the careful preparation of Contract documents, including the clauses that address schedule and time; developing Contractor interest in the Project; preparation of responsive bids; the Owner to evaluate Contractor bids; and most important, for the Project to be constructed under normal conditions.

The CM must manage all the Project parties to make sure that the Project stays on schedule. Just as the Contractor is typically responsible for means and methods of construction, the CM is responsible for the means and methods to monitor and manage the performance of all the parties including the evaluation of delays. Some Project parties may not be used to strict time management by the CM. The CM must ensure that the Contract language for all the parties includes time management provisions and procedures.

Even with the best planning, delays might occur. The CM must be able to foresee delays and take proactive measures to mitigate or recover from delays. Often, Project delays will be caused by the performance or lack of performance of one or more of the Project participants. The CM must keep detailed accurate records of the performance of all the parties so that it can evaluate liability for delays. Once again, the CM is expected to be the expert who has the responsibility to sort out Project delays for the benefit of all parties including the Owner.

CM Responsibility to Contractors and Subcontractors

If the CM is in an "at risk" arrangement, it is now responsible to the Owner for managing the construction and to the Contractors and Subcontractors for administering the contracts. In this case, again the CM is seen as the construction management expert that should be able to manage the performance of the Project parties. Detailed accurate performance records, clear Contract provisions, and dispute resolution procedures must be developed and maintained by the CM. However, a CM "at risk" functions much like a General Contractor. Consequently, the Owner must ensure that its Contract with the CM is appropriately structured so the Owner's risks are managed.

CM Responsibility for Managing Changes

Along with managing time and schedule, the CM must manage changes and the change order process. Change orders must address the additional cost of the changed work and the time required to perform the changed work. By properly maintaining a current CPM schedule and detailed performance records, and by seeking adequate information from the Contractor, the CM should be able to evaluate time extensions and additional costs. All too often, only additional costs resulting from changes are addressed, and additional time or time extension requests are left out of the change order. The "postponing" or "delaying" of the management and assessment of time is not recommended and usually results in unnecessary disputes. The CM should be proactive in preventing this common problem.

CM Responsibility for Delay Analysis

Construction projects involve many variables in terms of Project needs, the participants' motivations, and the need to build a Project under sometimes unpredictable circumstances. As a result, it is not uncommon for Project delays to be caused by a combination of actions or lack of actions by more than one Project party, changes, or unforeseen conditions. Therefore, even with the best records, it may be difficult to identify and evaluate responsibility for delays. Some CMs may be more experienced than others with delay analysis and evaluation. It may be necessary to engage the services of a scheduling and delay analysis specialist to augment the CM's services and to assist the Owner to evaluate the delays on a Project. If that is the case, it is recommended that a schedule/delay consultant be retained as soon as a problem is perceived.

CM Responsibility for Quality, Safety, and Environment

The CM may also be retained as the Owner's representative for ensuring quality Project work, Project safety, and/or compliance with environmental regulations. Again, because the CM is expected to be the expert in construction and representative of the Owner, the choice of CM should only be made after careful consideration of previous experience in these areas. Consideration should be given to the experience of the CM firm and the credentials of the individuals that the CM plans to assign to the Project.

GENERAL CONTRACTOR'S CONSIDERATIONS

Like the Owner, the General Contractor also must assess the risks of delays to the Contract completion, These considerations parallel those of the Owner, with a slightly different perspective.

Assess the Time Allowed in the Contract

The Contractor should assess the time allowed by the Contract to determine if enough time is provided to perform the work without the use of extraordinary resources. The Contractor must include in its bid the cost for additional effort (such as overtime) required to meet the Contract completion date.

Assess Exculpatory Language

If the Contract is rampant with exculpatory language, especially in the area of no-damages-for-delay, the Contractor should carefully consider accepting the risks involved. Some projects are not worth the risk of bidding. The Contractor should consult with qualified counsel before entering into a highly restrictive Contract with exculpatory language. Again, if the risk is too great, the Contractor may consider not bidding on the Project or pricing it accordingly.

Not only must Contractors assess the risk of exculpatory language in a Contract, but they must also read, understand, and comply with the Contract provisions, particularly with respect to changes and claims. For instance, the

Contract may specify a time limit for issuing notice of a change or for filing a claim. The Contractor must comply with these requirements. Also, the Contractor should make sure to submit all information and documentation required by the Contract.

CPM Schedules

CPM schedules required by Owners on projects can be effective tools for managing a Project. Contractors resisted using this management tool for many years, but most professional builders now realize the usefulness of these schedules. Aside from managing the work, the CPM schedule, properly updated, has become the most respected and reliable document for recording a Project's as-built history. Trying to recreate the progress on a Project, after it is completed, is far more laborious than contemporaneously updating the Project schedule.

CPM schedules consider not only time but resources. It is the efficient use of resources that will allow the Contractor to maximize its profits. If delays arise on the Project, the CPM schedule is probably the most effective tool that the Contractor has to demonstrate the delays that occurred to both critical and noncritical activities.

Risk to Subcontractors

The General Contractor may pass some requirements on to its Subcontractors. The amount of risk and responsibility is dictated to some extent by the terms of the Contract. Passing risk to the Subcontractors is not always as easy as including a general pass-through clause. This type of clause incorporates by reference all the conditions of the general Contract into the subcontract agreement. With a general pass-through clause, if a Subcontractor delays a Project, the damages assessed against the Subcontractor may be limited to the liquidated damages amount specified in the general Contract. Yet the General Contractor is liable for that same amount of liquidated damages to the Owner, plus its own additional costs. For example, a subcontract may include the following general language: "All of the conditions of the Contract between the Owner and the General Contractor are incorporated herein by reference and are binding upon the Subcontractor."

If the general Contract has a liquidated damages amount of $200 per calendar day, the incorporation by reference of the general Contract may effectively limit any Subcontractor's liability for delays to the $200 per day. A Contract should make it clear to Subcontractors that they can be held liable for delay costs from the General Contractor or from other Subcontractors, not just a share of liquidated damages to the Owner.

Consider Early Finish

General Contractors should try to complete projects earlier than the time allowed in the Contract. By reducing time on the site, the Contractor reduces general conditions costs and thereby realizes greater profit. At the planning stage, the

General Contractor should approach every Contract with the intent of early completion. If the Contractor plans to finish the Project early, the Project schedule should so state. There is no sense in having two schedules—one for the early completion (the actual schedule) and one that is shown to the Owner reflecting the full Contract duration—on the job.

SUBCONTRACTOR'S AND SUPPLIER'S CONSIDERATIONS

Subcontractor Considerations

Normally, the General Contractor dictates the Subcontractor's schedule. At times, the schedule requirements incorporated into subcontract agreements may be undefined or unreasonable. For example, subcontract agreements commonly state that the Subcontractor will perform its work in accordance with the General Contractor's schedule and will adjust accordingly so as not to delay the job. Since this statement leaves the work period undefined, the Subcontractor may have to accelerate its work for the entire duration of the job.

Subcontractors can insist on a lot of things regarding their contracts, but few Subcontractors can expect much success with this. What the Subcontractor does need to make clear is that it bid the Project for the Contract start and completion dates in the bid. If asked up front to bid an accelerated schedule, this request needs to be documented in the subcontract. The Subcontractor will have its own idea of how long it needs to perform the work, and if the General Contractor requires a shorter duration or a work sequence that the Subcontractor did not contemplate in its bid, the Subcontractor should carefully examine its ability to perform and request a Contract increase before agreeing to a change in the schedule.

Specific Schedule

To reduce risks associated with acceleration costs, the Subcontractor should clearly communicate to the General Contractor the time and schedule to which the bid applies. The Subcontractor should insist that schedules be included in the terms of the subcontract agreement.

Contract Language

Subcontractors should seriously consider whether to bid on contracts with extensive exculpatory (General Contractor protective) language. As noted before, some jobs are not worth bidding. This cannot be emphasized enough. Many General Contractors have their own specific contracts written to suit them, and they strongly resist any Subcontractor changes to the language. A Subcontractor should review the General Contractor's agreement before bidding and find out if the General Contractor is amenable to Contract changes. If not, then the Subcontractor needs to make a decision to live with the language and account for it in the bid or not bid the Project. The Subcontractor is better off knowing what it is getting into in advance before spending the time and money to estimate the Project, be the low Subcontractor bidder, and be faced with an onerous Contract.

Subcontractors must become thoroughly familiar with all clauses in their contracts. For instance, it is common to see clauses that state that for Subcontractor claims, the General Contractor will "pass through the claim to the Owner." The Subcontractor will accept whatever damages the General Contractor is able to collect and will also share in any costs for litigation or arbitration. Clauses of this nature may not sound fair, but they are common.

It is also common to see clauses that state that if the Subcontractor is delayed by another Subcontractor, it must seek compensation directly from that Subcontractor. Obviously, the clause is not desirable and creates legal problems for the Subcontractor who suffers the loss, vis-à-vis privity of Contract. Clearly, the General Contractor may not always be motivated to act in the Subcontractor's best interests. Therefore, the Subcontractor should attempt to include the following in its agreement with the General Contractor:

- An equitable breakdown of awards or settlements for claims involving more than just the respective Subcontractor.
- The right to pursue damages only against the general, and not the other Subcontractors.
- Proportional legal and administrative costs in claims actions.

DESIGN CONSULTANT'S CONSIDERATIONS

Designer Considerations

Normally, the Designer provides input on the duration of the Project. Such observations should not be formed or communicated casually. It is a subject that should be analyzed in a careful, detailed manner to determine the time required to perform the work considering the Project, the site, the weather, and so forth.

Designer as Owner's Representative

Designers who act as the Owner's representative during construction should require the Contractor to submit a detailed schedule for construction. The Designer then reviews the schedule submitted by the Contractor and monitors progress during construction. It is advisable to establish procedures ahead of time with the Contractor for schedule monitoring.

Changes

If the Designer as the Owner's representative makes a change to the Project because of Owner decisions or errors and omissions in the plans and specifications, it should assess the time effect of that change on the Project. Designers should resist the tendency to deny any time extension simply because they fear it might negatively reflect upon them. If a change necessitates a valid extension to the duration of the Project, the problem can be resolved much sooner and usually at far less cost if assessed (fairly and impartially) as soon as it arises.

Being the Project Designer and the Owner's representative is a difficult dual role because there is a perceived, if not an actual, conflict of interest. Virtually all projects have design problems, this is one of the reasons construction contracts are unique in having a changes provision. If a Contractor submits a change request because of a design problem, the Designer must take all necessary measures to ensure that its decisions are fair and impartial.

REAL-TIME CLAIMS MANAGEMENT

On larger time-sensitive construction programs, many forward looking public agencies, private Owners, and Contractors have instituted coordinated measures aimed at preempting and mitigating claims and disputes. These claims-focused risk management programs have proven to be extremely cost-effective.

For Owners, program goals are established and monitored. These goals can include initiatives to (1) make each Project "claim-resistant," (2) mitigate potential cross-Contract problems during construction, (3) remove Contract administration inconsistencies for multiple Project programs, and (4) facilitate the timely resolution of any claim issues. For Contractors, goals can include initiatives to (1) implement best practices for scheduling and change administration, (2) implement effective documentation practices, and (3) satisfy Contract requirements for changes, claims, and time extensions. In both cases, the effective handling of changes and claims calls for advanced preparation and a coordinated strategy in four areas: claim avoidance, claim mitigation, claim evaluation, and claim resolution.

The basic concept of Real-Time Claims Management is to establish, before the Project starts, a program that is aimed at identifying problems as early as possible and resolving them as quickly as possible. It requires that the Project participants "war game" the Project before it starts and identify as many potential problem areas as they can. Once the potential problem areas are identified, the participants then develop a game plan and specific approaches to prevent these problems from occurring and to manage them expeditiously if they do occur. The overall attitude is that the Project participants will resolve their problems rather than seek a resolution from a judge or jury.

In the context of delays, the Real-Time Claims Management approach focuses on time-related areas of possible problems. As an example, the program should include the following:

- Type of schedules to be required by Contract
- Independent review of all schedules and schedule updates
- Precise requirements for identifying time extensions associated with changes
- Procedure for independent review of time extension requests
- Scheduled meetings dedicated to review of Project progress and Project schedules
- Independent review of initial analysis of changes and associated time extension requests

The practice of Real-Time Claims Management has shown that having independent input, review, and analysis before and during the Project results in fewer problems and more expeditious resolution of problems. Most often, independent assistance is required when a delay dispute goes to arbitration or trial. But with that timing, the expertise applied becomes a determination of liability and a measure of the delays. Applying the same level of expertise much earlier becomes a tool for prevention and resolution. This is a far more cost-effective approach to the problem of construction delays.

Index

Note: Page numbers followed by *f* indicates figures and followed by *t* indicates table.

D

E

Z

Printed in the United States
By Bookmasters